BOSTON STUDIES IN THE PHILOSOPHY OF SCIENCE

VOLUME XIV

METHODOLOGICAL AND HISTORICAL ESSAYS

IN THE NATURAL AND SOCIAL SCIENCES

SYNTHESE LIBRARY

MONOGRAPHS ON EPISTEMOLOGY,

LOGIC, METHODOLOGY, PHILOSOPHY OF SCIENCE,

SOCIOLOGY OF SCIENCE AND OF KNOWLEDGE,

AND ON THE MATHEMATICAL METHODS OF

SOCIAL AND BEHAVIORAL SCIENCES

Editors:

VOLUME 60

BOSTON STUDIES IN THE PHILOSOPHY OF SCIENCE

EDITED BY ROBERT S. COHEN AND MARX W. WARTOFSKY

VOLUME XIV

METHODOLOGICAL AND HISTORICAL ESSAYS IN THE NATURAL AND SOCIAL SCIENCES

Edited by

ROBERT S. COHEN AND MARX W. WARTOFSKY

D. REIDEL PUBLISHING COMPANY

DORDRECHT-HOLLAND / BOSTON-U.S.A.

Library of Congress Catalog Card Number 73–83558

Cloth edition: ISBN 90 277 0392 2
Paperback edition: ISBN 90 277 0378 7

Published by D. Reidel Publishing Company,
P.O. Box 17, Dordrecht, Holland

Sold and distributed in the U.S.A., Canada and Mexico
by D. Reidel Publishing Company, Inc.
306 Dartmouth Street, Boston,
Mass. 02116, U.S.A.

TABLE OF CONTENTS

PREFACE

Modern philosophy of science has turned out to be a Pandora's box. Once opened, the puzzling monsters appeared: not only was the neat structure of classical physics radically changed, but a variety of broader questions were let loose, bearing on the nature of scientific inquiry and of human knowledge in general. Philosophy of science could not help becoming epistemological and historical, and could no longer avoid metaphysical questions, even when these were posed in disguise. Once the identification of scientific methodology with that of physics had been queried, not only did biology and psychology come under scrutiny as major modes of scientific inquiry, but so too did history and the social sciences – particularly economics, sociology and anthropology. And now, new 'monsters' are emerging – for example, medicine and political science as disciplined inquiries. This raises anew a much older question, namely whether the conception of science is to be distinguished from a wider conception of learning and inquiry? Or is science to be more deeply understood as the most adequate form of learning and inquiry, whose methods reach every domain of rational thought? Is modern science matured reason, or is it simply one historically adapted and limited species of western reason?

In our colloquia at Boston University, over the past fourteen years, we have been probing and testing the scope of philosophy of science. To be sure, our principal attention has been to the physical sciences, and the preceding volume of these *Studies* is devoted to some central questions in the understanding of contemporary physics. In this volume we have gathered a range of methodological and historical essays on the physical and social sciences, presented to the Boston Colloquium for the Philosophy of Science during 1969–72. There are three exceptions. Our friend and colleague, Azaria Polikarov, of the Bulgarian Academy of Sciences, spent several months as Research Associate at our Center and kindly contributed an essay. Also, by courtesy of Professors Edwin Levy and Howard Jackson of the University of British Columbia, we have included their translation of the little-known essay of Frege on inertia, his "single published writing in non-deductive science". Finally, we are pleased that we can include a brief paper by another friend and colleague, Władisław Krajewski of the University of Warsaw.

The essays fall into several main groups: first, there are those primarily concerned with methodological and epistemological questions (Feigl, McMullin, Polanyi, Quinn, Suchting, Mayr, Novakovic, Berkson, Polikarov, and Frege). Second, there are studies in the history and development of scientific ideas (Bartley, Elkana, McGuire, Popkin, Lauer, Krajewski and Zeleny). Finally, Shapere's paper addresses itself to the importance of contemporary physical science for metaphysics; and those of Sommers and Sosa to philosophy of logic.

Boston University Center for the   R. S. COHEN
Philosophy and History of Science   M. W. WARTOFSKY
1973

HERBERT FEIGL

EMPIRICISM AT BAY?

Revisions and a New Defense

At the risk of being regarded a 'square' and 'reactionary' I wish to defend, explicate, reformulate and 'salvage' whatever seems valuable to me in the empiricist tradition in the philosophy of science. I am fully aware of the almost hostile vogue of 'beating-up' on the empiricists, a fashion that is still in full swing. My only consolation is the remark of a well-known British philosopher. When he visited us at the University of Minnesota a few years ago, he said to me: "Don't worry Feigl, if you don't catch the bus of philosophy the first time, just wait a while, it'll come 'round again!'". Since I was lucky enough to catch the bus of philosophical fashion some forty years ago, I hope to be lucky once again during the rest of my 'natural life'.

Of course I know full well that some drastic revisions – in several respects – are in order. What I shall criticize are mainly some highly questionable exaggerations and distortions that the critics of empiricism have mistaken for genuine objections against the central tenets of classical as well as of modern (or 'logical') empiricism. My major aims then are criticism and clarification of misunderstandings. Since my topic is large and since it seems wise to limit my speech to – possibly – less than an hour, I am forced to restrict myself to a number of terse remarks. I trust you will excuse the somewhat dogmatic tone thereby necessitated. I regret that I can think of no alternative manner of presentation.

Modern empiricism has been, and is still being attacked on many fronts. I shall try to deal with what seem to me the most important and incisive issues.

The late, great, Hans Reichenbach[1] distinguished – I think most fruitfully and cogently – between two types of analyses, viz.: in the 'context of discovery' and in the 'context of justification'. Historians of science recognize something closely related to this distinction (in their own way) in that they speak of 'external' and 'internal' history of science. *External* history of science concerns the psychological, social-political-economic contexts and conditions of the developments in the sciences. *Internal* history attends to the reasoning at a given stage of science, i.e., the matters of evidence and inference, of presuppositions or basic assumptions (be they scientific or philosophical), the conceptual, definitional and postulational aspects.

This basic distinction has been attacked by a great many outstanding thinkers. I was astonished that such brilliant and knowledgeable scholars as N. R. Hanson, Thomas Kuhn, Michael Polanyi, Paul Feyerabend, Sigmund Koch *et al.* consider the distinction invalid or at least as misleading. N. R. Hanson complained[2] about the 'bifocal' views of science as presented by the logical empiricists (such as Carnap, Reichenbach, Hempel *et al.*). Hanson criticized the distinction (separation was intended by no one!) between the logical and the empirical aspects of scientific theories. But he (along with others) was attacking a straw man of his own making. None of the Logical Empiricists ever maintained that in an account of the origination and development of scientific theories those two aspects can be sharply kept apart. Hanson may have been quite right in his view that in many a scientific revolution, i.e. in many a drastic change of scientific *theories* something like a 'Gestalt-switch' has been (and still is) operative. But interesting as that is, it pertains to the psychology of productive, creative thinking and problem solving. It is in this respect that the affinities between great science and great art – stressed recently, for example by Bronowski, Polanyi (and should I mention Koestler?) – are especially significant. But this does not begin to cast any doubt on the need for the 'bifocal' approach. It is clearly one thing to appraise the logical validity of a mathematical derivation (be it in pure mathematics, or in the mathematically formulated empirical sciences) – and it is quite another thing to examine the evidential support of the premises of such derivations in the empirical siences. Even in the most elementary of Aristotelian syllogisms the distinction is glaringly obvious. "Is there a formal fallacy?", and "what is the evidence for the major and minor premises?" are wholly different questions and require entirely different considerations in order to be responsibly answered.

To be sure, W. V. O. Quine has raised serious doubts about the traditional Kantian (but also empiricist) distinction of analytic and synthetic proposition. It must be admitted that both in ordinary language, and especially in scientific theories the distinction can often be drawn only by arbitrary definitional decisions. Scientific theories, consisting of networks of nomological postulates (plus explicit definitions of derived concepts; derivations; and operational definitions of low-level empirical concepts) contain propositions which are difficult to classify as either analytic – or synthetic. Carnap[3] has, I think, fairly successfully dealt with this very question. In any case, outstanding logicians like Quine or Tarski (their occasional protestations to the contrary notwithstanding) should not have any difficulty with, or resistance to, acknowledging the radical difference between pure mathematics and applied mathematics, (as e.g. in physical theories). The observations, measurements and experiments that are indispensable in the confirmation or refutation of physical theories function in a role *toto coelo* different from whatever the observation of written symbols does for the appraisal of mathematical proofs. (I shall not digress here into a discussion of the role of electronic computers. Their philosophical significance, in this respect, is not different from that of algorithms in written or printed form.)

There is one way of reconstructing physical theories (take as a relatively simple example that of classical mechanics), in which I have – long before N. R. Hanson or Ernest Nagel – in my seminars on philosophy of physics for some thirty-five years suggested that the analytic-synthetic distinction can be retained in that a given formula, e.g. Hooke's law or Newton's second law of motion, may be given alternatively (and systematically interchangeably) the status of definitions or of laws, such that the total set of feasible alternative reconstructions helps in explicating what is at stake in various given contexts of scientific inquiry. I don't insist that this mode of reconstruction is always the most felicitous, faithful or fruitful one. We may have to get used to the idea that every single specific reconstruction has some demerits along with whatever merits it may possess. So much on the reproach of 'bifocality'.

Now I am turning to the core doctrine of classical empiricism (from Hume to Carnap and Nagel) according to which there are no synthetic *a priori* truths in the factual sciences, or in factual knowledge generally. I am not much impressed (though perhaps just a little bothered) by the examples of the phenomenologists (Husserl and his disciples) and those of Langford and others. It is claimed that once we have confronted a case of the phenomenological *a priori* we shall unfailingly know (by 'intuitive

induction') the universal and necessary truth of the proposition that
formulates such 'facts' as e.g. orange is between red and yellow; whatever
has shape, has size (and *vice versa*); that a three dimensional body with
six squares forming its surface (i.e. a cube) has twelve edges; that a given
spot cannot be red and green simultaneously etc. Even if these examples
are at first puzzling, they have been subjected to various analyses, completely consistent with the nonexistence of synthetic *a priori* knowledge.
If the worse comes to the worst, I would be willing, if the joke be permitted, to classify cases of the mentioned sort as *(horribile dictu)* analytic
a posteriori or as synthetic *a priori.* In any case, examples of this sort represent the 'puny *a priori*' compared with the 'grandiose' *a priori* of the
classical rationalists and of Kant, viz. regarding space, time, substance
and causality. The refutation of the grandiose *a priori* through the development and application (in physics and astronomy) of the non-Euclidean
geometries, the elimination of the notion of absolute substance (e.g. of
the ether by relativity theory) and the extremely serious doubts regarding
strictly deterministic causality in quantum physics (if not since 1900, then
certainly since 1926!); I say these refutations, to put it mildly, have punctured rationalistic philosophies in their most important contentions. If
historians and philosophers of science refuse to see this, then I wonder
whether it is not a waste of time even to discuss these matters with them
any more.

It should hardly be necessary to emphasize that the new empiricism is
neutral in regard to the nativism issue that agitates the psycholinguists.
Noam Chomsky and Jerry Fodor would be the first to admit that only
empirical evidence can ultimately decide what, if any, of the 'deep-lan-
guage structures' are both universal and innate. Hence the main contention of logical empiricism thus remains unscathed. To be sure, Chomsky's
work is of great interest not only psychologically and linguistically, but
also philosophically. But as a refutation of empiricism it can be taken as
relevant only to the older, seventeenth century empiricism (e.g. parts of
Locke's epistemology in contrast to doctrines of innate ideas such as
those of Plato, Descartes or Leibniz). The question of the *origin* of our
knowledge and competences (abilities) is the *old* issue of rationalism vs.
empiricism. Even though neither Hume nor Kant were consistent on the
salient points, the *new* issue of rationalims vs. empiricism was clearly seen
by them and has ever since been the battleground of the Kantians and the

modern empiricists. This new issue concerns exclusively the grounds of the validity of our knowledge claims. Hence, no amount of evidence advanced by Michael Polanyi in favor of the 'tacit dimension' is relevant to this issue. In order to prevent possible misunderstandings here, I gladly admit that as regards 'knowing how' (in Gilbert Ryle's sense), Polanyi and Gerald Holton may well be right in thinking that Einstein in 1905 did not know (or consider) the Michelson-Morley experiment. From my scanty knowledge of the historical data of the period I am inclined to believe that Einstein – through his acquaintance with the work of H. A. Lorentz – must have been at least dimly aware of the negative outcome of that famous experiment. But be that as it may, Einstein's genius consisted (in part) in his extraordinary capacity for correct conjectures. Six years before the remarkable observations on double stars by W. de Sitter, Einstein quite cavalierly assumed 'on general electromagnetic grounds' that the velocity of light coming from those stars is not influenced by their radial motions (velocities) relative to the terrestrial observers. Einstein's reliance on the general proportionality of the gravitational and inertial masses of physical bodies is another example of his ingenious intuition. A fact already clearly known to (but left uninterpreted by) Galileo and Newton thus became the empirical cornerstone of the highly speculative general theory of relativity.

Our new empiricism can thus fully agree with Leibniz' famous and brilliant saying *Nihil est in intellectu quod non prius fuerat in sensu, nisi intellectus ipse.* As we also realize from recent computer science, unless a computer has the requisite 'competence', it cannot be programmed for the corresponding tasks of calculation or problem-solving. The human brain (or, if you prefer, mind) has an amazingly rich organization which no doubt is the basis of our learning capacities and of our productive thinking. Questions of embryological development aside, surely there are innate features. They are beginning to be understood through neurophysiological and microbiological research.

Empiricism has often been accused of leading to skepticism, if not to solipsism. Philosophers should know that reproaching them for solipsism is psychologically akin (if you pardon a little psychoanalytic jargon) to the diagnosis of narcissism, if not an outright schizophrenia or paranoia. But most forms of classical as well as modern empiricism can easily be shown to be free of these vices or diseases. Skepticism and solipsism thought through consistently to the bitter end are self-refuting – in addition to being self-stultifying. As is well-known, the type of reasoning that leads to ordinary solipsism relentlessly pursued ends up with a solipsism of the present moment. This would condemn the philosopher to complete and permanent silence – and what worse fate could befall his loquacious profession?

More significant for the issues regarding the status of scientific theories is the fact that modern empiricism is no longer to be identified with positivism or operationism. In their search for a secure foundation of knowledge (really engendered by the rationalist Descartes) empiricists like Berkeley and Hume, positivists like Comte, Mill, and Mach, and operationalists like Bridgman attempted to ascribe this role to sense impressions, the immediately given, sense data (neutral elements in Mach, elementary mensurational and/or experimental procedures in Bridgman). Carnap in his early impressive *Logical Structure of the World*[4] (stimulated by Mach and B. Russell) chose a neutral basis of elementary direct experience and the relation of remembered similarity as the basis of his rational reconstruction of all empirical concepts.

After decades of criticism, these 'sense-data' and 'pointer-readings' doctrines have been largely abandoned. Some philosophers of science again flirt with Otto Neurath's suggestion that logical-reconstruction – from scratch – (i.e. from an ultimate and indubitable basis) is chimerical. Neurath himself, in order to illustrate the lines of scientific progress, used the analogy of rebuilding a ship on the high seas. Translated into what is common to Quine and Popper, this means that scientific progress (or if you prefer a more neutral designation, scientific change) consists in working from some tentatively accepted background knowledge, and by (often bold) conjectures introducing hypotheses or theories which then have to be severely scrutinized both for logical consistency and conclusiveness; as well as especially their evidential (observational-experimental) corroboration. But what precisely is to be taken as empirical evidence? Popper's basic propositions are perhaps still not so very different from Wittgenstein's elementary propositions or Carnap's experiential ones.

In the course of developments the 'incorrigibility' (i.e. infallibility) of these foundations was completely repudiated. Feyerabend, in a long sequence of essays[2], has thoroughly undermined all attempts at reconstruction on a phenomenal basis. And going far beyond this he has – buttressing his arguments with a multitude of examples from the history of astronomy and physics – energetically and indefatigably argued that all (or most) observation statements are 'theory laden', some of them even contaminated by the very theories which they are evidentially to support. Perhaps Goethe had something like that in mind when (in one of his aphorisms) he remarked: "The most important thing to remember is that whatever we call 'fact' already contains some theory."

In what follows I beg to differ sharply with this point of view. Of course, I grant – and would even insist – that there are hardly any important statements in the sciences that do not in some way presuppose a conceptual frame-work, if not some theoretical presuppositions. But I wish to point out that there are 'theories' and 'theories'. Our common sense assumptions of the relations of observer to observed events or processes, i.e. of the manner in which the observing subject (person) is embedded in the world observed are indeed taken for granted in the testing of scientific theories. But these common sense 'theories' are, as a rule, not under scrutiny when some high level theory is examined in regard to its empirical support.

By way of a brief digression I do want to state my opinion that in an *epistemological* reconstruction of the common sense view of the world (i.e. roughly what my friend Wilfrid Sellars calls the 'manifest image') something akin to a sense data analysis is indispensable. But sense data, far from being 'given', let alone being 'constructs', are 'reducts' (or 'destructs'), i.e. artificial reconstructions of the items of direct experience which can serve as substantiating evidence for our everyday knowledge of matters of fact. Consider, for example, the acquisition (i.e. the learning) of what the late, great psychologist Edward C. Tolman called a 'cognitive map' of our environment. My scanty knowledge of the topography of parts of Boston; the location of the University buildings, the surrounding streets, the River Charles, other public buildings like post offices, hotels, restaurants, banks, etc., is picked up by my visual perceptions – which in the serial order in which they came to me, form a chronological sequence, something like a movie film. A German philosopher, I remember only his last name, Gerhards (in *Die Naturwissensch.*, 1922) called such sequences 'phenograms'. Moreover it is not merely the temporal sequence of impressions (as represented in the 'phenograms') but rather – and more decisively – the repeated and repeatable observed con-

ditions → observed consequences among our impressions that form the
testing ground and the supporting (or refuting) evidence for our 'cognitive maps'. C. I. Lewis, I think, was emphasizing this in his conception
of 'terminating' judgments. It is the more or less lawful relationships
among our impressions, rather than the impressions (sense data) themselves that furnish the corroborating evidence for our knowledge of
things, topographies, and the regularity of events on the level of the
common life conception of the world.

Now, even if some critics of empiricism are not ready to accept the
foregoing sketch of an epistemology, I maintain – with much greater confidence – that the situation in the philosophy of science is largely analogous – (or perhaps 'homologous' is a more apt expression). To put it
briefly, I think that a relatively stable and approximately accurate basis –
in the sense of testing ground – for the theories of the factual sciences is
to be located not in individual observations, impressions, sense-data or
the like, but rather in the empirical, experimental *laws* (not necessarily in
metricized form). Infallibility ('incorrigibility') is *not* claimed for these
empirical laws. They, of course, are revisable not only in view of further
empirical evidence, but also in the light of theories. In the well-known
level structures of descriptions, empirical laws and theories (sometimes of
two or more layers of theories) corrections 'from above' are the rule, al
though there are exceptions as well. By 'corrections from above' I mean
such revisions as that of the Kepler laws in the light of Newtonian (and
Laplacean) perturbation theory, or the incisive re-interpretation of the
second law of thermodynamics in view of the Kinetic theory of heat and
of classical statistical mechanics, or the modification of Newton's laws of
mechanics and gravitation in Einstein's special and general theories of
relativity. These modifications, it is true, become significant obervational-
ly only for the rather extreme ranges of such relevant variables as velocities, masses, and (in cosmology) for the topology and metrics of the entire
universe. The development of the theories of the chemical bond, of the
genetic theory of evolution, and the corrections of Mendel's laws in the
light of the new molecular biology, furnish further examples.

Despite these admissions, I maintain, especially against Feyerabend,
Lakatos – and to some extent their teacher, Sir Karl Popper, that

(1) There is an important practical difference between empirical laws
and theories. This difference is epistemic, not ontological.

(2) While it may well be the case that all theories were (or are) 'born false' – i.e. that they all suffer from empirically demonstrable anomalies, there are thousands of empirical laws that – at least within a certain range of the relevant variables – have *not* required any revision or corrections for decades, – some even for centuries of scientific development.

(3) While I admit that 'theories come and go' (but nevertheless favor a realist over an instrumentalist philosophy of theories), I insist that the growth of scientific knowledge depends upon the relative (comparative) stability of empirical laws.

(4) That this is so, seems to follow from the methods by means of which scientific theories are successively 'secured' (Reichenbach's term) and often successfully established – the 'until further notice clause' of course always understood. Similarly understood is the promise of possible improvements in quantitative exactitude.

(5) The successive securing of theoretical knowledge-claims rests upon the (tentative!) reliance upon the (approximate) correctness, within the pertinent range of the relevant variables, of the empirical laws which characterize the functioning of the instruments of observation, experiment, measurement (or statistical designs).

To illustrate briefly by a few typical examples: Such theoretical principles of physical chemistry as the one of Guldberg and Waage according to which the rate of chemical reactions is proportional to the concentration of the reacting substances, may ultimately be tested by weighing the amounts of chemical elements or compounds before they are brought together in aqueous (or alcoholic, etc.) solution. The weighing itself is usually done by the use of chemical balances. The (nowadays very obvious) principle of the balance is the ancient law of the lever discovered by Archimedes. And while this law was much later recognized as a consequence of the conservation of energy principle, Archimedes' law has not been in need of any revisions for about 2200 years. The laws of light reflection and of refraction (precisely known at least since Snell's discoveries in 1621), and of geometrical optics generally (certain corrections on the 'fringes' of the relevant phenomena to the contrary notwithstanding), are still used in reflector and refractor telescopes, and in optical microscopes (and on a simpler everyday level in magnifying glasses, spectacles, etc.). Spectroscopes since Fraunhofer, Bunsen, and Kirchhoff exhibit the spectral lines of various chemical elements (or of compounds).

All those optical devices are utilized in modern astronomy (or biology); they are often of crucial importance in the testing of far-flung astrophysical and cosmological (and microbiological) theories. It should be clear that in the context of such testing the empirical ('low level') laws of optics are hardly ever questioned. Of course, I grant, that it is in principle conceivable that astrophysical theories may some day suggest revisions of optics. But I am not impressed with such purely speculative possibilities which the opponents of empiricism indefatigably keep inventing with shockingly abstruse super-sophistication! My point is very simply that thousands of physical and chemical ('low-level') constants figure in amazingly stable empirical laws. The refraction indices of countless transparent substances (such as the various types of glass, quartz, water, alcohols, etc., etc.), the specific weights, specific heats, thermal and electric conductivities of myriads of substances, the regularities of chemical compounding (or disassociation), the inverse square laws for the propagation of sound and light, similarly the Coulomb laws for magnetic and electric interaction and (after the experimental confirmations by Cavendish, Jolly, and Krigar-Menzel!) yes, even Newton's inverse square laws for gravitational forces, Ohm's, Ampère's, Biot-Savart's, Faraday's, Kirchhoff's, and so on, laws of electricity, etc., etc. are all intact, and are needed for the testing of higher lever theories. I could continue with the empirical regularities of the thermal expansion of mercury, alcohol or of various gases which have been used for a long time as thermometric substances – and (again with certain corrections 'coming from above') have served in the confirmation of important theoretical assumptions. Thus the (approximate) correctness of the empirical Boyle-Charles law – or for that matter, of Newton's law of cooling is well confirmed, and relied upon in the scrutiny of thermodynamic theories. Or I could mention that the variability of g – a consequence of Newton's theory that has been tested with the help of the empirical law of the (physical) pendulum; or that the empirically discovered formula of Balmer (in 1885) for the order of lines in the hydrogen spectrum still holds, and thirty years later became an important piece of empirical evidence for Bohr's quantum theory of the atom.

It is obvious that examples of this sort could be indefinitely multiplied. But I trust the tedium need not be prolonged. My main point, I repeat, is that while there are of course genuine revolutions in the realm of theory (shifts from one paradigm to a new one, if we speak Thomas Kuhn's language) or of research programs, be they 'progressive' or 'degenerative', to use the terminology introduced by Imre Lakatos, the empirical laws have remained amazingly stable. But I disagree with Kuhn and Feyerabend in that differences in the conceptual framework and the presuppositions of scientific theories, do not make them logically incommensurable. Not only is it always (in principle) possible to compare the logical structures and the evidential bases of differing theories, but we can also appraise their respective explanatory powers in the light of *which* and *how many* empirical laws (often qualitatively quite heterogeneous ones) can be derived from the postulates of the competing theories.

Sub specie eterni, I admit, of course, that Feyerabend has some excellent arguments for the 'theory-laden' character of (at least some) empirical laws. But since both he and I wish to do justice to the history and the practice of the scientific enterprise, it is imperative to give a faithful account of the actual procedures in the growth of scientific knowledge. Anyone who has worked in physical or chemical laboratories (I *have* for six years) can not help being impressed with the (relative) constancy and stability of the empirical laws, and their indispensability in the testing and stepwise corroboration of theories.

As I know only too well from my many discussions with Popper, and his disciples (most of them now renegades from the original position of Sir Karl), if confronted with my 'benighted' inductivism, they roll out the heavy artillery ('Big Bertha' I have come to call it) of Humean skepticism in regard to inductive inference or inductively based beliefs. Popper keeps saying that he does not 'believe in belief'[5]. But he has never deigned to answer *the* crucial question of his critics as to what justifies him to trust (or 'place his bets') on theories, hypotheses or laws that are – in *his* sense – highly corroborated. If he responds at all (and I have often pestered him with that query[6]) he says either – that is a matter of practical action (and I think this is quite compatible with Hume's psychological view of belief and 'custom'), or he says this is a piece of 'good metaphysics'. I wholeheartedly agree with the pragmatic answer and will elaborate on it shortly. But when it comes to metaphysics, Popper and I (along with most logical empiricists) have a common 'enemy' there which we both have vigorously and relentlessly criticized. I know of no criterion that Popper could use in order to demarcate good from bad metaphysics, except the one by which he distinguishes between science and non-science. In that case 'good metaphysics' would have to be continuous with science, perhaps mostly more general – but still in principle testable. But then the problem of induction returns with a vengeance to good metaphysics. As I know from many personal conversations with my good friend Karl Popper, he regards the general assumption of the lawfulness (not necessarily deterministic, according to Popper's opinion, actually indeterministic – but not chaotic, i.e. with essentially statistical regularities) of the universe as a 'good' metaphysical presupposition of the scientific quest. But this makes him vulnerable to the same sort of criticism which applies to the early 'world hypotheses' of C. D. Broad, J. M. Keynes, and the later (1948) B. Russell.

A side remark may be in order here. Popper's *modus tollendo tollens* view of falsification is highly persuasive in the case of simple syllogisms. I doubt that the much vaunted asymmetry of falsification and verification obtains in the case of scientific theories. With the late Bela Juhos I think that empirical laws in science are refuted only by better (inductively!) established alternative laws that are logically incompatible with the previous ones. Thus, for example, the regularities of the speed of light (discovered by Foucault in optically denser media) are incompatible with Newton's assumption regarding the speed of light (for instance in water or glass). This is the usual way in which assumed empirical laws are refuted, and this casts serious doubts on Poppers' asymmetry thesis.

Has Nelson Goodman (ever since his article on the 'Infirmities of Confirmation' and his book *Fact, Fiction and Forecast*) contributed to the solution of Hume's problem? I doubt it. The 'pathological' predicates like 'grue' and 'bleen' that have been discussed *ad nauseam* seem to be (after a suggestion by Paul Teller – in conversation) merely the analogon – on the qualitative-predicate level – of the well-known interpolation – extrapolation problems long familiar in connection with 'curve-fitting'. With a limited supply of discrete mensurational data there is a non-denumerably infinite number of curves (or functions) that fit the given array of points. Einstein and Popper are surely right in saying that there is no 'straight' (unique) path that would lead from the data of observation to the laws (let alone the theories!). Kurt Gödel, in an unpublished appendix to the doctoral thesis (Vienna 1928) of our dear late friend Marcel Natkin, proposed an elegant definition of the formally simplest curve (for interpolation purposes). It singles out (using variation calculus) that curve by the criterion of the minimum integral of the curvature of the curve within the given interval of the relevant variables. But, of course, as to whether 'nature' is kind enough to display that sort of simplicity is again a matter of empirical evidence. Hence this does not solve the problem of induction.

Nelson Goodman's own attempt at a constructive solution, recommends that we use only 'entrenched' predicates in our inductive projections. But 'entrenched' predicates in his sense are of the kind that have thus far proved fruitful in inductive generalizations. Hence, Goodman has at least re-stated the problem of induction but he has hardly advanced its solution.

Jerrold J. Katz, in his ambitious book *The Problem of Induction and Its Solution* (University of Chicago Press, 1962) tries to show by highly polemical – but partly defective – criticisms that there is no and there cannot be a justification of induction. In other words, Katz's solution is: there is no solution!

Rudolf Carnap in more than thirty years of impressive work on an inductive logic, had to revise, fix and mend various axiom systems in order to adapt them to the elementary requirement that they must be in keeping with the fact that we learn from experience and that 'probability is our guide in life'. Moreover, even if Carnap's inductive logic were otherwise impeccable, it is applicable to only extremely simple worlds of individuals and a finite number of predicates. Carnap hoped optimistically that his inductive logic could be generalized and extended to the quantitatively (metrically) formulated theories of science. But his much lamented death (at the age of 78) in September of 1970 terminated his work abruptly.

Imre Lakatos, perhaps Popper's keenest disciple (though nowadays a renegade) has in recent years advocated a solution in terms of what he calls 'research programs'.[7] As anyone even superficially familiar with the history of science knows very well, certain theoretical frames (I am tempted to say ideologies) have been extremely fruitful in solving problems, explaining a variety of phenomena (i.e. empirical regularities), and in suggesting new problems whose solution may be plausibly and auspiciously pursued within the same frame. On the other hand there have been theories (or ideologies) in the sciences which even despite some initial plausibility needed so much 'fixing' and 'mending' with *ad hoc* hypotheses, that they became suspect, and in the light of a new research program were displaced by a different more fruitful theory. Without entering into the otherwise important differences between Lakatos and Kuhn, such 'shifts' (of research programs) are what Kuhn characterizes as scientific revolutions or as transitions from one 'paradigm' to another. Now consider Lakatos' claims. He does not wish merely to chronicle these changes in scientific outlook. He presents his views as instruments of critique and advice to scientists. He encourages progressive problem shifts and research programs and warns against degenerative ones. But doesn't this make him *(horribile dictu!)* an inductivist after all? Doesn't he have to watch the development of scientific theories and place his bets, or (since probabilities may be unobtainable here) in accordance with how his 'race horses' (i.e. research programs) *run*? If Lakatos thus turns out to be a second level inductivist, I wonder whether he could not be persuaded to become as well a first level inductivist in regard to the validity of empirical laws.

To be sure, such justifications of induction (or inductivism) as have been presented by R. B. Braithwaite[8] and Max Black[9] are – despite their sophistication – clearly circular and thus beg the question at issue. Admittedly it is tempting to justify inductive inference by all its past successes, but David Hume, and most empiricists after him, have criticized this grievous fallacy. It is conceivable, I submit tentatively, that there is logically speaking, such a thing as 'virtuous circularity'. As C. I. Lewis so nicely put it (in *Mind and the World Order*) – a circle is the less vicious the bigger it is. So, I consider it possible, that in an all encompassing reconstruction of scientific theories together with their supporting evidence, a network of propositions may be so exhibited that the postulates justify the observational propositions *and vice versa*. A. S. Eddington may have had this sort of thing in mind in his masterly (and unjustifiably maligned) three philosophical books.[10] But no exact reconstruction of this type has thus far been worked out in any detail. ('The thing to do, is do it!') In any case certain constraints imposed by whatever be the observational evidence would seem indispensable for the adequacy of any such reconstruction.

Carnap had some limited sympathy with the approach of D. C. Williams in *The Ground of Induction* (Harvard University Press, Cambridge, Mass. 1947). The essential point here (originally suggested by C. S. Peirce) is the inference from sample to population. Williams thought he could derive inductive probabilities converging toward practical certainty in this manner. But he overlooked one essential precondition. Without some uniformity assumption (something like permanent kinds and limited variety à la Broad or Keynes) no probabilities can be derived. Nevertheless I think there is a kernel of truth in Williams' ideas. I think the much discussed pragmatic justification of the adoption of an inductive rule, the reasoning that Hans Reichenbach,[11] and I have advocated (my term 'vindication' of inductive inference has been widely used, e.g., by Wesley Salmon, Jerrold Katz et al.) can be given an alternative formulation in terms of 'sampling'. What I have in mind is the following simple thought: Consider observations, measurements and experimentation (and, of course, statistical designs) as a way of obtaining samples of the regularities of nature. Now of course without further assumptions (or prior probabilities – as used by the Bayesians and the 'personalists') there is absolutely no way of estimating the probability of the 'fairness' (representative character) of the sample. But consider how '*ir*rational' it would be to assume without any reason whatever that the samples are unrepresentative. (Of course they may well be unrepresentative in fact, in one way or another.) Yet, to *assert* that the samples *are un*representative opens the floodgates to limitless possibilities. It seems to me that assuming (until further notice!) that a given sample *is* representative is tantamount to one of the more important formulations of the requirement of simplicity. It secures the uniqueness – with respect to previous experience – which in many discussions has been seriously questioned.

Let me candidly state that this sort of vindication (just as in Reichenbach's or my own earlier formulations) must appear weak to anyone who expected a 'genuine solution' of Hume's 'grand problem' of the justification of induction. Compared with the subjective practical certainty that we attach to the inductive and analogical inferences on which we base the countless expectations of everyday life, and the best established empirical laws in the sciences, the pragmatic justification looks very pale indeed. But here I am inclined to use, for once, some 'Oxbridge' type philosophy: If this sort of justification (of the adoption of a principle of induction) seems disappointing to you, what else on earth (heaven or hell) would you possibly suggest? Even if we had a guarantee from God (or from Laplace's demon) that the world is strictly deterministic (and no such guarantee will be forthcoming!), scientists will always realize (or should at least be willing to admit in principle), that there may be as yet unknown parameters in our galactic region of space-time which are absent or completely different in other parts of the universe. No assurance can be given that the regularities of the world are unchanging. Entirely different empirical laws may have to be adopted tomorrow (or even beginning with the next second!).

If it is the major aim of the factual sciences to understand – in the sense of explaining the facts, events, processes (and their regularities) of nature, and as long as scientific explanation consists of the hypothetico – inferential reasoning (from deterministic or statistical) premises, then we had better hold on to whatever regularities have been empirically established.

The one concession I (still as an empiricist!) gladly make to Kant (and some of the neo-Kantians) is that our understandings of nature is predicated upon the well-known patterns of explanation, i.e., of derivation – deductive or probabilistic, as the situation may require – from (tentatively) accepted premises. We have not a glimmer of an idea what else understanding or explaining could consist of. It seems plausible that this basic feature of human reason (mind or brain) is thus a logically contingent, but extremely important feature of our world. The forms of intuition and of the understanding that Kant himself considered necessary conditions for *Erkenntnis überhaupt,* have had to be loosened up tremendously with the advances of recent science. Nevertheless, it is just conceivable that there are limits of our understanding, owing to the structure of our minds (or brains). There is nothing sacrosanct about the empirical laws of science. But I have tried to point out that very many of them have stood the test of time and of many searching criticisms. If the data of observation should surprise us (and of course I admit they have frequently done just that, think, for example, of the discovery of radioactivity or of the isotopes), then it is time to revise our laws and theories.

Popper, Feyerabend, Lakatos and many other brilliant philosophers of science, have understandably fixed their attention upon such dazzling and exciting high level theories as those of Newton, Einstein or of the quantum physicists (Bohr, Heisenberg, Schrödinger, Born, Dirac, von Neumann *et al.*). They have rarely condescended to the 'low' level of experimental laws. Of course, I agree with them, high-level theories are much more fascinating. But if Popper thinks that his critical approach helps us to understand how science can get 'nearer to the truth', the neglect of the empirical laws prevents him from giving a convincing account of the progress of knowledge. Popper relies on the well-known Tarski-Carnap semantic definition of truth. But this definition merely clarifies what we *mean* by the word 'truth'; as Popper admits, it does not provide him with a criterion of truth or of an explication of 'nearer to the truth'.

Closer attention to the actual procedures of the scientists does show that they do use very simple, almost homely, criteria for the truth and the approximation of the truth by laws or theories. A theory is 'near to the truth' if it enables us to derive a large number of well-established empirical laws. And empirical laws are well-established or 'close to the truth' if repeated experiments and measurements, if possible performed under varying conditions, yield practically convergent series of values for the natural constants that are pivotal in the empirical laws. To take at random one of thousands of examples: the speed of sound in air of normal temperature and pressure (on the surface of the earth) is approximately 1087 feet per second. Repeated measurements will show a practical convergence toward this (non-mathematical!) limit – once the theory of the errors of measurement has made the necessary corrections.

Finally, just a few words about Popper's criterion of demarcation. Empiricists can be quite happy with his distinction of scientific and non-scientific questions. The logical positivists took testability as a criterion of meaningfulness, but as Carnap has pointed out already in 1934 this was a mere proposal, to be judged by its fruitfulness. What is not so clear in Popper is an additional distinction that the positivists *did* make, namely between scientific and *un*scientific knowledge claims. Though I am extremely wary and skeptical of the claims of Psychical Research (ESP), I would certainly not reject them as meaningless. But I can afford to wait for more convincing evidence before I am persuaded that we have here a genuinely scientific enterprise. In other words, until further notice I go at least through the 'motions of an open mind'. Astrology, palmistry, alchemy, and phrenology, however, are, I think, rightly regarded as *un*scientific because all (or most) of the available evidence speaks harshly against these fake disciplines. In regard to some recent hypotheses in astrophysics, cosmology and of nuclear theory, however, the evidence is still too weak to decide whether or not they will end up on the scrap heap of failures of – however highly respectable – approaches toward honest science.

Psychology and the social sciences present problems and difficulties of their own. It is my considered opinion that nothing like a Newtonian or Einsteinian synthesis is here to be expected. I am inclined to agree with B. F. Skinner, whose search for reliable empirical laws (mostly statistical ones) of learning and motivation, and his studies of schedules of reinforcements, have been remarkably successful. He does not, like his predecessor in the behavioral sciences Clark L. Hull, look for all-embracing psychological theories in molar-behavioral terms. As Sigmund Koch has shown in his devastsating critique of Hull's work, it is both logically and empirically extremely questionable. I am, however, disappointed by B. F. Skinner's ignoring (if not denying) the possibility of progress in psychology through advances in neurophysiology. (Some of the contributions in that field already have considerable explanatory power.) And I am even more gravely disappointed and dismayed with S. Koch (my former brilliant student of ca. 1938 at the University of Iowa) who has lately adopted what I can only call the obscurantist stance of Polanyi, Kuhn and Feyerabend. An unprejudiced closer look at the actual research in psychology seems to me to show that we do not need any 'holistic' psycho-social-economic theories of the relation of observers to observed organisms, for fruitful theory construction. How a rat turns in a T-maze, and with what frequency how, and at what rate, it presses the lever on a Skinner box, or the words uttered by a person in psychoanalytic interviews, etc., etc. can be recorded by reliable machines. But this is a large topic and it is getting late.

Let me say by way of conclusion that empiricism, though in need of renovation, will remain a fruitful and adequate philosophy of science.[12]

Postscript. In order to forestall misunderstanding – such as involved in the reproach that I cannot 'have it both ways' – let me briefly clear up the essential point. Of course I *can* have it both ways: Thousands of empirical laws in physics and chemistry have *not* been in need of revision (except for minor adjustments in quantitative exactitude). Nevertheless, I admit, of course, that in quite a few (perhaps a few dozen) cases, empirical laws received incisive 'corrections from above', i.e. in the light of new and well-established theories. The empirical laws of Newtonian mechanics, and the empirical laws of classical electrodynamics *have* been drastically modified. But the thousands of physical or chemical constants that figure in the vast majority of the 'low-level' empirical laws, and are listed in such tables (as in Landoldt-Boernstein years ago) are pivotally characteristic of the empirical laws which remained essentially unchanged!

University of Minnesota

NOTES

[1] H. Reichenbach, *Experience and Prediction,* University of Chicago Press, 1938; and H. Feigl 'Beyond Peaceful Coexistence', in *Historical and Philosophical Perspectives of Science* (ed. by R. H. Stuewer) *Minnesota Studies in the Philosophy of Science* (H. Feigl and G. Maxwell, General Editors), University of Minnesota Press, Minneapolis, 1970.

[2] N. R. Hanson, 'Logical Positivism and the Interpretation of Scientific Theories', in *The Legacy of Logical Positivism* (ed. by P. Achinstein and S. F. Barker), Johns Hopkins Press, Baltimore, 1969.

Even more critical of the empiricist and positivist philosophy of science: P. K. Feyerabend, 'Problems of Empiricism', in *Beyond the Edge of Certainty* (ed. by R. G. Colodny), Prentice-Hall, Englewood Cliffs, N.J., 1965; and 'Problems of Empiricism, Part II', in *The Nature and Function of Scientific Theories* (ed. by R. G. Colodny), University of Pittsburgh Press, 1970; and 'Philosophy of Science, a Subject with a Great Past', in *Minnesota Studies in the Philosophy of Science,* Vol. V (ed. by R. H. Stuewer), University of Minnesota Press, Minneapolis, 1970. Equally challenging, P. K. Feyerabend, 'Against Method: Outline of an Anarchistic Theory of Knowledge', in *Analyses of Theories and Methods of Physics and Psychology* (ed. by M. Radner and S. Winokur), *Minnesota Studies in the Philosophy of Science,* Vol. IV (H. Feigl and G. Maxwell, General Editors).

[3] *The Philosophy of Rudolf Carnap,* Library of Living Philosophers (ed. by P. A. Schilpp), Open Court Publ. Comp., 1963, pp. 915–923; cf. also: R. Carnap, *Philosophical Foundations of Physics,* Basic Books, New York, 1966 (especially Chapters 27 and 28).

[4] R. Carnap, *The Logical Structure of the World* (English translation by R. A. George), University of California Press, Berkeley and Los Angeles, 1969, originally published as *Der Logische Aufbau der Welt,* Weltkreis Verlag, Berlin-Schlachtensee, 1928 (later republished by F. Meiner, Leipzig).

[5] K. R. Popper, 'Conjectural Knowledge: My Solution of the Problem of Induction', in *Revue Internationale de Philosophie* **95–96** (1971), 167–197.

[6] H. Feigl, 'What Hume Might Have Said to Kant', in Popper's *Festschrift, The Critical Approach to Science and Philosophy* (ed. by M. Bunge), The Free Press of Glencoe, and Macmillan, London, 1964.

[7] *The Problems of Inductive Logic* (ed. by Imre Lakatos), North-Holland Publishing Co., Amsterdam, 1968, *Criticism and the Growth of Knowledge* (ed. by Imre Lakatos and and Alan Musgrave), Cambridge University Press, 1970.

20   HERBERT FEIGL

[8] R. B. Braithwaite, *Scientific Explanation*, Harper & Row, New York, 1960.
[9] Max Black, *Models and Metaphors*, Cornell University Press, Ithaca, New York, 1962.
[10] A. S. Eddington, *The Nature of the Physical World*, Cambridge University Press, 1928. *New Pathways of Science*, Ann Arbor Paperbacks, University of Michigan Press, 1959. *The Philosophy of Physical Science*, Ann Arbor Paperbacks, University of Michigan Press, 1958; cf. also: Joh. Witt-Hansen, *Exposition and Critique of the Conceptions of Eddington Concerning the Philosophy of Physical Science*, G. E. C. Gads Verlag, Copenhagen, 1958; and the positivistically biased critique in L. S. Stebbing, *Philosophy and the Physicists*, Methuen, London, 1937; paperback, Dover 1958.
[11] H. Reichenbach, *The Theory of Probability*, University of California Press, Los Angeles, 1946 (see especially the exciting last chapter of this book). More recent statements regarding this 'vindication' of induction may be found in H. Feigl, 'De Principiis non Disputandum?', in *Philosophical Analysis* (ed. by Max Black), Prentice-Hall, 1963, and Wesley C. Salmon, 'Should We Attempt to Justify Induction?', in *Philosophical Studies* **8** (April 1957). But cf. also his later excellent and extensive, penetrating discussion and scrutiny in 'The Foundations of Scientific Inference', in *Mind and Cosmos* (ed. by R. G. Colodny), University of Pittsburgh Press, 1966 (also separately available as a slender book).
[12] For a modified statement of modern empiricism in the philosophy of science, cf. C. G. Hempel, 'On the "Standard Conception" of Scientific Theories', and a qualified argument in favor of traditional empiricism, H. Feigl, 'The Orthodox View of Scientific Theories: Remarks in Defense as well as Critique', in *Minnesota Studies in the Philosophy of Science* (ed. by M. Radner and S. Winokur), Vol. IV, University of Minnesota Press, Minneapolis, 1970. Also, H. Feigl, 'The Origin and Spirit of Logical Positivism' in the important book *The Legacy of Logical Positivism* (ed. by P. Achinstein and S. F. Barker), The Johns Hopkins Press, Baltimore, 1969.

For a brilliant and 'unorthodox' critique of traditional empiricism, as well as of some aspects of K. R. Popper's views on the logic of science, cf. Grover Maxwell's 'Corroboration without Demarcation', in *The Philosophy of Karl Popper* (ed. by P. A. Schilpp), *Library of Living Philosophers*, Open Court Publishing Co., La Salle, Illinois, forthcoming (soon).

A very lucid and judicious discussion (with ample references) is contained in J. J. C. Smart, *Between Science and Philosophy*, Random House, New York, 1968.

The literature on this subject is growing so vast that I must stop giving references. There are extremely sane and sound papers by Ernest Nagel, 'Theory and Observation', in *Observation and Theory in Science*, The Johns Hopkins Press, Baltimore, 1971, and by Dudley Shapere, 'Meaning and Scientific Change' in *Mind and Cosmos* (see above).

Quite recently (August 1972) a most brilliantly sound article came to my attention: Carl R. Kordig, 'The Theory-Ladenness of Observation' in *Review of Metaphysics* **24** (March 1971), 448–484. This, I think is the very best and concise critique of the ideas of Feyerabend, Hanson, Kuhn, and Toulmin. I wish I had known about this article sooner.

For an excellent and very thorough analysis and discussion of this and closely related issues cf. also: Adolf Grünbaum, 'Falsification and Rationality, forthcoming in the *Proceedings of the September 1971 International Colloquium on Issues in Contemporary Physics and Philosophy of Science* (held at Pennsylvania State University), (ed. by J. Kockelmans, G. Fleming and, S. S. Goldman).

ERNAN MCMULLIN

EMPIRICISM AT SEA

Ought one represent the Quine-Kuhn-Hanson-Feyerabend-Lakatos-Po-
lanyi rejection of the logical positivist account of science that dominated
the world of English-speaking philosophy in the decades between 1930
and 1960 as a rejection of empiricism itself? Is it permissible to group
these critics together as though the differences between them in this context were not significant? These are the first questions that are likely to
leap to the reader's mind as he peruses Professor Feigl's spirited defence
of a threatened orthodoxy. As he reads on, another question may begin
to nag him a little too. From the beginning, Professor Feigl concedes
that "drastic revisions are in order". Indeed, he admits so many of these
revisions as his essay progresses that one is not altogether sure finally
where to locate the substantial disagreement he believes to exist between
him and those whom he is attacking. His aim, he says, is to defend "the
central tenets of classical as well as of modern (or 'logical') empiricism"
against the mistaken objections of 'the critics of empiricism'. But as one
notes the concessions he makes to these same critics, one cannot but ask
whether many of the central tenets of these two forms of empiricism are
not in fact being abandoned, and whether the critics are *really* critics of
empiricism, as such, or only of two historical forms of it associated with
18th century Britain and 20th century Vienna respectively. In short, is it
really *empiricism* that is at bay? And has not Professor Feigl, in the light
of the concessions he makes, become one of the hunters himself, displaying (incidentally) an admirable openness of mind thereby?

To disentangle the complex of issues lying behind these questions, it
will be necessary to separate the strands of the noose that in the last
decade has tightened around the theory of science associated with logical
positivism in its heyday. It will be convenient to distinguish between three
facets of that theory which we can identify by the loose titles: logicism,
foundationalism and inductivism, or the L-thesis, the F-thesis and the
I-thesis, respectively. The 'critics' assailed by Feigl differ somewhat regarding these theses; Popper accepts the first, and rejects the second and

especially the third; Kuhn rejects the first and second, and has nothing specific to say about the third; Polanyi rejects the first and the third and not much to say about the second; Feyerabend and Toulmin reject all three. Feigl himself has been led to reject the second, but wishes to retain the first and third. Since the three theses are intricately interconnected, it will be important to determine whether if one is abandoned, the others *can* be maintained without substantial modification.

I. LOGICISM: THE L-THESIS

This thesis, which is central to the great 'classical' theories of science (whether Aristotelian, Cartesian or positivist), is to the effect that the theory of science itself *is* a logic, and that the goal of the philosopher of science is thus to disclose its structures (deductive according to Aristotle and Descartes, inductive for Hume and Carnap). Science is a *justified* form of knowledge; what marks it off is the kind of justification used. This justification must, in principle, be capable of expression in formal rules of inference (i.e. a logic) since there is question of moving from what is known to something in need of justification, in virtue of a procedure capable of specification, broad application and independent validation.

Associated with this thesis is the assumption that a sharp distinction can be drawn between the contexts of justification and of discovery, and that only the former is of concern to the philosopher. To the extent that recurrent patterns are noticed in the processes of discovery in science, these are either part also of the processes of justification (in which case the philosopher will study them) or else they are not (in which case he will leave them to psychologists, historians, sociologists...). Feigl equates this distinction with the one sometimes drawn between 'external' and 'internal' history of science. But they are not the same; those who drew the former distinction set all *historical* issues (whether external or internal) firmly on the side of discovery and encouraged philosophers to ignore them. The 'internalist' emphasis favored by historians of ideas is by no means equivalent to a logicist one; though logical factors are among those they regard as 'internal' to the issues studied, the historical development of concepts would never be reduced to formal terms of logical implication alone.

The discovery-justification dichotomy was challenged by Hanson, who claimed not only that discovery has its characteristic cognitive patterns, constituting a 'logic' in a broader sense, but also that justification cannot be reduced to any explicit set of formal rules. Polanyi carried this argument further by noting the 'tacit' components of knowing; he rejected the L-thesis on the grounds that both discovery and justification in science demand intuitive and patiently-acquired skills of recognition that are in principle irreducible to itemization in terms of explicit criteria or formal rules. Kuhn came at the problem from a different quarter; arguing both as historian and as sociologist, he claimed that the structure of scientific revolutions is not a logical one. The shift from one paradigm to another cannot be made rational (in the sense of coercive) by means of previously-agreed-upon methodological procedures. Toulmin is the most recent critic of what he calls 'logicality'. In his view, the rationality of science in no way derives from the presence within scientific procedures of an implicit logic; rather, it is disclosed by the ways in which conceptual change characteristically occurs, somewhat after the fashion in which genetic change occurs in a population of organisms.

Feigl wishes to maintain the L-thesis, from which the adjective in the label 'logical positivism' originally derived. To do so successfully, he would have to meet the wide variety of arguments brought against it by these writers. He alludes, in fact, to only one of these arguments, and in doing so concedes that something like a gestalt switch may occur in moments of theoretical change, but then proceeds to banish it from the consideration of the philosopher. For this to be justified, however, he would have to *show* that integrative acts of assessment play only a heuristic role in science and are completely replaceable at the level of 'justification' by an inductive or HD warrant. At the most general level of theory, this does not appear to be the case, crucial though these latter types of warrant are. The assurance of an Einstein or a Darwin does not, in fact, rest entirely upon his ability to justify his claim in terms of induction or of successful prediction. And there is more to it than a "capacity for correct conjectures" (Feigl's phrase, p. 5); there is a learnt (empirical) skill of appraisal at work, even after the conjecture has been validated. Feigl notes the 'ingenious intuition' of an Einstein who saw the significance of the proportionality of gravitational and inertial masses. But to see this (rather than something else) as significant was not just important as a clue suggesting the original conjecture, it was still part of the 'reason' (in the proper sense) why Einstein saw in his completed theory the best theory.

II. FOUNDATIONALISM: THE F-THESIS

A second constituent in all 'classical' philosophies of science, Aristotelian, Cartesian, Humean, Carnapian, is the assumption that science must begin from an indubitable foundation of some sort, whether first principles or sense-datum statements or whatever else. Without such a foundation, scepticism seemed unavoidable, or at the very least, progress in science (it was thought) could never be assured. The positivist reconstruction of science made it depend on protocol statements, themselves supposed to be in no danger of later modification of any sort.

The F-thesis had already been criticized by Peirce and the pragmatists long before the Vienna Circle assumed it as one of their basic principles. Yet, as Feigl notes, even among the members of the Circle there was not complete unanimity regarding it. Neurath, for instance, rejected it out of hand. Quine later continued his line of criticism. Hanson, and especially Feyerabend, developed new arguments against it, and (in Feigl's own words) "thoroughly undermined" it, so that it has been "largely abandoned" (p. 6). Indeed, Feigl so far agrees with Sellars' well-known critique of the "myth of the given" that he is willing to treat the sense-data supposedly reported in observation-statements as "artificial reconstructions of the items of direct experience" though they can nevertheless "serve as substantiating evidence for our everyday knowledge of matters of fact" (p. 7), the evidence here residing rather in the more or less law-like relationships between the impressions than in the impressions themselves (p. 8).

In the light of these concessions, it is puzzling that Feigl should also claim that he "differs sharply" with Feyerabend's view that "all (or most) observation-statements are 'theory-laden', some of them even contaminated by the very theories which they are evidentially to support" (p. 7). There are really two theses here. The former (that observation-statements are theory-laden) Feigl has already by implication accepted as his reason for abandoning foundationalism. But perhaps it is only the second (the 'vicious circularity' objection) that Feigl is rejecting, in which case he is in good company since none of those whom he is criticizing, except

Feyerabend, maintains this position. It is well to be clear that the 'critics' as a group do not subscribe to the subjectivist thrust of this and others of Feyerabend's arguments.

Feigl is obviously uneasy about the loss of epistemological security that the abandonment of the F-thesis entails. He keeps implying that in a way he hasn't *quite* given it up, using qualifiers like 'at least some', 'as a rule', 'hardly any', to underline that in his view one can still get pretty close to something that is ' perhaps still not so very different" from the foundational propositions of Wittgenstein and Carnap. But the F-thesis is one of those you can't give up just a teeny bit; if the 'foundation' is just the least bit 'tentative', 'corrigible' (p. 8), then an entirely new situation arises because the criteria for a reconstruction of the 'foundation' are not going to be either deductive or inductive, nor given in any formal way in advance. As the essay progresses, Feigl warms a little more to the "virtuous circularity" (p. 14) of the epistemology he is being forced to embrace, and concludes with an anti-foundationalist suggestion that even Feyerabend could applaud:

"I consider it possible that in an all-encompassing reconstruction of scientific theories together with their supporting evidence, a network of propositions may be so exhibited that the postulates justify the observational propositions and *vice versa*" (p. 14).

III. INDUCTIVISM: THE I-THESIS

This last suggestion is seen by Feigl as a possible solution to the problem of induction. He briefly reviews and rejects other solutions, and appears to assume that a proper solution will also serve as a justification of inductivism (p. 14). But this is by no means the case. Though philosophers may disagree about the significance of induction, none would query the propriety of the sort of generalizing from limited samples that goes on in science under the name of induction. Inductivism on the other hand is a theory of scientific method which proposes induction as the mode of inference proper to science, understanding it as a progression from non-problematic observational instances to problematic generalizations. The generalizations are regarded as revisable 'from below', as accuracy and range of observation improve. Hypothesis and deduction play secondary roles; an hypothesis is confirmed only if verified inductive laws can be deduced from it. Induction itself is assumed to follow an implicit logic which is in principle capable of formalization so as to yield numerical estimates of probability or likelihood.

The inductivist model was much attacked in the late nineteenth century but was adopted by many logical positivists, and became more or less official doctrine. From the beginning, it caused problems, notably in the context of explanation which does not reduce easily (most would now say: it does not reduce at all) to inductivist terms. Criteria that clearly play an important role in the justification of theory, such as simplicity and fertility, do not lend themselves to statement in inductivist terms. The goal of attaching a numerical probability estimate to laws and theories seems dubious, many would say misguided. These difficulties were well-known before the dawn of the Kuhnian era, and were quite enough on their own to render inductivism unacceptable.

Feigl still calls himself an "inductivist" but he is using the term in a broad sense, as the last quotation above shows. He admits that laws are frequently revised "from above", i.e. from the level of theory (p. 8). He admits that observation-statements depend in part for their warrant upon the theories presupposed in their formulation, so that if these theories change, the observation-statements may shift also. These kinds of revision cannot be treated in purely inductivist terms. In fact, by 'inductivist' he appears to mean nothing more than: maintaining the validity of empirical laws (i.e. that we can tentatively rely on their approximate correctness, pp. 9, 14), or even more broadly: requiring that research-programs be evaluated in terms of their performance in prediction (p. 14). In both of these senses, *all* philosophers of science are 'inductivist', so it ceases to be a helpful label.

Feigl wishes to shift his inductivism one level upwards: the testing-ground of theory "is to be located not in individual observations, impressions, sense-data or the like, but rather in the empirical, experimental *laws*" (p. 8). He reminds us at some length of the impressive stability of the thousands of laws of chemistry, biology, physics, and argues that the growth of scientific knowledge depends on this stability. I think he is right in this, and right too in suggesting that not enough attention has been paid to this quite central feature of science in recent discussions of scientific change. But he is surely taking Popper's well-known strictures on induction too literally if he thinks that Popper would disagree with him regarding the relative permanence of the myriad empirical laws closest to the level of observation and experiment in each branch of science. Popper may stress the 'conjectural' aspect of the laws and attack the view that they are arrived at by something called 'inductive inference'. But he is as emphatic as is Feigl himself about their existence and also (another point Feigl stresses) their relational character:

By choosing explanations in terms of universal laws of nature. we offer a solution to precisely this last (Platonic) problem. For we conceive all individual things, and all singular facts, to be subject to these laws. The laws (which in their turn *are* in need of further explanation) thus explain regularities or similarities of individual things or singular facts or events. And these laws are not inherent in the singular things .... Laws of nature are conceived, rather, as (conjectural) descriptions of the structural properties of nature.... Our 'modified essentialism'... suggests that our laws or theories must be universal that is to say, must make assertions about the world – about all spatio-temporal regions of the world. It suggests, moreover, that our theories make assertions about structural or relational properties of the world.[1]

Professor Feigl separates law and theory rather sharply, and takes law as his paradigm in characterizing the method and status of science; Popper sees law and theory in a spectrum, and characterizes science rather in terms of the paradigm of theory. This is not an unimportant difference; it is, however, largely a matter of emphasis, and certainly not an issue to be resolved by simple recourse to large-scale traits of science like the relative permanence of low-level empirical laws or the conjectural character of high-level theories.

IV. INCOMMENSURABILITY

Feigl appeals to the stability of laws once again in criticizing the incommensurability thesis of Kuhn and Feyerabend. Commensurability, i.e. the possibility of putting competing theories in direct confrontation with one another, was a direct consequence of the F-thesis, one that was normally taken for granted without discussion. In a well-known section of *The Structure of Scientific Revolutions*, Kuhn argued that during a scientific revolution, "proponents of competing paradigms must fail to make complete contact with each other's viewpoints." This "incommensurability of the pre- and post-revolutionary normal scientific traditions" derives from three main sources, disagreement "about the list of problems that any candidate for paradigm must resolve," lack of fit between the competing sets of concepts, and the fact that "the two groups of scientists see different things when they look from the same point in the same direction." Consequently:

before they can hope to communicate fully, one group or the other must experience the conversion that we have been calling a paradigm-shift. Just because it is a transition between incommensurables, the transition between competing paradigms cannot be made a step at a time, forced by logic and neutral experience. Like the gestalt switch, it must occur all at once.[2]

The abandonment of the F-thesis clearly poses a problem for commensurability; if there are no neutral and irrevisable 'basic' propositions (and Feigl himself admits this), then theories are no longer commensurable in the way in which it was thought they were. The real issue is, therefore, not commensurability as such, but whether one can define a new and a more complex sort of commensurability sufficient to allow and retain the notion of science as a rational enterprise. It is this issue more than any other that has embroiled philosophers of science in such passionate debate in the past decade. Feyerabend is, perhaps, the only one who would push incommensurability to the limit. In his most recent articles, he seems to be saying that there are no methodical ways in which one can go about resolving basic theoretical differences; one has to rely on all the variety of means of persuasion at one's disposal, propaganda and trickery not excluded.

Kuhn, on the other hand, in the Postscript to the new edition of his book, insists that the 'conversion' he spoke of is one based on 'good reasons', except that "such reasons function as values and...can thus be differently applied, individually and collectively, by men who concur in honoring them."[3] The competing theories are incommensurable in the basic sense that "there is no neutral algorithm for theory-choice, no systematic decision procedure which, properly applied, must lead each individual in the group to the same decision." What the philosopher has to come to grips with, then,

is the manner in which a particular set of shared values interacts with the particular experiences shared by a community of specialists to ensure that most members of the group will ultimately find one set of arguments rather than another decisive.[4]

Feigl would object to this formulation of the problem. He is willing to surrender the F-thesis, but argues that competing theories do not in consequence become 'logically incommensurable':

Not only is it always (in principle) possible to compare the logical structures and the evidential bases of differing theories, but we can also appraise their respective explanatory powers in the light of *which* and *how many* empirical laws...can be derived from the postulates of the competing theories (p. 11).

He does not provide us with any detail as to how this is to be done. And this is what is needed, because Kuhn and the others *have*, in fact, done very detailed analyses of historical instances to substantiate their point of view. Kuhn would say that the sort of comparison and appraisal of which Feigl speaks is possible only during the long periods of normal science, and that the plausibility of Feigl's case depends on excluding from consideration (or at least treating as exceptional) the 'revolutionary' moments of theoretical shift in science. The relative stability of low-level empirical laws would, of course, be conceded by Kuhn, but he would maintain that inspection of such laws could not of itself suffice to decide cases of paradigm-shift on the basis of some simple consideration of 'which and how many'.

V. VARIETIES OF EMPIRICISM

Professor Feigl sees himself as a defender of empiricism against a horde of attackers coming at that ancient and honorable doctrine from all sides. And indeed no one has better earned the right to be its defender. But there are two inescapable objections against that way of presenting his battle. First, empiricism is one of the most commodious of all philosophic mansions; so large is it, indeed, that few ever find their way outside it. None of the 'attackers' (save, perhaps, Feyerabend) would be content to have himself described as as anti-empiricist, *tout court*. Second, Feigl's own abandonment of the F-thesis means that he has rejected 'classical' empiricism. So that it is not at all clear that he is any more (or less) empiricist than those he is attacking.

He contrasts his 'new empiricism' with the 'older empiricism' whose doctrines on the origin of our knowledge have been refuted, in his view, by Chomsky (p. 4). The 'new empiricism' concerns "exclusively the grounds of validity of our knowledge claims" (p. 5); it is not a theory of sensation, but rather a view of the nature of scientific proof. It retains the 'core doctrine' of the older empiricism, however, namely that "there are no synthetic *a priori* truths in the factual sciences" (p. 3). A question immediately occurs to the reader. Who *would* hold that there are synthetic *a priori* truths in the physical sciences? Certainly not Kuhn, Polanyi, and the others criticized here. They have attacked this view even more effectively than did the logical positivists. Indeed, they would even challenge the analytic-synthetic distinction on which the formulation of the entire problem rests. Professor Feigl seems to have the phenomenologists in mind, and argues that 'rationalistic philosophies' have been punctured by recent developments in science, warning those "historians and philosophers of science (who) refuse to see this" (p.4).

But who *are* these? Surely empiricism can no longer be defined as it once was by opposition to 'rationalism'. There is no "new issue of rationalism vs. empiricism", as Feigl seems to suppose (p. 4). In particular, it would be altogether misleading to suggest that Polanyi, Kuhn *et al.* have somehow inherited the mantle of the rationalists. In a very real sense, Feigl himself is much closer in spirit to the rationalists because he shares with them the crucial L-thesis on which the 'critics' have made their strongest attack. The real dichotomy here is no longer between (classical) empiricism and rationalism, but between both of these doctrines and the newer post-Hegelian views which would challenge the basic assumption of *all* classical theories, Aristotelian, Cartesian, Humean, positivist, that the objectivity of science lies in a neutral methodology, capable of being made explicit in a set of logical rules.

To make 'empiricism' the banner-emblem is thus incorrect. It is the L-thesis that is at stake, not empiricism as such. Polanyi, for example, will insist that the intuitive skills of 'personal knowledge' on which scientific judgement (in his view) critically rests are *learned* skills. They are not innate; they are gained by the long and patient experience of the apprentice. It is true that he repudiates the phenomenalist doctrine of sensation of classical empiricism. But then so does Feigl, albeit with many a backward glance of regret. Polanyi is empiricist in very much the sense in which Aristotle was. He is not appealing to an *a priori*, or to some source of knowledge other than in expereince. Likewise, Kuhn (as we have seen) is arguing for an interaction of a set of shared values (such as accuracy, fertility, simplicity) with "the particular experiences shared by a community of specialists". His emphasis on the metaphor of conversion led many critics to claim an element of the arbitrary in the final option for or against a paradigm. In the Postscript, already quoted, he stresses the role played by 'good reasons' and successful efforts at translation, but still maintains that "neither good reasons nor translation constitute [i.e. are logically sufficient to account for] conversion."[5] What one must ask for here is, of course, the source of the 'residue' in decisions to adopt a new paradigm or theory. To my mind, Kuhn has still not quite declared himself in this regard. Does it lie in the common learning experience of the group (as Polanyi and Toulmin would, I think, say)? Does it involve arbitrary subjectivist elements that can, however, be continually reduced by a disciplined fidelity to experience? Or are such elements *inescapably* – and properly – present, as Feyerabend and some of the neo-Marxist theorists of science would say?

The 'new issue' (not really all that new) is thus not empiricism vs rationalism, but empiricism vs subjectivism, with most of the 'new critics' (despite appearances) on the side of empiricism, but with the issues so far insufficiently analyzed to allow clear-cut debate. What has occupied much of the attention of the critics of the new challenge up till now has been the defence of the L-thesis. But what has to be clearly grasped is that the L-thesis is *not* synonymous either with objectivity or with empiricism. It is possible to reject it and still maintain objectivist and empiricist views of science. The task that lies ahead for philosophers of science has got to be a more and more detailed analysis of the history of science with a view to determining just what sort of rationality it *does* exhibit, particularly in its moments of fundamental conceptual change. Abstract formulations of a normative logic of science prior to any consideration of actual scientific practice has to be considered the last vestige of an older rationalism (even though it may be found mainly among those who would call themselves empiricists). What we need, then, is a reformulation of empiricism in the light of the challenges thrown it by Feyerabend, Habermas, Kuhn, and the rest. It will have to be considerably more flexible than the older doctrine; it must be prepared above all to accept a far richer notion of experience, to be documented by appeal to historical materials.

Empiricism is thus not really 'at bay', as Professor Feigl's title suggests. It is at sea, rather, but still very much afloat, with the storm raging around it. But then, that is the way it should be, unless one is to hold that the notion of experience can at some point in time receive its definitive formulation. Judging by the singular lack of success in regard to such a formulation in the past, it would be wise not to foreclose the future too readily.

University of Notre Dame

NOTES

[1] 'The Aim of Science', in *Objective Knowledge*, Oxford 1972, pp. 196–7.
[2] *Op. cit.*, Chicago, 2nd edition, 1970, pp. 148–50.
[3] *Op. cit.*, p. 199.
[4] *Op. cit.*, p. 200.
[5] *Op. cit.*, p. 204.

PHILIP L. QUINN

WHAT DUHEM REALLY MEANT

> It is the nature of an hypothesis, when once
> a man has conceived it, that it assimilates
> every thing to itself, as proper nourishment;
> and from the first moment of your beget-
> ting it, it generally grows the stronger by
> every thing you see, hear, read, or under-
> stand. This is of great use.
>
> Laurence Sterne, *Tristram Shandy*

One of the central tasks of the philosophy of science is to give an account
of scientific progress. We have more reason than our ancestors to wonder
whether economic and political history show forth the increasing moral
perfection of mankind; skeptics and pessimists conclude that the 'Myth
of Progress' has been exploded once and for all. But myths do not die
easily, and the history of ideas seems to provide examples of the actual
growth of knowledge and not merely of changes in intellectual fashion.
In the history of science, paradigmatically, we can trace out both the
growing acceptance of theories of increasing scope, generality and explanatory power and the declining influence of theories which fail to square
with the facts of nature. Surely this is intellectual progress from ignorance
to knowledge! The philosophical question becomes: How is scientific
progress possible?

The obvious answer would be that the correct application of the scientific method guarantees this progress. But which feature of the scientific
method is it that provides the guarantee? It cannot be the conclusive
verifiability of all scientific statements, for in general theoretical statements are not conclusively verifiable. All interesting scientific theories go
beyond the observations which are their inductive bases in order to make
risky predictions; indeed, the content of typical scientific hypotheses is
not exhausted by any finite number of observations. Thus, even our best
confirmed theories may very well turn out to be false, as did the myths of
our primitive ancestors. In any case, we cannot prove that our theories are
true, for they are not deducible from the experimental results that follow

Boston Studies in the Philosophy of Science, XIV. All Rights Reserved.

from them. This we have all known since Hume shook us out of our dogmatic slumber.

If the comparison of scientific theories with experiment is never logically decisive in establishing their truth, then the scientific method cannot guarantee that science will finally succeed in attaining truth. But perhaps the scientific method can at least assure us that some falsehoods are avoidable. If scientific hypothesss are conclusively falsifiable *via* their observational consequences, then false hypotheses can, at least in principle, be eliminated from the scientific corpus by testing. In this case, the asymmetry with respect to conclusiveness between verifiability and falsifiability will guarantee that science can make progress in uprooting error. Scientific progress will consist in the rejection of falsified conjectures and the further testing of conjectures which have not yet been falsified. Thus, on such a Popperian account of scientific method, it is the falsifiability of scientific hypotheses which guarantees the rationality of scientific change and insures that science can make progress in eliminating error.

It would, therefore, be philosophically worrisome to be told that there is no asymmetry between the verifiability and the falsifiability of individual hypotheses in science, for the rationality of the scientific enterprise as well as the very possibility of scientific progress would seem to be threatened. But this is precisely what we seem to hear from the eminent physicist and historian of science Pierre Duhem; he says that "an experiment in physics can never condemn an isolated hypothesis" ([2], p. 183). Can Duhem demonstrate to us that there is no asymmetry between verification and falsification, that the one is as inconclusive as the other? In order to find out, some exegetical work on the Duhemian text is needed.

The problem of falsifiability in science is closely connected with the problem of conventionalism in science. Let us call a theoretical hypothesis true by convention just in case it is held to be true, it is not a logical truth and its truth or falsity is underdetermined by the facts. Since neither logical nor empirical considerations can serve as a basis for the truth of such an hypothesis, it must acquire its truth by stipulation, by our decision. And, if there are any scientific hypotheses which are neither verifiable nor falsifiable, then those hypotheses are potential conventional truths – we can decide that we wish them to be true and stipulate that they are true.

Many philosophers have thought that some conventional truths were to be found in physical geometry. Thus, Poincaré claimed that by comparison with Euclidean geometry

If Lobachevski's geometry is true, the parallax of a very distant star will be finite, *i.e.* positive non-zero; if Riemann's is true it will be negative. These are results which seem within the reach of experiment, and there have been hopes that astronomical observations might enable us to decide between the three geometries.

But in astronomy 'straight line' means simply 'path of a ray of light'.

If therefore negative parallaxes were found, or if it were demonstrated that all parallaxes are superior to a certain limit, two courses would be open to us: we might either renounce Euclidean geometry, or else modify the laws of optics and suppose that light does not travel rigorously in a straight line.

It is needless to add that all the world would regard the latter solution as the more advantageous.

The Euclidean geometry has, therefore, nothing to fear from fresh experiments. ([9], p. 81)

Poincaré's claim seems to amount to this. Given certain particular facts about the trajectories traversed by light rays traveling to the earth from distant stars, there are two courses open to the astronomer. He can either assert that the geometry of physical space is Euclidean but light rays do not travel along straight lines, or that the geometry of space is non-Euclidean and light rays travel along the geodesics (straight lines) of some particular non-Euclidean geometry. The optical facts are consistent with either assertion; they underdetermine the truth-value of any hypothesis about what the geometry of physical space is. Hence, the astronomer is free to stipulate what geometry physical space is to have, and suitable adjustments in his hypotheses about the paths traversed by light rays will suffice to account for all possible optical facts. Of course, Poincaré was mistaken in thinking that the Euclidean alternative would be universally regarded as the more advantageous. Einstein's General Theory of Relativity shows that, even though Euclidean geometry might be stipulated to be true, a non-Euclidean geometry which produces a simple theory overall, i.e. a simple geometry plus gravitational theory, has methodological advantages that outweigh simplicity in the geometrical portion of the theory alone. Recent philosophical work by Hans Reichenbach and Adolf Grünbaum has amplified and strengthened these conclusions.

So there is reason to believe that some theoretical hypotheses in physical geometry are true by convention. But, is there any reason to suppose that any theoretical hypothesis in science could be stipulated to be a conventional truth. The inductive leap from some geometrical hypotheses to all

scientific hypotheses seems formidable; nevertheless, some philosophers
have been willing to assert that any theoretical hypothesis in science is a
potential conventional truth. Quine goes so far as to claim that "any statement can be held true come what may, if we make drastic enough adjustments elsewhere in the system." ([11], p. 43), and among his reasons for
this claim is the doctrine that "our statements about the external world
face the tribunal of sense experience not individually but only as a cor-
porate body." a doctrine Quine thinks Duhem argued for successfully
([11], p. 41). More recently Grünbaum has formulated and attacked a
philosophical thesis, called the D-Thesis in honor of Duhem, which he
attributes to Quine and Duhem ([4], p. 276). The D-Thesis consists of the
conjunction of the following two subtheses:

> (D1)   No hypothesis which is a component of a scientific theory T
>   can ever be sufficiently isolated from some set of auxiliary
>   assumptions or other so as to be separately falsifiable by
>   observations;
> (D2)   Any hypothesis which is a component of a scientific theory T
>   can be held to be true come what may by way of observational
>   evidence and can be used to explain any observations which
>   refute T as a whole. ([5], pp. 1070–1071; [12], pp. 383–384)

Obviously, if the D-Thesis is correct, then any theoretical hypothesis in
science can be stipulated to be true by convention.

But does Duhem really endorse the D-Thesis? Does he accept the unrestricted conventionalism endorsed by Quine and attributed to him by
Grünbaum? If so, can Duhem demonstrate to us that unrestricted conventionalism in science is a philosophically tenable position? As before,
in order to find out, we must do some exegetical work on the Duhemian
texts.

Now that we have some idea of what philosophical issues are at stake
in Duhemian studies, let us try to clarify Duhem's own position.

I

In Chapter VI of Part II of *The Aim and Structure of Physical Theory*
Duhem discusses the relations between physical theory and experiment.

There he argues for four distinct theses concerning the testing of scientific theories:

(S)   The Separability Thesis – No single or individual theoretical hypothesis *by itself* has *any* observational consequences.

(F)   The Falsifiability Thesis – No single theoretical hypothesis can be conclusively falsified by any observations.

(D)   The Disjunction Thesis – No single theoretical hypothesis can be conclusively verified by performing one or more crucial experiments.

(I)   The Induction Thesis – No single theoretical hypothesis can be conclusively verified by induction from observations.

I shall consider each of these theses separately at this point, and later on I shall explore the logical relations among them. In order to simplify the discussion I shall assume without argument that some sense can be made of the distinction between a whole theory and an individual theoretical hypothesis which is a component of a theory, and of the distinction between theoretical hypotheses and observational statements.[1] Furthermore, I shall assume that the status of each and every statement with respect to these categories has been settled by some means or other prior to this discussion and will not, in this context, be a topic of controversy. The letter T, without subscripts, will be used as a symbol for whole theories; the letter T, with subscripts, will symbolize individual theoretical hypotheses; the letter O, with and without primes, will symbolize individual observational statements. It will be obvious from the discussion that the (unanalyzed) notion of entailment with which I operate is different from the standard conception of material implication.

One might think, says Duhem, that if a physicist disputes a certain theoretical hypothesis, then he can, at least in principle, test that hypothesis experimentally.

From the proposition under indictment he will derive the prediction of an experimental fact; he will bring into existence the conditions under which this fact should be produced; if the predicted fact is not produced, the proposition which served as the basis of the prediction will be irremediably condemned ([2], p. 184).

In other words one might suppose that

$$(T_i)\,(\exists O)\,(T_i \rightarrow O),$$

in which case each any every theoretical hypothesis would be falsifiable in isolation from all other theoretical hypotheses. This would amount to claiming that every theoretical hypothesis is *separable* from all others for purposes of testing.

But, continues Duhem, one would be mistaken to suppose that this is the case. For, in order to deduce an observational statement, the use of a conjunction of several theoretical hypotheses is necessary.

The prediction of the phenomenon whose nonproduction is to cut off debate does not derive from the proposition challenged if taken by itself, but from the proposition at issue joined to that whole group of theories ([2], p. 185).

Therefore, we must suppose that

$$- (\exists T_i) \, (\exists O) \, (T_i \rightarrow O),$$

or, equivalently,

$$(T_i) - (\exists O) \, (T_i \rightarrow O),$$

contrary to what it might seem natural to assume. In other words, no theoretical hypothesis is falsifiable by separating it from all other theoretical statements and testing it in splendid isolation, for no theoretical hypothesis has any observational consequences when so isolated. Theoretical hypotheses are inseparable for purposes of testing.

Duhem does not offer a proof that the thesis S is universally true; indeed, it is hard to see how he, or anyone else for that matter, could do so. However, he presents two examples from the history of physics which are confirming instances for the thesis: Wiener's attempted refutation of Neumann's hypothesis that a ray of polarized light vibrates parallel to the plane of polarization; and Arago's attempted refutation of the hypothesis that light consists of projectiles emitted at high velocities from the sun and other light sources (cf. [2], pp. 184–187). And, of course, it is easy to find other confirming instances for S. Hence, the thesis S is highly confirmed, even if it is not conclusively verified.

Duhem also maintains that no single theoretical hypothesis can be conclusively falsified by observations. He asserts that

... if the predicted phenomenon is not produced, not only is the proposition questioned at fault, but so is the whole theoretical scaffolding used by the physicist. *The only thing the experiment teaches us is that* among the propositions used to predict the phenomenon and to establish whether it would be produced, *there is at least one error; but where this error lies is just what it does not tell us* ([2], p. 185, my italics).

We should, I think, read the phrase 'not only is the proposition questioned at fault' as 'not only is the proposition in question called into question' to avoid the misleading suggestion that being at fault means for Duhem being false. The following passages lend additional support to the claim that Duhem was a proponent of the thesis F:

... when the experiment is in disagreement with his predictions, what he learns is that at least one of the hypotheses constituting this group is unacceptable and ought to be modified; *but the experiment does not designate which one should be changed* ([2], p. 187, my italics).

A disagreement between the concrete facts constituting an experiment and the symbolic representation which theory substitutes for this experiment proves that some part of this symbol is to be rejected. But which part? *This the experiment does not tell us; it leaves to our sagacity the burden of guessing* ([2], p. 211, my italics).

... experimental contradiction *always* bears as a whole on the *entire* group constituting a theory *without any possibility of designating which proposition in this group should be rejected* ([2], p. 216, my italics).

In other words, falsification is always both *holistic* in that it bears only on conjunctions of theoretical statements and *non-selective* in that it bears on all conjuncts equally. Therefore, with respect to single theoretical hypotheses falsification is necessarily inconclusive.

It is noteworthy that Duhem does not offer a separate argument in favor of the thesis F. I conjecture that this is because he takes the thesis F to be an obvious corollary of the thesis S. Given the thesis S, whenever

$$(T_i \ \& \ T_j) \to O,$$

and all that is observed is that the predicted phenomenon fails to materialize, i.e. $-O$, then one is entitled to conclude *only*

$$-T_i \lor -T_j,$$

and this conclusion *by itself* provides no grounds for singling out one or the other of T_i and T_j as the hypothesis to be rejected. Therefore, there is some initial plausibility associated with the claim that the thesis F is a corollary of the thesis S; nevertheless, the claim is false as we shall see later.

Duhem also maintains that crucial experiments are impossible in physics, where a crucial experiment is understood to be one that verifies some theoretical hypothesis by conclusively eliminating all of its competitors.

One might, says Duhem, suppose that observations could be used to verify conclusively a theoretical hypothesis is the following way:

Enumerate all the hypotheses that can be made to account for this group of phenomena; then, by experimental contradiction eliminate all except one; the latter will no longer be a hypothesis, but will become a certainty ([2], p. 188).

To facilitate discussion let us restrict our attention to the special case in which there are only two competing hypotheses.

Seek experimental conditions such that one of the hypotheses forecasts the production of one phenomenon and the other the production of quite a different effect; bring these conditions into existence and observe what happens; depending on whether you observe the first or the second of the predicted phenomena,[2] you will condemn the second or the first hypothesis; the hypothesis not condemned will be henceforth indisputable; debate will be cut off, and a new truth will be acquired by science ([2], p. 188).

This argument for the existence of crucial experiments can be easily schematized. Suppose we are given the disjunction

$$T_i \lor T_j$$

and the fact that

$$T_i \to O \qquad T_j \to O'$$

where $-(O \ \& \ O')$. If observations yield the result expressed by O', then we can deduce $-O$ and, hence, $-T_i$ by *modus tollens,* and this, in turn, allows us to deduce T_j by the rule of disjunctive syllogism. On the other hand, if the observations yield O, then we can deduce $-O'$ and, hence, $-T_j$, which yields the conclusion T_i. In either case, it looks as if one of the two hypotheses is conclusively verified.

What flaw does Duhem find in this procedure? We might expect Duhem's objection to amount to the claim that the supposition that $T_i \to O$ and $T_j \to O'$ is false in the light of the thesis S, and that, hence, the hope of establishing either $-T_i$ or $-T_j$ by any observations is doomed to frustration. And, if $-T_i$, for instance, cannot be established, then the deduction of T_j from $(T_i \lor T_j) \ \& \ -T_i$ will fail to verify T_j conclusively even if $T_i \lor T_j$ is known to be true, and the thesis D will be demonstrated. But, this is not, in fact, Duhem's objection. His argument is completely in-dependent of both the theses S and F; he allows us the (presumably coun-terfactual) assumption that a single theoretical hypothesis can be isolated from all but logically necessary statements.

But let us admit for a moment that in each of these systems everything is compelled to be necessary by strict logic, except a single hypothesis; consequently let us admit that the facts, in condemning one of the two systems, condemn once and for all the single doubtful assumption that it contains ([2], p. 189).

Hence, we are allowed to suppose for the purposes of this argument that, in effect, a single theoretical hypothesis has observational import, i.e. that $-O$ (or $-O'$) will conclusively falsify T_i (or T_j). Why then are we not al-lowed to conclude that T_j (or T_i) has been conclusively verified ?

Duhem's answer to this question consists in pointing out that in order for T_i to be transformed into an indubitable truth by the falsification of T_j, the disjunction $T_i \vee T_j$ must be true. This means the disjunction must be exhaustive of the possible explanatory premises.

... it would be necessary to enumerate completely the various hypotheses which may cover a determinate group of phenomena; but the physicist is never sure he has ex-hausted all the imaginable assumptions ([2], p. 190).

In order that the deduction of T_i from the premiss $(T_i \vee T_j)$ & $-T_j$ in accordance with the rule of disjunctive syllogism count as putting the truth of T_i beyond all possible doubt, the disjunction $T_i \vee T_j$ must be exhaustive, i.e. there must not be a T_k distinct from both T_i and T_j which could cover the same group of phenomena they do. In the trivial case in which, say, $T_j = -T_i$ logic alone will guarantee that $T_i \vee T_j$ is true; in scientifically interesting cases, e.g., light is either undular or it is corpuscular, there is no such guarantee. Thus, even if single theoretical hypotheses were isola-ble for purposes of testing and, hence, separately falsifiable, this would not entitle us to conclude that a crucial experiment is possible in physics. We would, in addition, have to be able to imagine and eliminate all possible alternatives to the hypothesis to be verified. But since the class of all pos-sible alternatives to a given hypothesis is, at least potentially, infinite, there is no reason to suppose that we would ever be able to accomplish this task, nor is there reason to suppose we could ever know whether we had accomplished it. Therefore, Duhem seems justified in concluding that,

independent of the difficulties surrounding the falsification of single
theoretical hypotheses, the verification of a theoretical hypothesis by a
crucial experiment is, apart from possible cases in which the alternatives
are somehow fortunately limited, impossible.

On this interpretation of the argument in favor of the thesis D, Duhem's
purpose is to show the impossibility of crucial experiments even if the
thesis S should be false; he wants to prove the thesis D in a way that does
not presuppose the truth of the thesis S. Of course, given the thesis S, the
thesis D can be proved, as has been acknowledged above. Thus, the fact
that Duhem does not avail himself of the argument from the thesis S to
the thesis D does not tell us anything about Duhem's views on falsifiability. In particular, it does not support the claim that, if Duhem had
believed that any theoretical hypothesis T_i can be held true and used to
explain nontrivially *the very observation* which refutes the conjunction
T_i & T_j for some T_j, then he would have denied the possibility of a crucial
experiment on the grounds that T_i "cannot be falsified but can rather
always be upheld in the face of any evidence whatever" – a claim made by
Laurens Laudan in private correspondence with Grünbaum (cf. [4], p.
282). For Duhem might have believed that any T_i can be upheld in the
face of any evidence whatever and believed that the impossibility of a
crucial experiment is a consequence of this claim, and yet have chosen to
argue for the thesis D from other premises. And, on my interpretation,
Duhem chose to argue for the thesis D from the impossibility of establishing
that a disjunction of alternative hypotheses is true in virtue of being exhaustive *not because* this was the only line of argument open to him *but
because* this argument, if sound, establishes the truth of the thesis D without making use of any claims about the impossibility of falsification.

In the light of this discussion, it is evident that the following remark by
K. R. Popper is highly misleading: "Duhem, in his famous criticism of
crucial experiments (in his *Aim and Structure of Physical Theory*), succeeds
in showing that crucial experiments can never *establish* a theory. He fails
to show that they cannot *refute* it" ([10], p. 112). This suggests that
Duhem's discussion of crucial experiments contains an unsuccessful attempt to show that theories cannot be refuted. But Popper overlooks
several relevant considerations. In the first place, Duhem's argument is
designed to show that a so-called crucial experiment cannot verify conclusively a single theoretical hypothesis; it does not, as stated by Duhem,

even apply to whole theories. Now, doubtless the argument that the elimination of all possible alternatives is not feasible can be extended to cover the case of whole theories, but Duhem himself does not make this extension in his discussion of crucial experiments. In the second place, Duhem never doubted that experiments can refute whole theoretical systems; he was only concerned to argue that no single theoretical hypothesis can be falsified by any observations. It is decidedly odd to speak of Duhem 'failing to show' something which he explicitly denies. Finally, the arguments Duhem offers in behalf of his claim that no single theoretical hypothesis is falsifiable by observations do not occur in his discussion of crucial experiments at all, but elsewhere in Chapter VI of Part II of *The Aim and Structure of Physical Theory*. In the discussion of crucial experiments Duhem even assumes for argument's sake that individual theoretical hypotheses can be conclusively falsified by experiment. Thus, in his criticism of crucial experiments Duhem does not even try to show that crucial experiments *cannot* refute a theory; rather, he assumes in this discussion that both whole theories and single theoretical hypotheses *can* be refuted by experiment. Elsewhere, however, Duhem does contend that no single theoretical hypothesis can be conclusively falsified by any observations. The interesting question is whether or not Duhem can demonstrate this thesis, for, if he (or anyone else for that matter) could prove that single hypotheses are not falsifiable, then the application of the Popperian method of conjecture and refutation would be of necessity limited to whole theoretical systems. This situation would not please Popper. And, if the only theoretical system large enough to be falsifiable as a whole is the totality of our knowledge and beliefs, as Quine contends, then the Popperian method will be incapable of deciding which belief or beliefs are to be rejected when the total system is falsified – the method will have no discriminatory power.

Duhem claims that the conclusive verification of a theoretical hypothesis by 'direct demonstration' is impossible. As opposed to the indirect demonstration of a hypothesis by a *reductio* argument in a series of crucial experiments, a direct demonstration of a theoretical hypothesis would consist in drawing it "from observation by the sole use of those two intellectual operations called induction and generalization" ([2], p. 190). The intellectual operations Duhem seems to have had in mind reduce, as his subsequent remarks indicate, to simple and straightforward generalization of empirical laws taken to be established by observation or of the consequences of such laws. According to Duhem, direct demonstration is the method espoused by Newton when he rejected "any hypothesis that induction did not extract from experiment; when he asserted that in a sound physics every proposition should be drawn from phenomena and generalized by induction" ([2], p. 191).

What is Duhem's objection to direct demonstration as a method of verification? Interestingly enough, he does not avail himself of the standard argument that the conclusions of inductive inferences do not follow logically from their observational premises and, hence, that inductive arguments from experience are incapable of conclusively verifying theoretical hypotheses. Instead, he develops the objection that a theoretical hypothesis is often inconsistent with the empirical generalizations which are supposed to be part of its inductive basis. This happens because no empirical generalization is perfectly precise; all experimental laws are merely approximate. Every observational regularity

... is therefore susceptible to an infinity of distinct symbolic translations; and among all these translations the physicist has to choose one which will provide him with a fruitful hypothesis without his choice being guided by experiment at all ([2], p. 199).

Thus it can happen that an accepted theoretical hypothesis, rather than being a generalization of experimental laws, formally contradicts them, even though both the experimental laws and the theoretical hypothesis yield approximately the same predictions or agree to within the limits of experimental error. For instance,

The principle of universal gravity, very far from being derivable by generalization and induction from the observational laws of Kepler, formally contradicts those laws. If Newton's theory is correct, Kepler's laws are necessarily false ([2], p. 193).

Or, to cite an example elaborated by Feyerabend in the course of his attack on the orthodox doctrine of the reduction of one theory to another, Newton's law of universal gravitation is inconsistent with Galileo's law for freely falling bodies near the surface of the earth (cf. [3], pp. 46–47ff.). Hence, since Newton's principle of universal gravitation, which might be supposed to arise by a generalization of the experimental laws of Kepler and Galileo, is actually inconsistent with those laws, it could hardly have been derived from them by induction, generalization, or any other method.

It is worth emphasizing that Duhem's objection is not that verification by direct demonstration fails to be decisive by invariably falling short of logical conclusiveness, as the standard objection has it; rather, he maintains that, since the choice of theoretical hypotheses is not even constrained to be formally consistent with the experimental laws in a given domain, these laws could not possibly serve as premises for the derivation of a fruitful theoretical hypothesis. In Duhem's opinion, the results of experiment are more like the scaffolding for theory construction than the foundations upon which a theory rests.

This point of view has interesting consequences for the attempt to construct a logic of discovery, as the work of Hanson indicates (cf. [6], Chapter IV). However, construed as an objection to the possibility of conclusive verification, it faces two difficulties. In the first place, Duhem never demonstrates that this objection can be lodged with equal justice against all putative direct demonstrations. He argues, convincingly enough, that it applies to the cases of Newton's work in gravitational theory and Ampère's research in electrodynamics, but this does not suffice to establish the thesis I in full generality since, after all, two cases constitute a rather small basis for an inductive argument. In the second place, it is not at all clear that Duhem is not conflating the logic of discovery with the logic of justification. It may well be that, in the context of discovery, the results of experiment always fail to pick out a uniquely fruitful hypothesis for precisely the reason Duhem gives, and yet, in the context of justification, it might sometimes happen that, once a fruitful theoretical hypothesis is discovered and the observational statements and experimental laws are reformulated to take account of their character as approximations, a direct demonstration of the theoretical hypothesis in question would provide a conclusive verification. We, of course, are skeptical about this possibility because of what we know about the problems of induction and confirmation, but the question is whether Duhem's argument gives us any independent grounds for our skepticism.

Therefore, there are some reasons for doubting that Duhem's objection to verification by direct demonstration is sufficient by itself to show that conclusive verification by this procedure is impossible. Nevertheless, since the standard objections to conclusive verification by induction are decisive, we may take it that Duhem is correct in claiming that conclusive verification by direct demonstration is impossible, even though the argument he offers in support of this claim may not be sufficient to prove it.

These considerations are the basis for my conclusion that Duhem does, indeed, propound and argue for the theses S, F, D and I enunciated above.

II

If my interpretation is faithful to Duhem's intentions, it is clear that Duhem accepts and argues for all the theses attributed to him above. What is not yet established is how the logical relations among the several theses are to be conceived. I shall now attempt to give an account of some of the logical relations among the four theses, but it is to be understood that, due to the lack of textual evidence bearing on theses matters, I cannot be certain that Duhem would have agreed with everything I attribute to him.

Duhem believes that conclusive verification of a single theoretical hypothesis by means of a series of crucial experiments or by means of direct demonstration is impossible. Does he also believe, as we do today, that conclusive verification of a theoretical hypothesis is impossible *by any means*?

The question is whether or not Duhem accepts the following thesis:

(V)   The Verifiability Thesis – No single theoretical hypothesis can be conclusively verified by any means whatever.

Now, it is clear that

$$V \rightarrow (D \ \& \ I),$$

and, on the standard twentieth-century accounts of the logic of confirmation, the thesis V is true; however, these facts do not show that Duhem must have accepted the thesis V. Moreover, as far as I have been able to discover, Duhem never explicitly endorses or argues for the thesis V. He may have tacitly accepted the thesis V on the strength of the argument that, since conclusive verification by direct demonstration is impossible and conclusive verification by indirect demonstration is impossible, and since there is no third alternative, conclusive verification by any means is impossible. In other words, Duhem may have thought

$$(D \ \& \ I) \rightarrow V.$$

The attribution of this line of thought to Duhem must, however, remain tentative because it is not certain that these two alternative modes of verification are, in fact, exhaustive of all possibilities. The expressions 'verification by direct demonstration' and 'verification by indirect demonstration' suggest that exhaustiveness is assured on logical grounds alone, but when the expressions 'verification by crucial experiment' and 'verification by induction from experiment' are used to describe that alternatives envisaged in theses D and I, respectively, this assurance evaporates. There may be means of indirect demonstration other than the crucial experiment just as there may be means of direct demonstration other than simple induction; as long as the possibility of as yet unconsidered means of conclusive verification remains open, the alternatives envisaged in theses D and I have not been shown to be exhaustive. And, since Duhem himself is so clear about this problem in his discussion of the thesis D, we should, I think, be hesitant about imputing to him an argument for the thesis V of this sort. Nevertheless, in my subsequent discussion of Duhem's views on falsifiability, I shall find it convenient to entertain the hypothesis that Duhem did accept the thesis V. This will do no harm, since, if he had accepted the thesis V, he would have been perfectly justified in doing so.[3]

It is evident that if no single theoretical hypothesis can be conclusively falsified by observations, then no single theoretical hypothesis can *by itself* have observational consequences, for if any theoretical hypothesis by itself had observational consequences, then it would be conclusively falsified by the failure of the states of affairs expressed by these consequences to be produced in appropriate circumstances. Thus, Duhem is entitled to accept the entailment

$$F \to S.$$

The unresolved question is whether or not he would also be justified in accepting the entailment

$$S \to F.$$

That Duhem did accept the latter entailment has been suggested above. There I conjectured that Duhem may have regarded the thesis F as a trivial corollary of the thesis S. Duhem may have noticed that since, given the truth of the thesis S, an observational statement O can only be deduced from some conjunction of two (or more) theoretical hypotheses T_i & T_j, the failure of the state of affairs expressed by O to obtain warrants only the conclusion $-T_i \lor -T_j$. And, he may have thought this fact alone entitled him to conclude that it is never possible to be forced to give up one of the two hypotheses rather than the other. I shall now demonstrate that this conclusion is a *non-sequitur*, that Duhem, if he accepted this piece of reasoning, was not justified in doing so.

Suppose that the thesis V is false, i.e. that there is some method for conclusively verifying at least some theoretical hypotheses, and suppose that the thesis S is true, i.e. that no theoretical hypothesis by itself has any observational consequences. The only way to deduce an observational consequence is to make use of some conjunction of theoretical hypotheses. Suppose, without loss of generality, that some conjuction of theoretical hypotheses with exactly two conjuncts permits the deduction of an observational statement, i.e. that

$$(T_i \& T_j) \to O.$$

Only the conjunction $T_i \& T_j$ will be falsified by $-O$; hence, if $-O$ is established by observation, then only $-T_i \lor -T_j$ will be deducible. Now, assume that either T_i or T_j is one of those theoretical hypotheses which is conclusively verifiable, i.e. one of those hypotheses in virtue of which the thesis V is false, and that the conclusive verification of the hypothesis in question, say T_i, does not involve the other hypothesis T_j. In this situation, since

$$(((T_i \& T_j) \to O) \& -O \& T_i) \to -T_j,$$

the falsification of T_j by $-O$ and whatever observations suffice to verify conclusively T_i will be conclusive. In this case, the thesis F will be false. Therefore, the thesis F does not follow from the thesis S alone. (Note, however, that this example shows that the truth of V is a necessary condition of the truth of F.) For if the thesis V is false, then if any refuted two-termed conjunction has one conjunct which is a counterexample to V its other conjunct will be a counterexample to F. The schema above illustrates this situation.

The fact that this counterexample to the claim that the thesis S entials the thesis F makes use of the possibility of conclusive verification suggests that Duhem might be justified in asserting that

$$(S \& V) \to F.$$

Can we also show that this entailment does not hold?

One way of doing this would be to construct a counterexample in which the refuting observations, though *holistic* in that they bear only on conjunctions of theoretical hypotheses, are *selective* in that they bear with different force on different conjuncts so that cumulatively they decisively refute one of the conjuncts. And one might suppose that such a situation can be envisaged with little trouble in the following way. Suppose that the theses S and V are both true. Further, suppose that for some T_i and T_j it happens that both $(T_i \& T_j) \to O'$ and $(T_i \& -T_j) \to O''$, and suppose that what is observed is expressed by some observational statement O such that both $-(O \& O')$ and $-(O \& O'')$. Since from $((T_i \& T_j) \to O') \&$ $\& ((T_i \& -T_j) \to O'') \& -O' \& -O''$ one can deduce $-(T_i \& T_j) \&$ $\& -(T_i \& -T_j)$ by *modus tollens* and, hence, $(-(T_i \& T_j) \& -T_i) \vee$ $\vee (-(T_i \& T_j) \& T_j)$ by DeMorgan's laws and the laws of distribution and double negation, and, therefore, $-T_i$, the observation or observations expressed by O are sufficient to falsify T_i conclusively. Unfortunately, this elaborate reasoning is of little avail, for the situation described is self-contradictory. As John Winnie has pointed out to me, from $((T_i \&$ $\& T_j) \to O') \& ((T_i \& -T_j) \to O'')$ we can deduce $T_i \to (O' \vee O'')$ which is a counterexample to the thesis S. Hence, the supposition that the thesis S is true contradicts the further supposition that both $(T_i \& T_j) \to O'$ and $(T_i \& -T_j) \to O''$.

Is there another way of providing a counterexample to the entailment under discussion? It may well be impossible since any such counterexample would have to show that some single theoretical hypothesis is incompatible with some set of observational statements, but in this case the entailment of some truth-functional compound of observational statements incompatible with those in the set by that single theoretical hypothesis would provide a counterexample to the thesis S. My only reservation about this argument is that it is not clear that there will always be an entailment of the sort required (as opposed to a material implication). This lack of clarity shows that the notion of entailment appropriate for science, and the associated problem of non-triviality in scientific explanation, deserve more careful attention, but that is the subject for another paper.

The thesis F has, therefore, a peculiar position in Duhem's philosophy. Although it is not a consequence of the thesis S, it may follow from the theses S and V together. But, since the thesis V may well be stronger than

the conjunction of the theses D and I, we have not conclusively demonstrated that the thesis F follows from the conjunction of the theses S, D
and I. Hence, it has not been established that

$$(S \ \& \ D \ \& \ I) \rightarrow F.$$

On the other hand, it is clear from the discussion above that $F \rightarrow S$, and
also that the truth of V is a necessary condition for the truth of F, and
that $V \rightarrow (D \ \& \ I)$. Since the thesis F is stronger than the conjunction of
the theses S, D and I in this sense, it is not surprising that it should have
attracted special attention from Duhem's readers. What is surprising is
that Duhem himself may have supposed it to be a trivial corollary of the
thesis S and, on that account, devoted so little labor to the attempt to
establish its truth. It is this anomaly, the fact that the thesis F is both
far-reaching and controversial, that has made it an interesting topic for
debate among Duhem's philosophical heirs.

III

The argument so far, if correct, has established the claim that Duhem
endorsed the thesis that no single theoretical hypothesis can be conclusively falsified by observations, but, since Duhem offered no special
argument in support of this thesis and since it may not follow from the
other, less controversial theses about the testing of theoretical hypotheses
which he did accept, the question of its truth remains unanswered. There
is, however, another claim which has sometimes been attributed to Duhem
that has not yet been discussed. This is the following thesis:

(E)   The Explainability Thesis – Any single theoretical hypothesis
        can continue to be held true whenever a total theory T of
        which it is a component is refuted, and can be used as a
        component of some theory T' which non-trivially explains[4]
        the very observations that refuted T.

There are two questions worth asking about the thesis E in the context of
this discussion: Did Duhem believe the thesis E to be true? Is the thesis
E a consequence of any other thesis that Duhem accepted; in particular,
does the thesis E follow from the thesis F?

To the first question, the textual evidence does not, as far as I can see, provide a clear answer. When in the course of scientific testing it becomes necessary to revise a theory, physicists can usually legitimately disagree about how to proceed.

... one may be obliged to safeguard certain fundamental hypotheses while he *tries* to reestablish harmony between the consequences of the theory and the facts by complicating the schematism in which these hypotheses are applied, by invoking various causes of error, and by multiplying corrections. The next physicist, disdainful of these complicated artificial procedures, may decide to change some one of the essential assumptions supporting the entire system. The first physicist does not have the right to condemn *in advance* the boldness of the second one, nor does the latter have the right to treat the timidity of the first physicist as absurd. The methods they follow are justifiable only by experiment, and *if they both succeed in satisfying the requirements of experiment* each is logically permitted to declare himself content with the work that he has accomplished ([2], p. 217, my italics).

Does this passage show that Duhem is a proponent of the thesis E? I think not. Given the truth of the thesis F, a physicist is always free to *try* to use his favorite hypothesis in reestablishing harmony between the consequences of the theory and the facts. But the freedom to make an attempt does not guarantee that the effort can or will succeed, for there may not be any theory containing the physicist's favorite hypothesis which non-trivially explains the facts. And Duhem does *not* say that a physicist who tries to save his favorite hypothesis can always succeed in this enterprise. The physicist who tries to use his favorite hypothesis in reestablishing harmony between theory and experiment is not doing something *absurd,* nor can he be condemned *in advance*; but this does not imply that his trials are bound to be successful in every case, or that he cannot be condemned *in retrospect*. Duhem is surely correct in asserting that two physicists, one preserving and the other abandoning a given hypothesis, can both be minimally content with their work *if* each succeeds in satisfying the requirements of experiment. However, Duhem refrains from claiming that in all possible circumstances both physicists *can or will* succeed in satisfying these requirements; he does not assert that each physicist is justified in hoping for contentment with his work. Duhem's contention (cf. [2], pp. 216–218) that good sense, in fact, decides which hypotheses ought to be abandoned on grounds such as the simplicity and elegance of the total theoretical system does not, of course, *by itself* imply that good sense must decide this question.[5] Hence, this claim is compatible with the thesis E. It may, as a matter of historical fact, be generally true that

... the day arrives when good sense comes out so clearly in favor of one of the two sides that the other side gives up the struggle even though pure logic would not forbid its continuation ([2], p. 218).

This conclusion, however, does not support the thesis E, for it does not mean that, if good sense were to fail in its office as judge of scientific controversies, then any theoretical hypothesis which is a component of a refuted theory T can be successfully incorporated into some theory T' which non-trivially explains, among other things, the refuting facts. Therefore, the textual evidence does not support the conjecture that Duhem himself was a proponent of the thesis E. Although Duhem does not reject this thesis explicitly, he does not endorse it either.

But, to take up the second question, perhaps Duhem ought to have accepted the thesis E. If the thesis E were a logical consequenc of the thesis F, then, since Duhem did accept the thesis F, he would have been obliged for the sake of consistency to endorse the thesis E also. However, it is not difficult to show that the thesis E is *not* a logical consequence of the thesis F.

Suppose that the thesis F is true, i.e. that no single theoretical hypothesis can be conclusively falsified by observations. From this assumption it follows that any given theoretical hypothesis T_i can continue to be held true whenever a total theory T of which it is a component is refuted by some observation expressed by the statement O. But, further suppose that every theory T' which non-trivially explains O implies some statement incompatible with T_i. In such circumstances no theory which explains O can contain T_i as a component on pain of inconsistency. Hence, T_i cannot be used as a component of any theory T' which explains the observations that refuted T, and the thesis E is false because its second conjunct is false. Notice, however, that the thesis F remains true. It is not that T_i has been falsified by observations; T_i can consistently continue to be held true, but *only if* the attempt to explain O is abandoned. On the other hand, T_i can be abandoned in order that an explanation of O be consistently possible, even though no observation has conclusively falsified T_i.[6] Therefore, since in such circumstances the thesis F would be true and the thesis E would be false, the thesis E is not a logical consequence of the thesis F alone. Thus I conclude that Duhem need not have been committed to the thesis E simply because he was a proponent of the thesis F.

In sum, since textual evidence is insufficient to show that Duhem accepted the thesis E, and since the denial of the thesis E is consistent with the affirmation of the thesis F (and *a fortiori* with the affirmation of the theses S, D and I), the claim that Duhem himself was a proponent of the thesis E rests on insecure foundations.[7]

IV

In this paper I have attempted to give a coherent interpretation of Duhem's doctrine of the testability of scientific hypotheses as it appears in Chapter VI of Part II of *The Aim and Structure of Physical Theory*. Perhaps a summary of the results of this investigation will prove helpful to the reader by way of a conclusion to the discussion.

On my interpretation Duhem was a proponent of the theses S, F, D and I. It is likely that he also accepted the thesis V, and he may have thought that thesis to be a consequence of the conjunction of the theses D and I. The thesis F, the most interesting of the lot in connection with the problem of the falsifiability of individual theoretical hypotheses, has a peculiar status, for although the thesis F entails the thesis S and has as a necessary condition the thesis V, which in turn entails the conjunction D & I, we have discovered no conclusive demonstration that the conjunction S & D & I entails F. Thus it is clear that the thesis F is the strongest of Duhem's claims, and it is, therefore, somewhat astonishing that he seems to regard it as an obvious corollary of the thesis S. Finally, there is not sufficient evidence of either a logical or a textual nature to attribute to Duhem the thesis E with much confidence.

If this interpretation is basically sound, it means that Duhem himself cannot be claimed as a defender of the D-Thesis *without qualification*. Although Duhem would accept subthesis D1 of the D-Thesis (which is a variant of the thesis F) since he accepts the thesis F, there is no convincing reason for supposing that he would accept subthesis D2 (which is a variant of the thesis E) because there is no conclusive reason for supposing that he accepts the thesis E. Hence, I find myself in *partial* agreement with Laudan's suggestion that attacks on the D-Thesis are wrongly directed against Duhem (cf. [7], p. 295). My own view of the matter is that attacks on the D-Thesis which are specifically directed against subthesis D2 are, almost certainly, not attacks on Duhem, whereas assaults aimed speci-

fically at subthesis D1 are quite properly taken to be directed against a
position Duhem himself held.[8]

Brown University

NOTES

[1] The latter distinction has, of course, been under attack in the recent philosophical
literature. The most persuasive criticsms of the theoretical-observational distinction
relate to its bearing on problems of the reference of theoretical terms (realism, fiction-
alism) and questions concerning the meaning of theoretical terms (Are theoretical
terms operationally definable? Are observational terms theory-laden?). In discussions
of testability alone, however, there seems to be general agreement that some scientific
statements are (more or less) directly testable while others are only indirectly testable.
Since in this paper only issues relating to testability are to be discussed, a minimal
theoretical-observational distinction seems feasible. Observational statements are all
those statements which are taken to be directly testable, i.e. whose truth or falsity is
taken to be directly establishable by making observations; theoretical hypotheses are
those statements taken not to be directly testable but at best indirectly testable *via* the
testing of their observational consequences. This distinction will, I claim, be of assistance in elucidating Duhem's views on testability because for Duhem testability and
meaning are separable issues. It would not, I think, help very much in illuminating
Quine's views; since Quine insists that factual and linguistic components of truth cannot be factored out at the level of individal statements, problems of testability and
questions of meaning cannot be neatly separated in a discussion of his defense of conventionalism (cf. [11], p. 42).
[2] For Duhem, the supposition that we will not observe something incompatible with
both follows from the assumption in the case under discussion that there are only two
possible competing hypotheses.
[3] In his comments on the oral presentation of this paper Professor Imre Lakatos
argued that Duhem would have vehemently rejected the thesis (V) because he believed
that scientific theory can arrive at a natural classification of phenomena and that the
scientist can know when such a natural classification has been achieved. This interpretation of Duhem is quite common; D. G. Miller, for example, attributes to Duhem
the belief that "as physical theories evolve by successive adjustments to conform to
experiment, they approach asymptotically a sort of 'natural classification' that somehow reflects an underlying reality" ([8], p. 47). But this is not the whole story. When
speaking cautiously about the physicist's belief that a theory reflects an underlying
reality, Duhem asserts that "while the physicist is powerless to justify this conviction,he
is nonetheless powerless to rid his reason of it... yielding to an intuition which Pascal
would have recognized as one of those reasons of the heart 'that reason does not know'
he asserts his faith in a real order reflected in his theories more clearly and more faith-
fully as time goes on" ([2], p. 27). In other words, physicists have an incorrigible but
unjustifiable belief that their theories tend toward reflecting an underlying reality. But
this is faith not knowledge and is hardly firm ground for rejecting the thesis V. Furthermore, the only sort of evidence Duhem allows in support of the claim that a theory involves a natural classification is predictive power, i.e. "anticipating experience and
stimulating the discovery of new laws" ([2], p. 28), but predictive power is not a suffi-

cient condition for truth or verification. Moreover, it would seem to be a mistake to take the scientist's judgments about the naturalness of a classificatory system as evidence for or a criterion of the truth of theories employing that system. In practice, scientists will tend to regard as natural what is familiar to them, and their judgments of naturalness will tend to reflect the conceptual *status quo*. Far from contributing to the verification of theoretical hypotheses, serious reliance on naturalness as a criterion for the acceptability of scientific hypotheses would impede scientific progress by discouraging novelty. It may not be unreasonable to suggest that Duhem's well-known rejection of relativity because he believed it violated common sense illustrates the dangers of taking the scientist's opinions about what is natural too seriously (cf. [8], p. 52).

4 In the thesis E the notion of explanation presupposed is the standard one, namely, subsumption under covering laws, with a single restriction. Explanations are required to be non-trivial to prevent O from being explained by T_i & $(- T_i \vee O)$ for an arbitrary T_i and to exclude other similar cases. The terminological convention involved in calling such non-trivial subsumptions "explanation" is a bit anachronistic in a discussion of Duhem, for many things that are explanations in this sense Duhem would have called descriptions. He reserves the term 'explanation' for metaphysical explanations in terms of hidden realities 'behind' or 'underneath' the sensible appearances (cf. [2], pp. 7–18).

5 Mario Bunge has recently spelled out in some detail how good sense might come to reach a decision (cf. [1], pp. 160–161). According to Bunge, to deduce testable consequences from a theory T we need additional information including subsidiary assumptions A, e.g., a theoretical model of the system being studied plus simplifying assumptions such as linearizations, background information I and initial conditions C. When the complex consisting of T, A, I and C confronts adverse empirical evidence, good sense specifies proceeding in one of two ways. If T has stood up to tests in the past, the subsidiary assumptions A should be modified; only after unsuccessful attempts to explain the adverse evidence with such alternative subsidiary assumptions should serious doubt be cast on T. If, on the other hand, T has not stood up to tests in the past then both T and A should be subjected to exacting examination. In pragmatic terms Bunge's advice sounds reasonable and, hence, deserves serious consideration as a candidate for a satisfactory explication of Duhem's concept of good sense's procedures. However, Bunge offers no argument to show that these procedures will succeed in isolating the component or components of T or A which are actually false in every case or in any particular case. Nor does Bunge show that applying these procedures will prohibit any particular component of T from satisfying the thesis E in any case. Indeed, it would seem to be impossible to demonstrate this point. Thus, Bunge's procedures, reasonable though they may be, do not guarantee success when applied and cast no doubt on the theses S, F or E. Therefore, if the task Bunge's procedures are supposed to perform is to 'find the subset of premises responsible for the failure' in explanation, there is no assurance that 'a solution may be possible in every case' ([1], p. 161).

6 It is interesting to speculate on whether or not scientists could reasonably tolerate O as a permanent anomaly. I suspect that, if O expressed a rather unimportant fact and the alternatives to T_i were vastly more complicated than T_i, considerations of theoretical simplicity could induce scientists to learn to live with an unexplained O. If every theory is born refuted, absolute explanatory completeness may not be a necessary condition even for ultimate scientific success.

7 In his comments on the oral version of this paper Professor Hilary Putnam suggested that the thesis E might be strengthened so as to be interestingly false by imposing constraints, other than nontriviality, on the explanations allowable in a revised theory T'.

Although the suggestion may have some merit, I am unable to think of any restrictions on permissible explanations, apart from non-triviality, which would clearly be generally plausible. Besides, the thesis E may be interestingly false as it stands, as G. J. Massey has pointed out, if it is interpreted as tacitly asserting the existence of a deductive explanation for any observation. There may, after all, be individual observable events which cannot be explained deductively due to the irreducibly statistical character of the laws of the relevant domain (cf. revised version of [5], p. 89).

[8] This paper is based in part on material included in my Ph. D. dissertation, University of Pittsburgh, 1970. I am grateful to Adolf Grünbaum and Laurens Laudan for helpful criticism.

BIBLIOGRAPHY

[1] Bunge, Mario, 'Theory Meets Experience', *Mind, Science, and History*, SUNY Press, Albany, 1970.
[2] Duhem, Pierre, *The Aim and Structure of Physical Theory*, Atheneum, New York, 1962.
[3] Feyerabend, P. K., 'Explanation, Reduction, and Empiricism', *Minnesota Studies in the Philosophy of Science, Vol. III*, University of Minnesota Press, Minneapolis, 1962.
[4] Grünbaum, Adolf, 'The Falsifiability of a Component of a Theoretical System', *Mind, Matter, and Method*, University of Minnesota Press, Minneapolis, 1966.
[5] Grünbaum, Adolf, 'Can We Ascertain the Falsity of a Scientific Hypothesis ?', *Studium Generale* **22** (1969), 1061–1093; revised version in *Observation and Theory in Science*, Johns Hopkins Press, Baltimore, 1971.
[6] Hanson, N. R., *Patterns of Discovery*, Cambridge University Press, Cambridge, 1958.
[7] Laudan, Laurens, 'Grünbaum on 'The Duhemian Argument' ', *Philosophy of Science* **32** (1965), 295–299.
[8] Miller, D. G., 'Ignored Intellect: Pierre Duhem', *Physics Today* **19** (1966), 47–53.
[9] Poincaré, H., *The Foundations of Science*, Science Press, Lancaster, 1946.
[10] Popper, K. R., *Conjectures and Refutations*, Harper & Row, New York, 1968.
[11] Quine, W. V. O., *From a Logical Point of View*, Harper & Row, New York, 1963.
[12] Quinn, P. L., 'The Status of the D-Thesis', *Philosophy of Science* **36** (1969), 381–399.

MICHAEL POLANYI

GENIUS IN SCIENCE

I. INEXACTITUDE OF SCIENCE AND THE WORK OF GENIUS

We accept the results of science, and we must accept them, without having
any strict proof that they are true. Strictly speaking all natural sciences
are inexact. They could all conceivably be false, but we accept them as
true because we consider doubts that may be raised against them to be
unreasonable. Juries base their findings on the distinction between reasonable doubts which they must accept, and unreasonable doubts which
they must disregard. They are instructed to make this distinction and to
do it without having any set rules to rely upon. For it is precisely because
there are no rules for deciding certain factual questions of supreme importance that these questions are assigned to the jury to decide them by
their personal judgment. The scientist combines the functions of judge
and jury. Having applied to his findings a number of specifiable criteria,
he must ultimately decide in the light of his own personal judgment whether the remaining conceivable doubts should be set aside as unreasonable.

Once it is recognised that all scientific discoveries ultimately rest on the
scientist's personal judgment, the path seems open for unifying the whole
sequence of personal decisions, beginning with sighting a problem and
then pursuing the problem throughout an enquiry, all the way to the
discovery of a new fact of nature.

We shall meet the main features of the principle that controls scientific enquiry from the dawn of a problem to the finding of its solution, by looking first at its highest actions in the work of genius.

Genius is known for two faculties which may seem incompatible.
Genius is a gift of inspiration, poets back to Homer have asked their
Muse for inspiration, and scientists back to Archimedes have acknowl-
edged the coming of a bright idea as an event that suddenly visited them.
But we have also ample evidence of an opposite kind; genius has been

Boston Studies in the Philosophy of Science, XIV. All Rights Reserved.

said to consist in taking infinite pains, and all kinds of creative pursuits are in fact extremely strenuous.

How can these two aspects of genius hang together? Is there any hard work, which will induce an inspiration to visit us? How can we possibly conjure up an inspiration without even knowing from what corner it may come to us? And since it is ourselves who shall eventually produce the inspiration, how can it come to us as a surprise?

Yet this is what our creative work actually does. It is precisely what scientific discovery does: We make a discovery and yet it comes as a surprise to us. The first task of a theory of creativity, and of scientific discovery in particular, must be to resolve this paradox.

The solution can be found on a biological level, if we identify inspiration with 'spontaneous integration' and look out for the effort that induces such integration. Suppose I move an arm to reach for an object: my intention sets in motion a complex integration of my muscles, an integration that carries out my purpose. My intention is about something that does not yet exist: in other words it is a project, a project conceived by my imagination. So it seems that it is the imagination that induces a muscular integration to implement a project that I form in my imagination.

Could we say that this integration is spontaneous? I think that in an important sense we can call it spontaneous, for we have no direct control over it. Suppose a physiologist were to demonstrate to us all the muscular operations by which we have carried out our action, we would be amazed at the wonderful mechanism that we had contrived in achieving our project. We would find that we had done something that profoundly surprises us.

This exemplifies a principle that controls all our deliberate bodily actions. Our imagination, thrusting towards a desired result induces in us an integration of parts over which we have no direct control. We do not *perform* this integration: we *cause it to happen*. The effort of our imagination evokes its own implementation.

And the way we evoke here a desired event by the action of our body offers in a nutshell a solution of the paradox of genius. It suggests that inspiration is evoked by the labours of the thrusting imagination and that it is this kind of imaginative labour that evokes the new ideas by which scientific discoveries are made.

These conclusions may seem too fast, but they will be confirmed and enlarged by passing on from voluntary action to visual perception. The constancy of objects seen is achieved by an integration of clues which takes place beyond our direct control. We see objects and their surroundings coherently by integrating two to three snapshots per second, which present to us overlapping images ranging over the area before us. The intelligent scanning of these consecutive snapshots shows that our imagination is at work guiding our integration. And we can add that in case of any difficulty in recognising what it is that we see, the imagination explores alternative possibilities by letting our eyes move round to look for such possibilities.

Different branches of science are based on different ways of seeing. When an object is composed of parts that function jointly, our vision integrates the sight of these parts to the appearance of a coherent functioning entity. This is how the engineer, who knows the way a machine works, sees the machine as a working whole. Such integration underlies all biology and psychology. The view of an organism, the sight of an intelligent animal, the image of a human person, are all based on such integrations. We may call these visual integrations spontaneous because their parts are not directly controlled and often cannot even be directly noticed. The process of scientific discovery consists generally of such integrations evoked by the work of the imagination.

II. POWERS OF ANTICIPATION

The progress of discovery falls into three main periods. The *first* is the sighting of a problem and the decision to pursue it; the *second*, the quest for a solution and the drawing of a conclusion; the *third*, the holding of the conclusion to be an established fact.

I have spoken of the way our eyesight organises consecutive snapshots by scanning them in an intelligent way, and how, in case of difficulty, the imagination explores alternative possibilities to find out what it is that we are seeing. These efforts of our eyesight are based on the assumption that any curious things before us are likely to have some hidden significance. Scientists speculating about strange things in nature act on a similar assumption. They try to interpret the facts they know, and go on collecting more facts, in the hope that these will reveal a coherence that is of interest to science. Such is the act of seeing a problem and pursuing it.

But here we meet a strange fact. In accepting the task of pursuing a problem, the scientist assumes it to be a *good* problem, a problem that he can solve by his own gifts and equipment and that it is worth undertaking in comparison with other available possibilities. He must estimate this; and such estimates are guesses.

But such guesses have proved sufficiently good to secure the progress of scientific enquiry with a reasonable degree of efficiency. It is rare to come across years of futile efforts wasted, or else to find that major opportunities were patently missed. Indeed, the opportunities for discovery are so effectively exploited that the same discovery is often made simultaneously by two or three different scientists. There is no doubt therefore of the scientist's capacity to assess in outline the course of an enquiry that will lead to a result which, at the time he makes his assessment, is essentially indeterminate.

How can we explain this capacity? I have said that scientific discovery is in essence an extension of perception. Remember how the different images of an object presented to our eyes from various distances, at different angles and in changing light, are all seen jointly as one single object, and that it is in terms of this coherence that our eyes perceive real things. This bears deeply on science. Copernicus laid the foundations of modern science by claiming his discovery of the heliocentric system in these very terms. He showed that his system included a parallelism between the solar distances of the planets and their orbital periods, and on this coherence he based his insistence that his system was no mere computing device, but a real fact.

But such claims to know reality are questioned by our antimetaphysical age. Can we define what it means to claim that an object is real? I think we can.

To say that an object is real is to anticipate that it will manifest its existence indefinitely hereafter. This is what Copernicus meant by insisting that his system was real. Copernicus anticipated the coming of future manifestations of his system, and these were in fact discovered by later astronomers who had accepted his claim that his system was real. *We can conclude then that, in nature, the coherence of an aggregate shows that it is real and that the knowledge of this reality foretells the coming of yet unknown future manifestations of such reality.* This concept of reality will now be extended to include all the phases of a scientific enquiry. It will explain the way discovery is anticipated, from the sighting of a problem to its final solution.

But let me stop first to recall the antecedents which led to this theory. I began my work on the nature of science twenty-five years ago guided by the idea that we make scientific discoveries in the same way we strain our eyesight to perceive an obscure object; and that in this effort we are guided by anticipating to some extent the direction which will prove most fruitful. "A potential discovery", (I wrote, "may be thought to attract the mind which will reveal it – inflaming the scientist with creative desire and imparting to him intimations that guide him from clue to clue and from surmise to surmise."[1]

For years I have written about this kind of anticipation.[2] At one stage I was joined in this idea by George Polya,[3] whose observations of mathematical discovery I had relied on from the start of my enquiry. And more recently I met with a brilliant description of anticipation in the postumous work of C.F.A. Pantin, who writes that: "(Intuition) does not only suddenly present solutions to our conscious mind, it also includes the uncanny power that somehow we know that a particular set of phenomena or a particular set of notions are highly significant: and we are aware of that long before we can say what that signification is."[4]

But only now can I see an explanation for such anticipations. I see that the anticipations offered to us by good problems should be understood in the same way as the anticipations aroused by all true facts of nature. Thus, when a coherent set of clues presents us with the sense of a hidden reality in nature, we are visited by an anticipation similar to that which we feel in seeing any object already recognised to be real. The expectations attached to a good problem differ only in their dynamic intensity from the expectations that will be attached to any facts eventually to be discovered in the end, once the problem has been solved. Of course, the sense of reality implied in adopting a problem, is pointing more clearly in a particular direction. And also the results anticipated in this kind of reality are expected to appear more soon than are the prospects implied in affirming the reality of an established fact; but I regard this difference as a mere matter of degree.

The whole of science, as it is known to us, has come into existence by virtue of good problems that have led to the discovery of their solution. The fact that scientists can espy good problems is therefore a faculty as essential to science, as is the capacity to solve problems and to prove such solutions to be right. In other words, the capacity rightly to choose a line of thought the end of which is vastly indeterminate, is as much part of the scientific method as the power of assuring the exactitude of the conclusions eventually arrived at. And both faculties consist in recognising real coherence in nature and sensing its indeterminate implications for the future.

This conclusion fulfills in substance my hopes of finding the same principles of personal judgment at work at all stages of a scientific enquiry, from the sighting of a problem to the discovery of its solution. Problems are discovered by a roaming speculative imagination, and once a problem is adopted, the imagination is thrust in the direction of the problem's expectations. This evokes new ideas of coherence which, if true, reduce the indeterminacy of the enquiry. The speculative or experimental examination of these ideas directs yet new thrusts of the imagination that evoke yet further surmises; and so the pursuit is narrowed down ever further, until eventually an idea turns up which can claim to solve the problem.

This rough sketch must suffice for the moment to outline the sequence of 'infinite pains' that finally evoke a surprise claiming to be the solution of the problem.

III. RATIONALITY TO THE RESCUE?

But scientific opinion has been reluctant to accept the fact that the scientist is guided essentially by a vague sense of still unrevealed facts. Hence textbooks of physics have taught for decades that Einstein dscovered relativity as an explanation of Michelson's observation that the earth's rotation causes no flow in the surrounding ether, and so, when I pointed out about twelve years ago that this was a pure invention,[5] the only response I evoked was from Professor Adolf Grünbaum of Pittsburgh[6] who said that my description of Einstein's way to discovery was like Schiller's story that his poetic inspiration came to him by smelling rotten apples. Fortunately, a study recently published by Professor Gerald Holton has shown at last in great detail, that I had been right.[7]

I have mentioned this story to illustrate the temper of our age which prefers a tangible explanation to one relying on more personal powers of the mind, even though the plain facts do show these less tangible forces at work. The theory of scientific discovery – most influential today – expresses this preference by dividing the process of discovery sharply into the choice of a hypothesis and the testing of the chosen hypothesis. The first part (the choice of a hypothesis) is deemed to be inexplicable by any rational procedure, while the second (the testing of the chosen hypothesis) is recognised as a strict procedure forming the scientist's essential task.

This theory of scientific discovery would save the strictness of science by declaring that scientific discoveries are merely tentative hypotheses which can be strictly tasted by confronting their implications with experience. And that if any of the implications of a hypothesis conflicts with experience, the hypothesis must be instantly abandoned; that indeed, even if the hypothesis is accepted on the grounds of having been confirmed in its predictions, it will ever remain on trial ready to be abandoned if any experience turns up that contradicts one of its claims. We are told that unless a hypothesis produces testable conclusions it should be disregarded as lacking any substantial significance.[8]

Let me test this theory. There may be cases where a scientific discovery was made and only claimed as such after some additional implications of it had been tested; but there is plenty of evidence to show that this is not necessary and, indeed, is often impracticable. On November 11, 1572 Tycho Brahe observed a new star in Cassiopeia, and this discovery refuted the Aristotelian doctrine of an unchangeable empyrean. This happened before the days of the telescope, and indeed the same observation was made also in China. The discovery was complete without producing testable implications, exactly as the eruption of Vesuvius on August 24, A.D. 79 was established as a fact, without any tests of its vast implications. Or take Kepler's discovery that for the six then known planets the square of the orbital period was proportional to the cube of the solar distance. The figures underlying this discovery had been known for eighty years or more; I happened to test the relative solar distances of the planets made available by the work of Copernicus and found that they agree with Kepler's Third Law within two per cent. All that Kepler did was to recognise this relationship, which is his Third Law. Yet Kepler hailed his discovery as the crowning of his search for celestial harmony, even though no testable implications of it were known at the time and indeed for a long time after his death.

Admittedly, many discoveries were not made at one stroke. But of these too, many fail to exemplify the orthodox 'hypothetico-deductive model'. On March 13, 1781. William Herschel observed a slowly moving nebulous disk which he first took to be a comet, but, after a few weeks of watching its motion, recognized as a new planet, to be named Uranus. Later on, Leverrier and Adams, basing themselves on the irregularities of Uranus, derived the existence of one more planet, and the prediction of its position promptly led to its discovery. It was named Neptune. Thus the existence of Uranus and Neptune were claimed the moment they were observed and this observation completed their discovery, without regard to particular testable implications.

Turning to physics, we can take Max von Laue's discovery of the diffraction of X-rays in 1912 as a parallel case to this. A conversation with P. P. Ewald aroused in Laue the idea that X-rays would show optical diffraction when passing through a crystal. His attempt to find experimental help to test his idea met with opposition from the director of the laboratory, but when his request finally prevailed, the result confirmed his expectation, and he announced his discovery, which was accepted on this evidence.

Further, we sometimes find examples of beautiful discoveries neither based on any new observation nor predicting anything which would confirm or refute them – e.g., in theoretical work in physics and physical chemistry. Van't Hoff's derivation of the chemical mass action law from the Second Law of Thermodynamics was a fundamental discovery based only on known facts and predicting nothing. In a period extending close on half a century, no one has been able to find a test for the statistical interpretation of quantum mechanics that we owe to Max Born. Its radically new conception of physical laws as predicting only the probability and not the actual course of events controlled by the law is generally accepted today, though it was originally grounded on no new facts and has never offered factual implications that could test it.

Such unempirical theories can be of supreme importance in all the experimental sciences, including biology. Darwinism is an example of it, and indeed in two senses. First, for seventy years Darwinism was accepted by science, even though its evolutionary mechanism could not be understood in terms of known facts, and second, up to this day no such empirical implications of it are known which, if experimentally tested, could disprove the theory.

The second point is widely recognised, so I shall only demonstrate the first. During the first forty years following on the publication of the *Origin of Species* in 1859, it became increasingly clear that the kind of variations known at that time were not sufficiently hereditary to form the basis of a selective process producing evolutionary transformation.[9] Yet the authority of scientific opinion continued to support the theory of evolution by natural selection and spread its deep influence on the world view of humanity. After the re-discovery of Mendelian mutations in 1900 the opposite difficulty arose. These variations were hereditary, but they were much too massive for producing a process of gradual adaptation. Yet the acceptance of Darwinism as our world view, supported by science, remained unshaken, while the new contradictions remained unexplained for another three decades. This difficulty may have been overcome since 1930 through the rise of Neo-Darwinism, and if this new theory holds, the previous disregard of the fact that the theory of natural selection conflicted with the known laws of nature, may turn out to have been justified.[10]

To sum up, we have seen examples to show that important scientific discoveries can be made at a glance and established without any subsequent tests; and that there have been great theoretical discoveries which had no testable experimental content at all. We have seen also that a theory interpreting in a novel way a vast range of experience was accepted by science, and then firmly held for many years by science, though its assumptions contradicted the laws of nature as known at the time, and also that it continues to be held by science, as other important theories are, though it has never been testable by predictions that could be empirically refuted.

IV. PERSONAL JUDGMENT IN SCIENCE

My own theory of scientific knowledge is, and has been from my the start twenty-five years ago, that science is an extension of perception. It is a kind of integration of parts to wholes, as *Gestalt* psychology has described, but in contrast to *Gestalt*, which is a mere equilibration of certain pieces to form a coherent shape, it is the outcome of deliberate integration revealing a hitherto hidden real entity. There are no strict rules for discovering things that hang together in nature, nor even for telling whether we should accept or reject an apparent coherence as a fact.[11] There is always a residue of personal judgment in deciding whether to accept or reject any particular piece of evidence, be it as a proof of a true regularity or, on the contrary, as a refutation of apparent rules. This is how I saw and accepted the fact that, strictly speaking, all empirical science is inexact. And as I came to realise that all such integration is largely based, like perception itself, on tacit elements of which we have only a vague knowledge, I concluded that science too was grounded on an act of personal judgment.

To show this, I became for many years a scandal-monger, collecting cases where the most generally accepted rules of scientific procedure had been flaunted and flaunted to the advantage of science. My first such case showed that even though a new idea conflicts from the start with experience, it may be generally accepted by science. The periodic system of elements shows that the sequence of rising atomic weights produces a striking pattern of the elements in respect of their chemical character. But two pairs of elements fit into the pattern only the wrong way round, that is, the direction of decreasing weights. Yet at no time has this caused the system to be called in question, let alone to be abandoned.

Another example: The idea that light is composed of particles was proposed by Einstein – and upheld, still unexplained, for twenty years – in spite of its being in sharp conflict with the well-established wave nature of light. Commenting on the later history of these cases, (which were among my first scandals) I concluded that any exception to a rule may involve not its refutation but the indication of its deeper meaning.

And I went on to declaring that the process of explaining away observed deviations from accepted teachings of science is in fact indispensable to the daily routine of research. In my laboratory – I said – I find the laws of nature formally contradicted every day, but I explain these events away by the assumption of experimental error. I know that this may cause me one day to explain away a fundamentally new phenomenon and to miss a great discovery; such things have happened in the history of science. Yet I shall continue to explain away my odd results, for if every anomaly observed in a laboratory were taken at its face value, research would degenerate into a wild-goose chase after fundamental novelties.

But these products of my early scandalmongering were surpassed by a statement of Einstein which recently came to my notice.[12] Werner Heisenberg has told the story how, in the course of shaping his quantum theory, he told Einstein that he proposed to go back from Bohr's theory to quantities that could be really observed. To which Einstein replied that the truth lay the other way round. He said "whether you can observe a thing or not depends on the theory which you use. It is the theory which decides what can be observed". Max Planck has also rejected Heisenberg's claim to deal with observables, on the grounds that science is a theory bearing on observations, and never including observations.[13]

The position of observations in the face of prevailing theories is of course precarious. Take once more the famous experiment of Michelson and Morley demonstrating the absence of the ether drift corresponding to the rotation of the earth. Far from rejoicing at this great discovery, which was eventually to form the main experimental support for Einstein's relativity, Michelson called his result a failure. Professor Holton has told in the paper that I quoted before, how both Kelvin and Rayleigh spoke of Michelson's result as 'a real disappointment' and Sir Oliver Lodge even said that this experiment might have to be explained away. Thus the ether theory, which was firmly supported by the current interpretation of physics, caused the experiment to be distrusted. But when some thirty-five years later the same experiment was repeated, with improved instruments by D. C. Miller, and this time did show the presence of an ether drift, this result was rejected in its turn. By this time relativity had overthrown the ether theory. And of course, this time, the theory was rightly preferred to the experiment.[14]

I have no space here to tell in detail the story that I picked up at the very beginning of my scandalmongering[15], how the way scientists of the first rank came out with experiments showing a transformation of elements, because they were encouraged by the radio-active transmutations discovered by Rutherford. There was one epidemic of such publications from 1907 to 1913 that was evoked by Rutherford's discovery (in 1903) that radioactivity involves a transformation of elements; and a second epidemic spread from 1922 to 1928, in response to Rutherford's discovery (in 1919) of an artificial transformation of elements. The observations published during these epidemics would otherwise of course have been cast aside as mere 'dirt effect'.[16]

Let me add a counter example, where plausibility justly triumphed over observation. I have in mind Eddington's derivation from his theory of the universe, developed in the mid-nineteen twenties, that the reciprocal of the fine-structure constant usually denoted by $h\,c/2\pi e^2$ is equal to the integer 137. The theory was widely rejected and this was facilitated by the fact that the experimental value for Eddington's figure was at the time 137.307 with a probable error of ±0.048. However, by the passing of twenty years, new experiments gave a value of 137.009, which brilliantly confirmed Eddington's theory. But this agreement was rejected as fortuitous by the overwhelming majority of scientists; and they were right.[17]

To sum up: Science is the result of an integration, similar to that of common perception. It establishes hitherto unknown coherences in nature. Our recognition of these coherences is largely based, like perception is, on clues of which we are often not focally aware and which are indeed often unidentifiable. Current conceptions of science about the nature of things always affect our recognition of coherence in nature. From the sighting of a problem to the ultimate decision of rejecting still conceivable doubts, factors of plausibility are ever in our minds. This is what is meant by saying that, strictly speaking, all natural science is an expression of personal judgment.

The machinery of genius, which I have described before, is at work all the way from the start to the finish of an enquiry. And once we have recognised this mechanism we can see that we are ourselves the ultimate masters of its workings. Exactitude is recognised then to be always a matter of degree and ceases to be a surpassing ideal. The supremacy of the exact sciences is rejected and psychology, sociology, and the humanities are thus set free from the vain and misleading efforts of emulating mathematical rules.

I started on this way many years ago in a short paper entitled 'The Value of the Inexact'.[18] I pointed out that if we insisted on exactitude of procedure, we would have no chemistry, or at least none to speak of. For chemistry relies for its guidance on judgments of 'stability', 'affinity', 'tendency' as descriptions of chemical processes and also on the skilful application of rules of thumb as guides for acting on such judgments. But the value of the inexact goes much further. It alone makes possible the science of biology. For the structure of living things can be recognised only if we allow our vision to integrate the sight of its parts to the view of a coherently functioning entity, an entity which vanishes if analysed in terms of physics and chemistry.

Hence I have defined scientific value as the joint produce of three virtues, namely (1) accuracy, (2) range of theory, and (3) interest of subject matter.[19] This triad of values distributes our appreciation evenly over the whole range of sciences. We have then greater exactitude and elegance being balanced by a lesser intrinsic interest of subject matter – or else the other way round. For example, most subjects of modern physics are interesting only to the scientists, while the horizon of biology ranges over our experience of animals and plants, and of our own lives as human beings; so the glory of mathematical precision and elegance in which physics far surpasses biology, is balanced in biology by the much greater interest of its subject matter.

Once science is appraised by a three fold grading, all scholarship is elevated to the same pride: as a pride free of pangs about not being a 'real science'. The foolish hierarchy of Auguste Comte is smashed and flattened out.

I am not making excuses for the inexactitude of science and for our personal actions that ultimately decide what to accept to be the truth in science. I do not see our intervention as a regrettable necessity, nor regard its result as a second-rate knowledge. It appears second-rate only in the light of a fallacy which systematically corrupts our conception of knowledge and distorts thereby wide regions of our culture.

Oxford

NOTES

[1] Michael Polanyi, *Science, Faith and Society,* published in 1946 by Oxford University Press; Phoenix Edition, 1964, p. 33.
[2] Michael Polanyi, 'Problemsolving', *Brit. Journ. Philos. Science* **8** (1957), 89–103.
[3] George Polya, *Mathematical Discovery,* John Wiley & Sons, New York, London, Sydney, 1965, Vol. II, p. 63.
[4] C. F. A. Pantin, *The Relation between the Sciences* (ed. by A. M. Pantin and W. H. Thorpe), Cambridge Univ. Press, 1968, pp. 121–122.
   We may look also at other creative work. Kant has described in *The Critique of Pure Reason* the part of anticipation in the pursuit of philosophic problems. He wrote: "It is unfortunate that not until we have unsystematically collected observations for a long

time to serve as building materials, following the guidance of an idea which lies concealed in our minds, and indeed only after we have spent much time in the technical disposition of these materials, do we first become capable of viewing the idea in a clearer light and of outlining it architectonically as one whole according to the intentions of reason".

H. W. Janson (*History of Art*, New York 1962, p. 11) describes anticipation in making a painting: "It is a strange and risky business in which the maker never quite knows what he is making until he has actually made it, or to put it another way, it is a game of find-and-seek in which the seeker is not sure what he is looking for, until he has found it".

Northrop Frye (*T. S. Eliot*, 1963, p. 28) speaks of Eliot's account of anticipation: "The poet has no idea of what he wants to say until he has found the words of his poem... [He] may not know what is coming up, but whatever it is, his whole being is bent on realising it".

Anticipations of this kind resolve the problem of Meno in which Plato questions the possibility of pursuing an enquiry in our inevitable ignorance of what we are looking for.

[5] Michael Polanyi, *Personal Knowledge*, Routledge and Chicago, 1958, pp. 9–13.

[6] Adolf Grünbaum, *Philosophical Problems of Space and Time*, New York 1963, pp. 385–386.

[7] Gerald Holton, 'Einstein, Michelson and the 'Crucial Experiment' ', *Isis* **60** (1969), 133–197.

[8] Clearly, the position to which I am referring is that Sir Karl Popper stated in *Logik der Forschung* (1934), translated into English as *The Logic of Scientific Discovery* (1946). It is this statement that has been most widely influential and though it was modified in some parts in Poppers *Conjectures and Refutations* (1963), the changes do not substantially affect the principles of 'refutationalism' which I shall test here.

[9] C. D. Darlington, *Darwin's Place in History*, Basil Blackwell, Oxford, 1960, p. 40.

Professor Darlington has described in Chapter 8 entitled 'The Retreat from Natural Selection' how in the successive editions of Darwin's *Origin of Species* natural selection is gradually abandoned and evolution "shored up with Lamackian inheritance".

[10] The present situation was described as follows by Julian Huxley, *Evolution the Modern Synthesis*, Allen and Unwin, London, 1942, p. 116 and repeated in the same words in its revised edition in 1963.

It must be admitted that the direct and complete proof of the utilization of mutations in evolution under natural conditions has not yet been given... Thus it is inevitable that for the present we must rely mainly on the convergence of a number of separate lines of evidence each partial and indirect or incomplete, but severally cumulative and demonstrative.

J. Maynard Smith in an article entitled 'The Status of Neo-Darwinism', pp. 82–89 in *Towards a Theoretical Biology* (ed. by C. H. Waddington), Aldine Publishing Co., Chicago, 1969, has listed some evidence as proving that Neo-Darwinism is not 'tautological'. But he merely shows, as Huxley does, that the evidence so far supports the theory.

[11] The view that discovery is an informal process was anticipated by William Whewell in his brilliant critique of the mechanistic scheme of induction effectively propounded at that time by John Stuart Mill. The way intuition suddenly follows on a time of strenuous searching has been described by Poincaré in his famous account of an experience of his own. These examples have supported me in my enterprise.

[12] W. Heisenberg, 'Theory, Criticism and a Philosophy', 71 in *From a Life of Physics*, special supplement of the *Bulletin of the International Atomic Energy Agency*, Vienna, pp. 36–37. This publication is not dated.

[13] Max Planck in *Positivismus und Reale Aussenwelt*, Akademische Verlagsgesellschaft, Leipzig, 1931, p. 21, says, "... there exists absolutely no physical magnitude which can be measured in itself" (my translation).

[14] For details of this event see my *Personal Knowledge*, pp. 12–13.

[15] *Op. cit., Science, Faith and Society*, pp. 91–92.

[16] For details see my *Science, Faith and Society*, 1946, Appendix 2.

[17] For further details see *Personal Knowledge*, 1958, pp. 43, 151, 158, 160.

[18] Michael Polanyi, *Philosophy of Science* 3 (1936), 233.

[19] See *Science, Faith and Society*, 1946, Chapter II, Section II.

W. A. SUCHTING

REGULARITY AND LAW*

I. INTRODUCTION

The concept of a law of nature has a central place in most comprehensive reflections on the history and philosophy of science. To confine attention here only to the systematic, philosophical aspect, the very aim of science (at least as the latter has been understood since roughly the Renaissance) has been widely held to be the discovery of empirical laws.[1] Again, the concept of law has been appealed to from a wide variety of viewpoints in the analysis of notions such as explanation, confirmation, and subjunctive conditionals.

In particular, the notion of law is fundamental to most Humean ('constant conjunction', 'regularity') accounts of cause. Hume himself did not, of course, speak much about laws; rather, he stated his famous doctrine of cause in terms simply of constant conjunctions of similar sorts of items. But the account in just the terms in which he stated it is subject to very obvious and decisive objections,[2] and later philosophers working in the spirit of Hume have generally formulated their theories in terms of the notion of law. Briefly (and therefore roughly) it has been held by such philosophers that a singular causal statement of the from 'a caused b' (e.g. 'the striking of the match caused it to light') is to be taken as equivalent (in some sense) to the claim that there is a set of statements which contains essentially at least one law of sequence and the statement of the occurrence of a, and which entails a statement of the occurrence of b.[3] This is the basic idea of the analysis not only of 'factual' conditionals of the above sort, but also of subjunctive and counterfactual conditionals.[4] And with these means other, connected concepts are accounted for, e.g. (causal) dispositions.[5]

Despite this, and despite the fact that an Humean view on causation and related matters has tended to be part of the orthodoxy of empiricism amongst both professional philosophers and scientists, Hempel was right in summing up the situation in 1964,[6] in the course of a postscript to a reprint of his classical 1948 paper on explanation with Oppenheim, by saying that "it remains an important desideratum to find a satisfactory version" of a definition of law "which requires more of a law-like sentence than that it must be essentially universal."[7]

Nevertheless, Hempel pretty clearly did not think that the problem is anything but just a very difficult one. And this attitude is fairly common. For example, in his book *Philosophical Foundations of Physics,* published in 1966, Carnap assured his readers at the end of the chapter entitled 'Does Causality Imply Necessity?' that "it should be kept clearly in mind that, when a scientist speaks of a law, he is simply referring to a description of an observed regularity,"[8] and the reader is left in no doubt that this is Carnap's view of laws too. Furthermore, his remarks in the succeeding chapter suggest that he is confident that further work will yield an adequate formal account of laws conceived in this way.[9] Again, Quine uses essentially the law/non-law distinction in the Humean account contained in his paper 'Necessary Truth', reprinted also in 1966, without any manifest worries about it.[10]

In this paper I want to give some reasons for thinking that the problem of giving an adequate account of law along Humean lines is an insoluble one. I do not have any strictly demonstrative arguments. But I think that there are basic and pervasive difficulties in attempts to develop such an account.

I shall first distinguish two general Humean strategies for marking off the class of law statements, and then comment critically on some typical accounts of these types. This method of making the object of analysis accounts which have actually been put forward rather than one's own formulations of what are claimed to be general trends, runs the risk of producing a scrappy result and a concentration on the trees rather than the wood. However, it has the advantage of largely protecting the discussion from being an examination of straw-men; moreover the comments on specific accounts are, by and large, meant to be of general import. At any rate the door is left open for someone to produce a view which is not subject to objections similar to those levelled at the particular doctrines examined here.

II. TWO HUMEAN APPROACHES TO THE PROBLEM OF LAW

I shall simply assume here, what is implicit in the citation from Hempel above, namely, that the class of true, empirical, universally quantified statements not logically equivalent to any singular statement is wider than the class of law-statements. I shall also assume that this is so even if the conditions are strengthened by adding the requirement that a law state-- ment should not contain any essential occurrences of proper names.[11] In a nutshell, the Humean problem of law is to find a way of marking off the narrower class of law-statements in a way consistent with the general Humean program – in particular, without using non-truth-functional sentential connectives, modal operators (other than 'contingently' or ones reducible to the latter), or irreducible 'causal' (or 'nomic') relational predicates.

For an Humean, a law-statement cannot be, in itself, more than a true universal statement of fact (subject to the above-mentioned restrictions).[12] Thus it would seem, *a priori,* that law-statements must be marked off by means of certain specific relations that they bear to other items. It would also seem *a priori* that the possible candidates for the other term of the relation must be either people who entertain statements, or other statements.

In fact, consideration of the Humean literature on this problem of law reveals two broad directions of attack on the problem of specifying the requisite further conditions. I shall call them, for want of better names, the *pragmatic* and the *systematic.* Philosophers pursuing the first line try to mark off universal statements which are laws from those which are not in terms of certain attitudes which people have to the former but not to the latter, attitudes which may be expressed, for example, in the use or uses which the one kind of universal but not the other have in guiding human action. Philosophers pursuing the second, 'systematic' line try to mark off universal statements which are laws from those which are not in terms of certain characteristic logical relations which law-statements have to other statements. The statements marked out as laws on this second approach may well be typically the objects of certain characteristic human attitudes, expressed, for example, in the uses made of them and not others, in guiding human action. But these attitudes are taken not to constitute, but rather to be a consequence of, their being law-statements.

I have said that theses two approaches may be distinguished in the various attempts which have been made to give an Humean definition of law. This is not to say that the two are not sometimes run together.[13] Furthermore, whilst I have not met with any approaches which cannot be fitted into one or other of these two categories, and am tempted to claim that they are the two fundamental ones, I certainly have no demonstrative argument for such a claim.

III. THE PRAGMATIC APPROACH TO THE HUMEAN PROBLEM OF LAW

III.1. Hume himself might be taken to be an early precursor of the pragmatic approach insofar as he held, firstly, that the idea of necessary connection is the central component of the idea of the causal relation, and, secondly, that the idea of necessary connection is to be explicated in terms of certain facts about human beings and other animals, viz. propensities to associate in a reflex-like manner certain 'impressions' with certain 'ideas'. But be that as it may, it will be best to consider some rather later examples of what I take to be this approach.

Such an approach is to be found at least as far back in the literature as a now generally forgotten paper in *Mind* for 1927 by R. B. Braithwaite, in which he defended the view that a universal of law as distinct from a universal of fact is, briefly, one which is believed on non-demonstrative grounds, i.e., on grounds other than a complete enumeration of instances.[14]

Similarly, F. P. Ramsey, though he regarded "variable hypotheticals or causal laws" as "not judgments but rules for judging 'If I meet a ϕ, I shall regard it as a ψ'",[15] distinguished such rules from others by saying that the conjunction of the evidence on the basis of which we adopt them is, in the case of laws, such that "we trust it to guide us in a new instance."[16]

Again, in the final section of his well known paper 'The Problem of Counterfactual Conditionals', Nelson Goodman poses the question of "what... distinguish(es) a law like 'All butter melts at 150 °F' from a true and general non-law like 'All the coins in my pocket are silver'?" His first answer to this question is subsequently refined later in the paper. But this initial statement may be taken for purposes of analysis, because it contains the core idea, and it is only with this that I am concerned. Goodman says then that the difference between the law and the non-law consists 'primarily' in

... the fact that the first is accepted as true while many cases of it remain to be determined, and further, unexamined cases being predicted to conform with it. The second sentence, on the contrary, is asserted as a description of contingent fact *after* the determination of all cases, no prediction of any of its instances being based upon it.

He goes on:

As a first approximation then, a law is a true sentence used for making predictions...
I want to emphasize the Humean idea that rather than a sentence being used for prediction because it is a law, it is called a law because it is used for prediction; and that
rather than the law being used for prediction because it describes a causal connection,
the meaning of the causal connection is to be interpreted in terms of predictively used
laws.[17]

Further, P. F. Strawson begins by characterizing a law-statement as a
true, empirical general statement which contains no words the reference
of which depends in any way upon the situation in which they are uttered.[18]
He then adds[19] as a final condition (à propos of one of William Kneale's
objections to the regularity theory) that the evidence for the truth of a
law, our grounds for accepting it, shall not contain, essentially, evidence
that the members of the subject-class of the law "is limited in a very definite temporal sense, i.e., evidence that there will be no more members,
and that there never were more than the limited number of which observations have been recorded." Thus, in sum, to call certain true, empirical, general statement law-statements "is both to emphasize the un-
restrictedness of their application and to say something about the nature
of the evidence by which they are supported."[20]

As a final example I shall cite the view expounded in A. J. Ayer's paper
'What is a Law of Nature?'.[21] In this paper he begins the positive part of
the treatment by suggesting that the difference between 'generalizations
of law' and 'generalizations of fact' "lies not so much on the side of the
facts which make them true or false, as in the attitude of those who put
them forward." Thus he proposes to explain the distinction between the
two sorts of generalization "by the indirect method of analysing the distinction between treating a generalization as a statement of law and treating it as a statement of fact."[22]

His summary of his account occurs in a passage beginning thus:

... for someone to treat a statement of the form 'if anything has Φ it has Ψ' as expressing a law of nature it is sufficient (i) that, subject to a willingness to explain away exceptions, he believes that in a non-trivial sense everything that in fact has Φ has Ψ (ii)
that his belief that something which has Φ has Ψ is not liable to be weakened by the
discovery that the object in question also has some other property χ...[23]

There follows some qualifications to the second condition which it is not
necessary to add, as I do not wish to criticize the details of his account.[24]

III.2.   My queries about this species of Humean account are very simple ones. But they also concern fundamentals. And I have not seen anything in the literature that resolves my doubts on these scores.

My first general query concerns the possibility of unknown laws.

We ordinarily think that it is possible that there should be unknown laws. Thus we think that, if there are laws of nature at all, at least some of them obtained before the existence of human beings and that, should the human race disappear, at least some laws would continue to obtain. We think it possible (and indeed very likely) that at least some laws which obtain, will never be known or even formulated by human beings. Indeed, I think we consider it unnecessary for the obtaining of laws that any law-statement ever be known or believed by any human being.

And indeed, that there can be unknown laws is explicitly assumed by some of the philosophers I have cited above, implicitly assumed by others, and probably believed by all of them.[25]

But an at least *prima facie* difficulty faces a pragmatic account of law here. For if the definition of a law (in terms of a statement which expresses it) makes an essential reference to people's attitudes, it is not immediately obvious how there could be laws in the absence of people to have attitudes.

Ayer is one of the proponents of a pragmatic approach who expressly affirm that there can be unknown laws. It might be expected that he would appeal, in this regard, to the fact that he presents his account as at most a sufficient and not a necessary condition for a statement's being a law-statement.[26] But he simply asserts that the possibility of unknown laws "is not inconsistent with holding that the notion is to be explained in terms of people's attitudes."[27]

Unfortunately Ayer does not spell out this remark. But it may be conjectured that what he has in mind is something like this: to say that there is an unknown law expressible by sentence S is to say that what would normally be asserted by S is true, though nobody in fact asserts S, and that there are certain situations which would be the subject-matter of evidence-statements E relevant to S, such that if someone were to be in possession of E, and also consider S, then that person would have an attitude to S satisfying conditions on laws (partly cited above).

An immediate objection to this way out is that it uses subjunctive conditional language (concerning people's tendencies to have beliefs of certain sorts, etc.) which is at least as problematic for a Humean as the notion of law. Indeed, of course, Humeans (like many others) have typically construed subjunctive conditionals in terms of law-statements. Thus the view in question reduces talk of unknown laws to subjunctive conditional talk about human behavior, and this is in turn construed, presumably, in terms of laws about human behavior. Now these laws must also be construed on Humean lines. That is, in the final analysis, they must also be held simply to describe certain regularities. But suppose there were no human beings to manifest regularities of behaviour. Then, in the logically possible world in which the regularities of human behaviour are not reducible to other, actual regularities, there would not, on the Humean view, be any laws of the appropriate sort. Hence it would seem that there could be no unknown laws in a world which was completely (omnitemporally and omnispatially) knowerless. But in such a world there are no known laws either. Therefore in such a world there are no laws.

The view that I have ascribed to Ayer regarding unknown laws is, I think, similar to one which was held by F. P. Ramsey. However, I have preferred to bring Ramsey's views into the discussion only after the above discussion had been completed because I do not find them entirely clear.

Ramsey says[28] that to speak of an unknown causal law

must mean that there are such singular facts in some limited sphere (a disjunction) as would lead us, did we know them, to assert a variable hypothetical. But this is not enough, for there must not merely be facts leading to the generalization, but this when made must not mislead us. (Or we could not call it a true causal law.) It must therefore be also asserted to hold within a certain limited region taken to be the scope of our possible experience.

He goes on to say that

To this account there are two possible objections on the score of circularity. We are trying to explain the meaning of asserting the existence of an unknown causal law, and our explanation may be said to be in terms of the assertion of such laws, and that in two different ways. We say it means that there are facts which would lead us to assert a variable hypothetical; and here it may be urged that this means that they would lead us in virtue of one possibly unknown causal law to form a habit which would be constituted by another unknown causal law.

His answer to this is

first, that the causal law in virtue of which the facts would lead us to the generalization must not be any unknown law, e.g. one by which knowledge of the facts would first drive us mad and so to the mad generalization, but the known laws expressing our methods of inductive reasoning; and, secondly, that the unknown variable hypothetical must here be taken to mean an unknown statement (whose syntax will of course be known but not its terms or their meanings), which would, of course, lead to a habit in virtue of a known psychological law.

This reply would seem to be open to the earlier objection.

III.3.   My second difficulty can be introduced in connection with what Ramsey says at the beginning of the last quotation from him above. He says here that "the causal law in virtue of which the facts would lead us to the generalization" viz. the generalization accepted as another causal law "must... be... the known laws expressing our methods of inductive reasoning." This seems to imply that the statements to which we take up or would take up the relevant attitudes such that those statements are laws, and hence the members of the class of law-statements, depend on *de facto* modes of functioning of human organisms. And nothing that the other exponents of the pragmatic approach quoted say is inconsistent with this. For example, Braithwaite defines laws simply in terms of our beliefs, Goodman in terms of our predictive-inductive practices, and so on.

My difficulty is this. It seems *prima facie* possible that the world should be just the same as it actually is except that the ways in which people actually form their beliefs, the ways in which they guide their actions, etc., should be different. Even supposing that there is uniform behavior in these respects now, it seems possible that there should be a different uniform behavior. Or, we seem to be able to imagine that in one and the same world, some people should act in one way and others in other ways. In general it seems to be possible that two possible worlds should differ only with respect to the attitudes, inductive practices, etc., of the people in them.

But all this seems to be logically impossible on the pragmatic view of law. For on this view, what laws there are depends, in the final analysis, on what attitudes, inductive practices, etc., people have. Thus the membership of the class of law-statements would be, logically, a matter of convention, depending on which set of attitudes (etc.) we decide on as that defining lawfulness. (This conventionality does not mean, to use Poincaré's distinction,[29] that the decision is arbitrary, for some choices may be preferable to others in terms of simplicity, elegance, and so on.) In brief, to say, e.g., that a statement is not used for predictive purposes because it is a law, but a law because it is used for predictive purposes, seems to have the consequence that there are as many possible sets of law-statements (asserted of one and the same natural world) as there are possible predictive practices; and this certainly conflicts with our pre-analytic understanding of the concept of law. In particular, insofar as we cash subjunctive conditionals in terms of laws, it makes it a logically conventional matter what subjunctive conditionals are true (or warrantedly assertable, or whatever).

I shall conclude with mention of another difficulty which I find with this view, connected but not identical with the one just mentioned. I shall put it in relation to Goodman's account, though I think that it could apply to the others too, *mutatis mutandis*.

It is said that 'All the coins in my pocket are silver' is 'accepted as a description of contingent fact' only after the determination of all cases, hence is not usable for predictive purposes, and hence is not a law (even if true). In all this it is said to contrast with 'All butter melts at 150 °F'.

Now suppose that the number of coins in my pocket were very large indeed; suppose even further that the temporal reference were omnitemporal, so that 'all the coins in my pocket' referred to all the 'coins ever in my pocket'. Then it is not clear to me that it would not ordinarily be thought to be inductively reasonable to predict, on the basis of a knowledge of the fact that such and such a proper subset of the coins was silver, that the next instance is. If it would be, then the universal in question seems to satisfy the requirements for being a law (assuming that all the coins were in fact silver), even if it were not explainable in terms of wider generalisations (e.g. about my habits, or the habits of people like me in relevant respects).

Or, consider the other case. Suppose that, as a matter of fact, the world ever contained only very few instances of butter, and all of these were contained in one storehouse. Would this make it that 'All butter melts at 150 °F' is 'accepted as a statement of contingent fact' only after the determination of all cases, 'no prediction of any of its instances being based upon it' ? If so, then the universal in question seems to satisfy the requirements for being a law. If not, what distinguishes this universal from the other ?

Now we would ordinarily think that what the above thought-experiments do is simply to alter, in imagination, the number and distribution of instances of the universals in question. And the actual number and distribution of instances of a universal seem to be, with respect to that universal at least, a completely accidental matter which could not affect the lawlikeness or otherwise of the universal.

But even if the conclusions that I have drawn from the account of law in question are validly drawn, I realize that it may be objected that I have begged one or more of the fundamental questions at issue. For, after all, it may be replied that the view being discussed is one which construes laws as being statements about what is omnitemporally actually the case and not about possible worlds, and that if the supposition about what the world is actually like is changed then we should not be surprised if what we say in the laws of the suppositious world are judged to be different. In reply to this I can only repeat my claim that the consequences in question are inconsistent with central features of the ordinary understanding of the notion of law, and that insofar as a criterion of adequacy of the philosophical account of law is consistency with central features of that ordinary understanding, the inconsistency in question constitutes an objection.

IV. THE SYSTEMATIC APPROACH TO THE HUMEAN PROBLEM OF LAW

IV.1. I turn now to a consideration of what I have called earlier the 'systematic' approach to the Humean problem of law, viz., the attempt to mark off law-statements in terms of certain logical relations which specifically law-statements have to other statements. The main lines of such an approach were sketched by J. S. Mill in his *Logic*. It was developed, in the renaissance of the problem of law in the second half of the 1940's by, for example, Reichenbach and Hempel and Oppenheim. It happens that the two main Humean accounts of law that have been published in the last five years are both examples of this approach. One of these – Popper's[30] – I have considered elsewhere.[31] In this section I shall consider whether the other is any more trouble-free than previous attempts along this line have been shown to be.

This second of the two recent attempts to state a comprehensive definition of law on a regularity basis is that by Bernard Berofsky in *Noûs*, November 1968. I shall leave aside the mainly polemical parts,[32] with which, as a matter of act, I agree. I shall also leave undiscussed his widely-held view that the most viable variety of the main alternative to a regularity theory of law, viz. the necessitarian view, is that which holds the alleged necessity of natural laws to be of a unique type ('nomic', 'causal', 'physical') rather than logical, though, as a matter of fact I do not agree with this.

Berofsky's thesis is that a sentence is a law-statement if and only if it satisfies eleven conditions, which are "selected in accordance with criteria actually used by scientists and laymen."[33]

Of these conditions the first seven[34] are less open to contention than the others and in the present context it will suffice simply to list them. A sentence is a law then only if it is: true; logically contingent; of universal conditional form; and only if the predicates occurring in it are: not self-contradictory; 'open-ended', where " 'P' is not open-ended if and only if 'Px' entails 'there are a finite number of individuals that satisfy 'Px' ' "; 'qualitative', where a 'qualitative' predicate is 'one that can logically apply to an indefinite number of instances'; 'non-positional', i.e., not of the sort exemplified by 'grue'.

According to the next condition, the *Deductive Systematization Requirement*[35] (DS) a sentence is a law only if it is 'deductively systematized'. And a sentence L_1 is 'deductively systematized' if and only if there exists a sentence L_2 which satisfies the first seven conditions, and is either higher or lower on the 'scale of deductive systematization' than L_1. Berofsky has a rather elaborate definition of 'higher (lower) on the scale of deductive systematization than'. Essentially, the idea is the same as that of one universal's being 'more general than' another. For example, 'All human beings are mortal' is 'higher on the scale of deductive systematization' (or 'more general') than 'All women are mortal'.

I shall not stop to expound in detail or discuss this definition, though it is of interest on its own account. For, as Berofsky explains, (DS) is not by itself a significant condition on laws, since any generalization at all can be deductively systematized in accordance with it. "Given any true sentence of the form 'All P is Q', there will always be some predicate which is true of all or some P's, say 'R', and [making it possible to] deduce 'All PR is Q', thereby satisfying the definition of deductive systematization. Moreover, in the same way, an indefinite number of sentences can be derived from 'All P is Q'. From 'All the men in room A are bald,' and 'All the tall men in room A are bald'."[36] Its importance lies in its being a necessary condition for the non-trivial applicability of the remaining three conditions.[37]

The next condition is the *Comprehensiveness-Requirement* (C)[38]. Berofsky states that this is "the most important requirement." This is correct because, as already mentioned, the first seven conditions are less controversial than the others and the previous condition (DS) is a presupposition of the non-trivial application of the remaining conditions, whilst the final two tidy up loose ends left by (C).

(C) states that L is a law only if (1) it appears in a deductive systematization which 'accounts for' each sentence of a set S_1, each member of which entails some sentences of the form Px (or Pxy, etc.), and (2) there is no other deductive systematization that 'accounts for' (a) the members of S_1 and (b) other sentences entailing sentences of the form Px (etc.).

A set of sentences S is 'accounted for by a deductive systematization' (DS) if and only if it is a set 'each member of which is entailed by the DS plus any legitimate state description' (SD). (There is a qualification here to block a possible objection but this will not be relevant to my further remarks.) Thus, 'All men in room A are bald' *accounts for* the set S_1: *Ba, Bb, ... Bn* (where 'B' is short for 'bald' and *a, b, ... n* designate all and only the men in room A), because in conjunction with the legitimate SDs: *Ma, Mb, ... Mn* ('M' short for 'man in room A') it entails S_1. 'All men in room A are bald' passes condition (1) of (C). But it fails condition (2) of (C) insofar as, let us assume, there is another universal, say, 'All men with scalp condition P are bald', which accounts for all the cases of baldness in room A, plus others.

But (C) allows unacceptably excessive width to the concept of law. Simply consider the example already cited. Suppose that in fact there were no alternative universal in terms of P or some other condition, accounting for the baldness of the men in room A. Then 'All men in room A are bald' would count as a law (assuming that it passes the final two conditions, which it may easily do). (I also assume that it passes the first seven conditions. In doing so I take myself to be following Berofsky, who appears to exclude it by virtue of its failing the Comprehensiveness Condition.) But it may be that people would be at least very unwilling to count this as a law. And this unwillingness would by no means necessarily depend on the assumption of any special insight into what is or is not lawlike. On the contrary, it might be accounted for in terms of Berofsky's own remark somewhat further on,[39] that "our belief that certain sentences are laws, while others are accidental generalizations ... hinges very often ... on a belief that certain general theories are true." In the above case, the very general theory may be that systematic similarities or differences between the properties of individuals do not depend merely on similarities or differences in spatial position (being in room A or elsewhere, for example).

Let me try to put the point a little differently and more generally. Consider an individual a which is (omnitemporally) uniquely F, and which also has a property G (which may or may not be instantiated by other individuals); and suppose that F and G satisfy the conditions restricting the sorts of predicates which may appear in laws. 'All F are G' is true, and if we further suppose it contingent, it then satisfies the first seven conditions. (DS) can always be satisfied, and satisfaction of the last two conditions may again be assumed. As regards (C) it is clear that 'All F are G' accounts, formally, for Ga. Now either there is an alternative deductive systematization which accounts for Ga or there is not. If there is, then Ga is explained by it. If there is not, then the conjunction of 'All F are G' and 'Fa' explains 'Ga'. Therefore, whatever is the case, 'Ga' is subsumable under a law. But to be subsumable under a law is, roughly, what is generally meant by saying that some state of affairs is determined (rather than 'accidental'). However, intuition suggests that the G-ness of that a might be quite accidental.

In fact, nothing essential hangs on the assumed uniqueness of a in being F. If there is any trouble with 'All F are G' where the subject-class is unique, then it is not clear how the trouble could be removed simply by increasing the number of instances. The general point in question is my claim that we should not ordinarily think that 'All F are G' is a law just because it is the only generalization that, in conjunction with Fa, entails Ga (even where 'All F are G' satisfies Berofsky's other conditions); for this underivability may be due to its being purely accidental as well as for its being naturally necessary.

The above remarks apply, *mutatis mutandis*, to Berofsky's comment on his example 'All Presidents of the United States from Missouri order an atomic bomb attack on Hiroshima'. He says that this is "easily eliminated because there is a competing account in terms of beliefs and desires held by Truman that is far more comprehensive."[40] Quite apart from the fact that the universal in question entails nothing about Truman, what if there were not such a competing account available? I think that we would be more likely to consider the universal purely accidental, rather than lawful; certainly it seems to be possible that it should be purely accidental, and this Berofsky's account does not allow for.

Berofsky remarks towards the end of the paper that his account may be criticized insofar as "the insistence upon systematization seems to rule out laws that do not belong to a comprehensive theory". As he says, quite correctly, "this criticism is easy to rebut. First of all, any generalization can be trivially systematized by making it an axiom and deducing lower-order generalizations from it. Then we consider whether or not a more comprehensive theory is in the offing. If there is none, a generalization that looks relatively isolated will be a law on the present approach."[41] The criticism made in this section is not that (C) unduly restricts the class of laws but rather that it allows too wide an extension to the class of law-statements. Some systematically isolated generalizations would, we should expect, be laws, and others not. Berofsky's conditions do not allow us to make the distinction.

I shall not comment specially on Berofsky's final two conditions as no new points of principle would be involved.

V. CONCLUSION

Even if all of the claims in the main parts of this paper were true, they would obviously not constitute a demonstration that a regularity account of law is not possible: as I said in the introductory remarks I do not have any *a priori* arguments against the possibility of such an account. It would thus always be on the cards that somebody even more imaginative and ingenious than those whose views I have discussed would come up with a relatively trouble-free account. But if all of my claims were true, I would consider this to be some evidence for the unlikelihood of this happening.

My claims may be disputed in various general ways. Firstly, it may be argued in the various special cases that what I have said follows from the account in question does not follow. Secondly, it may be agreed in such and such a case that the consequence does not follow but argued that it is not inconsistent with pre-analytical intuitions on the matter, where such conflict is the burden of the criticism. Thirdly, the conflict may be admitted, but its thrust may be sought to be avoided. And this may occur in two main ways. On the one hand, those who make it a criterion of adequacy of their analyses that the latter should fulfill the main intentions involved, obscurely perhaps, in the ordinary notions, may argue that the conflicts with intuition are merely peripheral, and part of the price that any precisification must pay. On the other hand, some may be prepared to ignore any such conflicts, provided only that the account be adequate for a language concerned with actual and foreseeable ongoing scientific work. These various types of rejoinder are progressively more and more general in their assumptions, and make it progressively more and more certain that no final solution to a problem like the present will be reached solely on the field of battle defined by this problem alone.

University of Sydney

NOTES

* The text of a paper given to the Boston Colloquium in March, 1970, with the omission of a section on Popper's account of natural necessity which did not add anything essential to what I had already argued elsewhere.

(I am indebted to remarks by Professor L. S. Carrier on a first draft for stimulus to a rewriting of various parts of it.)

[1] For changes in the notion of law see Zilsel (1942).

[2] See e.g. Taylor (1966), 22f.

[3] See e.g. Popper (1945), 342f, and Berofsky (1966) to cite just two of numberless examples.

[4] E.g. Nagel (1961), 68ff, Mackie (1962), etc.

[5] E.g. Hochberg (1967), Wilson (1969), etc.

[6] It has not been till comparatively recently (roughly in the last twenty years) that much attention has been given to formulating an exact account of law along Humean lines. There is practically a total gap between Hume and J. S. Mill's *System of Logic*. And if we except some work by the early Russell and Broad (Russell, 1914, Lecture VIII; Broad, 1914, Chapter II) there is pretty much a gap between Mill and the work of Goodman (1947), Reichenbach (1947), Hempel and Oppenheim (1948) and others about another century later. (The treatments in e.g. Mach, 1908; Pearson, 1911; and Peirce, 1903, added little if anything.) Other Humean accounts in the period between 1947 and 1964 include Braithwaite (1954, 1st ed. 1952), Reichenbach (1954), Ayer (1956), Popper (1968a), Nagel (1961), Pap (1962). For criticisms of these accounts see e.g. Ayer (1961), Cohen (1966, Chapter X), Hempel (1955, 1965), Jobe (1967), Madden (1963, 1969), Molnar (1969), Nerlich and Suchting (1967), Pap (1958, 1963). Scriven (1961, 1964),

[7] Hempel (1965), 293.

[8] Carnap (1966), 207.

[9] Carnap (1966), 211, 212f, etc. Cf. also Carnap (1963), 207.

[10] Quine (1966), 50f.

88 W. A. SUCHTING

[11] For relevant argument on this point see e.g. Hempel (1965), Kneale (1950, 1952, 1961, 1962).

[12] I am assuming throughout that laws are true/false statements and not prescriptions, rules or inference, etc. For arguments against such a view of laws and detailed references to the literature see Hempel (1965).

[13] For example, consider Walters (1968), 413.

[14] Braithwaite (1927), 470, 473. As regards the justice of this summing up of Braithwaite's account, cf. Ramsey (1929), 242.

[15] Ramsey (1929), 241.

[16] Ramsey (1929), 243.

[17] Goodman (1947) as reprinted in Goodman (1954), 26. (With this prediction – criterion of laws, cf. Schlick (1931).)

[18] Strawson (1952), 198.

[19] Strawson (1952), 199f.

[20] Strawson (1952), 200.

[21] Ayer (1956). References are to the reprint in Ayer (1963).

[22] Ayer (1963), 230, 231.

[23] Ayer (1963), 233.

[24] Though it can in fact easily be shown that, even accepting Ayer's pragmatic framework, the conditions are not sufficient, i.e., that statements which people would not treat as laws of nature are not excluded by the conditions which he lays down. On this see Jobe (1967), 368f.

[52] See Ramsey (1929), 243, Goodman (1954), 27 (bottom of page), Ayer (1963), 234.

[26] In an earlier passage on the same page he refers to the fact that he has not claimed necessity rather than sufficiency for his account. He says that this is "both because I think it possible that they could be simplified and because they do not cover this whole field. For instance, no provision has been made for functional laws, where the reference to possible instances does not at present seem to be eliminable". On the second claim see Suchting (1968).

[27] Ayer (1963), 234.

[28] Ramsey (1929), 244f. See further 251f. Cf. also Pap (1948), 220.

[29] Poincaré. I am indebted to Richard S. Taylor (University of South Florida) for bringing what is essentially Poincaré's distinction to my notice in this context.

[30] Popper (1968b). The 'revised' in the title of Popper's paper is an allusion to an earlier account published in (1968a) for the first time in the first edition of 1959. For criticism of this see Nerlich and Suchting (1967).

[31] See Suchting (1969).

[32] See especially 320–26, 331–34.

[33] Berofsky (1968), 316.

[34] Berofsky (1968), 318ff.

[35] Berofsky (1968), 327ff.

[36] Berofsky (1968), 330. The words in square brackets are mine: the original text is incomplete and I conjecture that my addition conveys the intended meaning. Cf. also 338.

[37] Berofsky (1968), 330, 335.

[38] Berofsky (1968), 334f.

[39] Berofsky (1968), 338.

[40] Berofsky (1968), 338.

[41] Berofsky (1968), 338f.

BIBLIOGRAPHY

Ayer, A. J., 'What is a Law of Nature?', *Revue Internationale de Philosophie* **2** (1956) No. 36. Reprinted in Ayer (1963).
Ayer, A. J., 'Review of Nagel', *Scientific American* **204** (1961) 197–203.
Ayer, A. J., *The Concept of a Person*, Macmillan, London, 1963.
Berofsky, B., 'Causality and General Laws', *J. Phil.* **63** (1966) 148–57.
Berofsky, B., 'The Regularity Theory', *Noûs* **2** (1968) 315–40.
Braithwaite, R. B., 'The Idea of Necessary Connection' *Mind* **36** (1927) 467–77; **37** (1928) 62–72.
Braithwaite, R. B., *Scientific Explanation*, 2nd ed., Univ. Press, Cambridge, 1954.
Broad, C. D., *Perception, Physics & Reality*, Univ. Press, Cambridge, 1914.
Carnap, R., 'Replies and Expositions' in *The Philosophy of Rudolf Carnap* (ed. by P. A. Schilpp), Open Court Publ. Co., LaSalle, Ill., 1963.
Carnap, R., *Philosophical Foundations of Physics*, Basic Books, New York, 1966.
Cohen, L. J., *The Diversity of Meaning*, 2nd ed., Methuen, London, 1966.
Goodman, N., 'The Problem of Counterfactual Conditionals', *J. Phil* **44** (1947). Reprinted in Goodman (1954).
Goodman, N., *Fact, Fiction & Forecast*, Athlone Press, London, 1954.
Hempel, C. G., 'Review of H. Reichenbach (1954)', *J. Symb. Logic* **20** (1955) 50–54.
Hempel, C. G. and Oppenheim, P., 'Studies in the Logic of Explanation', *Phil. Sci.* **15** (1948) 135–78. Reprinted in Hempel (1965).
Hempel, C. G., *Aspects of Scientific Explanation*, Free Press, New York, 1965.
Hochberg, H., 'Dispositional Properties', *Phil. Sci.* **34** (1967) 10–17.
Jobe, E. K., 'Some Recent Work on the Problem of Law', *Phil. Sci.* **34** (1967) 363–81.
Kneale, W., 'Natural Laws and Contrary to Fact Conditionals', *Analysis* **10** (1950). Reprinted in *Philosophy and Analysis* (ed. by M. MacDonald), Blackwell, Oxford.
Kneale, W., *Probability and Induction*, 2nd ed., Univ. Press, Oxford, 1952.
Kneale, W., 'Universality and Necessity', *Brit. J. Phil. Sci.* **12** (1961) 89–102.
Kneale, W., *The Development of Logic*, Univ. Press, Oxford, 1962.
Mach, E., 'Sinn und Wert der Naturgesetze', in *Erkenntnis und Irrtum*, Barth, Leipzig, 1908, pp. 449–63. (Abridged translation as 'The Significance and Purpose of Natural Laws', in A. Danto and S. Morgenbesser (eds.), *Philosophy of Science*, Meridian Books, New York, 1960.)
Mackie, J. L., 'Counterfactuals and Causal Laws', in R. J. Butler (ed.), *Analytical Philosophy. First Series*, Blackwell, Oxford, 1962.
Madden, E. H., 'Discussion: Ernest Nagel's *The Structure of Science*', *Phil. Sci.* **30** (1963) 64–70.
Madden, E. H., 'A Third View of Causality', *Rev. of Metaphysics* **23** (1969) 67–84.
Mill, J. S., *A System of Logic* (1843); 8th ed., London, 1872.
Molnar, G., 'Kneale's Argument Revisited', *Phil. Rev.* **78** (1969) 79–89.
Nagel, E., *The Structure of Science*, Routledge & Kegan Paul, London, 1961.
Nerlich, G. C. and Suchting, W. A., 'Popper on Law and Natural Necessity', *Brit. J. Phil. Sci.* **17** (1967) 233–35.
Pap, A., 'Disposition Concepts and Extensional Logic', in *Minnesota Studies in the Philosophy of Science* (ed. by H. Feigl, M. Scriven and G. Maxwell), University of Minnesota Press, Minneapolis, 1958.
Pap, A., *An Introduction to the Philosophy of Science*, Free Press, New York, 1962.

Pap, A., 'Reduction Sentences and Disposition Concepts', in *The Philosophy of Rudolf Carnap* (ed. by P. A. Schilpp), Open Court, LaSalle, Ill., 1963.

Pearson, K., *The Grammar of Science* (1892); 3rd ed. Everyman, 1937.

Peirce, C. S., 'The Laws of Nature and Hume's Argument Against Miracles' (1903), in P. P. Wiener (ed.), *Values in a Universe of Chance*, Stanford Univ. Press, 1958.

Poincaré, H., *Science and Method*, Dover, New York, (n.d.).

Popper, K. R., *The Open Society and Its Enemies*, Vol. II, Routledge & Kegan Paul, London, 1945.

Popper, K. R., *The Logic of Scientific Discovery*, 2nd (English) ed., Hutchinsons, London, 1968a.

Popper, K. R., 'A Revised Definition of Natural Necessity', *Brit. J. Phil. Sci.* **18** (1968b) 316–21.

Quine, W. van O., 'Necessary Truth', *The Voice of America Forum Lectures. Philosophy of Science Series*. No. 7, 1963. Reprinted in Quine (1966).

Quine, W. van O., *The Ways of Paradox*, Random House, New York, 1966.

Ramsey, F. P., 'General Propositions and Causality' (1929). Posthumous paper first published in Ramsey (1931).

Ramsey, F. P., *The Foundations of Mathematics*, Routledge & Kegan Paul, London, 1931.

Reichenbach, H., *Elements of Symbolic Logic*, Macmillan, New York, 1947.

Reichenbach, H., *Nomological Statements and Admissible Operations*, North-Holland Publ. Co., Amsterdam, 1954.

Russell, B., *Our Knowledge of the External World*, Open Court, LaSalle, Ill. 1914.

Schlick, M., 'Causality in Contemporary Physics', *Brit. J. Phil. Sci.* **12** (1962) 177–93, 281–98. (Transl. of a paper which first appeared in *Die Naturwissenschaften*, 1931.)

Scriven, M., 'The Key Property of Physical Laws-Inaccuracy', in H. Feigl and G. Maxwell (eds.), *Current Issues in the Philosophy of Science*, Holt Rinehart & Winston, New York, 1961.

Scriven, M., 'The Structure of Science', *Rev. Metaphysics* **17** (1964) 403–24.

Strawson, P. F., *An Introduction to Logical Theory*, Methuen, London, 1952.

Suchting, W. A., 'Functional Laws and the Regularity Theory', *Analysis* **29** (1968) 50–51.

Suchting, W. A., 'Popper's Revised Definition of Natural Necessity', *Brit. J. Phil. Sci.* **20** (1969) 349–52.

Taylor, R., *Action and Purpose*, Prentice-Hall, Englewood Cliffs, N. J., 1966.

Walters, R. S., 'Laws of Science and Lawlike Statements', in P. Edwards (ed.), *Encyclopaedia of Philosophy*, Vol. 4, Macmillan, New York 1968.

Wilson, F., 'Dispositions: Defined or Reduced?', *Australas. J. of Philos.* **47** (1969) 184–204.

Zilsel, E., 'The Genesis of the Concept of Physical Law', *Phil. Rev.* **51** (1942) 245–79.

ERNST MAYR

TELEOLOGICAL AND TELEONOMIC,
A NEW ANALYSIS

Teleological language is frequently used in biology in order to make statements about the functions of organs, about physiological processes, and about the behavior and actions of species and individuals. Such language is characterized by the use of the words 'function', 'purpose', and 'goal', as well as by statements that something exists or is done 'in order to'. Typical statements of this sort are 'It is one of the functions of the kidneys to eliminate the end products of protein metabolism', or 'Birds migrate to warm climates in order to escape the low temperatures and food shortages of winter'. In spite of the long-standing misgivings of physical scientists, philosophers, and logicians, many biologists have continued to insist not only that such teleological statements are objective and free of metaphysical content, but also that they express something important which is lost when teleological language is eliminated from such statements. Recent reviews of the problem in the philosophical literature (Nagel, 1961; Beckner, 1969; Hull, 1973; to cite only a few of a large selection of such publications), concede the legitimacy of some teleological statements but still display considerable divergence of opinion as to the actual meaning of the word 'teleological' and the relations between teleology and causality.

This confusion is nothing new and goes back at least as far as Aristotle, who invoked final causes not only for individual life processes (such as development from the egg to the adult) but also for the universe as a whole. To him, as a biologist, the form-giving of the specific life process was the primary paradigm of a finalistic process, but for his epigones the order of the universe and the trend toward its perfection became completely dominant. The existence of a form-giving, finalistic principle in the universe was rightly rejected by Bacon and Descartes, but this, they thought, necessitated the eradication of any and all teleological language, even for biological processes, such as growth and behavior, or in the discussion of adaptive structures.

The history of the biological sciences from the 17th to the 19th centuries

Boston Studies in the Philosophy of Science, XIV. All Rights Reserved.

is characterized by a constant battle between extreme mechanists, who explained everything purely in terms of movements and forces, and their opponents who often went to the opposite extreme of vitalism. After vitalism had been completely routed by the beginning of the 20th century, biologists could afford to be less self-conscious in their language and, as Pittendrigh (1958) has expressed it, were again willing to say 'a turtle came ashore to lay her eggs', instead of saying 'she came ashore and laid her eggs'. There is now complete consensus among biologists that the teleological phrasing of such a statement does not imply any conflict with physico-chemical causality.

Yet, the very fact that teleological statements have again become respectable has helped to bring out uncertainties. The vast literature on teleology is eloquent evidence for the unusual difficulties connected with this subject. This impression is reenforced when one finds how often various authors dealing with this subject have reached opposite conclusions (e.g. Braithwaite, 1954; Beckner, 1969; Canfield, 1966; Hull, 1973; Nagel, 1961). They differ from each other in multiple ways, but most importantly in answering the question: What kind of teleological statements are legitimate and what others are not? Or, what is the relation between Darwin and teleology? David Hull (1973) has recently stated "evolutionary theory did away with teleology, and that is that", yet, a few years earlier McLeod (1957) had pronounced "what is most challenging about Darwin, is his reintroduction of purpose into the natural world". Obviously the two authors must mean very different things.

Purely logical analysis helped remarkably little to clear up the confusion. What finally produced a breakthrough in our thinking about teleology was the introduction of new concepts from the fields of cybernetics and new terminologies from the language of information theory. The result was the development of a new teleological language, which claims to be able to take advantage of the heuristic merits of teleological phraseology without being vulnerable to the traditional objections.

1. TRADITIONAL OBJECTIONS TO THE USE OF TELEOLOGICAL LANGUAGE

Criticism of the use of teleological language is traditionally based on one or several of the following objections. In order to be acceptable teleological language must be immune to these objections.

1. *Teleological Statements and Explanations Imply the Endorsement of unverifiable Theological or Metaphysical Doctrines in Science*

This criticism was indeed valid in former times, as for instance when natural theology operated extensively with a strictly metaphysical teleology. Physiological processes, adaptations to the environment, and all forms of seemingly purposive behavior tended to be interpreted as being due to non-material vital forces. This interpretation was widely accepted among Greek philosophers, including Aristotle, who discerned an active soul everywhere in nature. Bergson's (1910) *élan vital* and Driesch's (1909) *Entelechie* are relatively recent examples of such metaphysical teleology. Contemporary philosophers reject such teleology almost unanimously. Likewise, the employment of teleological language among modern biologists does not imply adoption of such metaphysical concepts (see below).

2. *The Belief that Acceptance of Explanations for Biological Phenomena that Are not Equally Applicable to Inanimate Nature Constitutes Rejection of a Physico-Chemical Explanation*

Ever since the age of Galileo and Newton it has been the endeavor of the 'natural scientists' to explain everything in nature in terms of the laws of physics. To accept special explanations for teleological phenomena in living organisms implied for these critics a capitulation to mysticism and a belief in the supernatural. They ignored the fact that nothing exists in inanimate nature (except for man-made machines) which corresponds to DNA programs or to goal-directed activities. As a matter of fact, the acceptance of a teleonomic explanation (see below) is in no way in conflict with the laws of physics and chemistry. It is neither in opposition to a causal interpretation, nor does it imply an acceptance of supernatural forces in any way whatsoever.

3. *The Assumption that Future Goals were the Cause of Current Events Seemed in Complete Conflict with Any Concept of Causality*

Braithwaite (1954) stated the conflict as follows:

In a [normal] causal explanation the explicandum is explained in terms of a cause which either precedes it or is simultaneous with it; in a teleological explanation the explicandum is explained as being causally related either to a particular goal in the future or to a biological end which is as much future as present or past.

This is why some logicians up to the present distinguish between causal explanations and teleological explanations.

4. *Teleological Language Seemed to Represent Objectionable Anthropomorphism*

The use of words like 'purposive' or 'goal directed' seemed to imply the transfer of human qualities, such as intent, purpose, planning, deliberation, or consciousness to organic structures and to sub-human forms of life.

Intentional, purposeful human behavior, is, almost by definition, teleological. Yet I shall exclude it from further discussion because the words 'intentional' or 'consciously premeditated', which are usually used in connection with such behavior, endanger us with getting involved in complex controversies over psychological theory, even though much of human behavior does not differ in kind from animal behavior. The latter, although usually described in terms of stimulus and response, is also highly 'intentional', as when a predator stalks his prey or when the prey flees from the pursuing predator. Yet, seemingly 'purposive', that is goal-directed behavior in animals can be discussed and analyzed in operationally definable terms, without recourse to anthropomorphic terms like 'intentional' or 'consciously'.

As a result of these and other objections teleological explanations were widely believed to be a form of obscurantism, an evasion of the need for a causal explanation. Indeed some authors went so far as to make statements such as 'Teleological notions are among the main obstacles to theory formation in biology' (Lagerspetz, 1959:65). Yet, biologists insisted in continuing to use teleological language.

The teleological dilemma, then, consists in the fact that numerous and seemingly weighty objections against the use of teleological language have been raised by various critics, and yet biologists have insisted that they would lose a great deal, methodologically and heuristically, if they were prevented from using such language. It is my endeavor to resolve this dilemma by a new analysis, and particularly by a new classification of the various phenomena that have been traditionally designated as 'teleological'.

2. THE HETEROGENEITY OF TELEOLOGICAL PHENOMENA

One of the greatest shortcomings of most recent discussions of the teleology problem has been the heterogeneity of the phenomena designated as
'teleological' by different authors. To me it would seem quite futile to
arrive at rigorous definitions until the medley of phenomena, designated
as 'teleological', is separated into more or less homogeneous classes. To
accomplish this objective, will be my first task.

Furthermore, it only confuses the issue, when a discussion of teleology
is mingled with that of such of extraneous problems, as vitalism, holism,
or reductionism. Teleological statements and phenomena can be analyzed
without reference to major philosophical systems.

By and large all the phenomena that have been designated in the
literature as teleological, can be grouped into three classes:

(1) Unidirectional evolutionary sequences (progressionism, orthogenesis).

(2) Seemingly or genuinely goal-directed processes.

(3) Teleological systems.

The ensuing discussion will serve to bring out the great differences between these three classes of phenomena.

1. *Unidirectional Evolutionary Sequences (Progressionism, orthogenesis,
etc.)*

Already with Aristotle and other Greek philosophers, but increasingly
so in the 18th century, there was a belief in an upward or forward progression in the arrangement of natural objects. This was expressed most
concretely in the concept of the *scala naturae*, the scale of perfection
(Lovejoy, 1936). Originally conceived as something static (or even des-
cending owing to a process of degradation), the Ladder of Perfection was
temporalized in the 18th century and merged almost unnoticeably into
evolutionary theories such as that of Lamarck. Progressionist theories
were proposed in two somewhat different forms. The steady advance toward perfection was either directed by a supernatural force (a wise
creator) or, rather vaguely, by a built-in drive toward perfection. During
the flowering of Natural Theology the 'interventionist' concept dominated
but after 1859 it was replaced by so-called orthogenetic theories widely
held by biologists and philosophers (see Lagerspetz, 1959; 11–12 for a

short survey). Simpson (1949) refuted the possibility of orthogenesis with particularly decisive arguments. Actually, as Weismann had said long ago (1909), the principle of natural selection solves the origin of progressive adaptation without any recourse to goal-determining forces.

It is somewhat surprising how many philosophers, physical scientists, and occasionally even biologists, still flirt with the concept of a teleological determination of evolution. Teilhard de Chardin's (1955) entire dogma is built on such a teleology and so are, as Monod (1971) has stressed quite rightly, almost all of the more important ideologies of the past and present. Even some serious evolutionists play, in my opinion rather dangerously, with teleological language. For instance Ayala (1970:11) says,

the overall process of evolution cannot be said to be teleological in the sense of directed towards the production of specified DNA codes of information, i.e. organisms. But it is my contention that it can be said to be teleological in the sense of being directed toward the production of DNA codes of information which improve the reproductive fitness of a population in the environments where it lives. The process of evolution can also be said to be teleological in that it has the potentiality of producing end-directed DNA codes of information, and has in fact resulted in teleologically oriented structures, patterns of behavior, and regulated mechanisms.

To me this seems a serious misinterpretation. If 'teleological' means anything it means 'goal-directed'. Yet, natural selection is strictly an *a posteriori* process which rewards current success but never sets up future goals. No one realized this better than Darwin who reminded himself "never to use the words higher or lower." Natural selection rewards past events, that is the production of successful recombinations of genes, but it does not plan for the future. This is, precisely, what gives evolution by natural selection its flexibility. With the environment changing incessantly, natural selection – in contradistinction to orthogenesis – never commits itself to a future goal. Natural selection is never goal oriented. It is misleading and quite inadmissible to designate such broadly generalized concepts as survival or reproductive success as definite and specified goals.

The same objection can be raised against similar arguments presented by Waddington (1968: 55–56). Like so many other developmental biologists, he is forever looking for analogies between ontogeny and evolution. "I have for some years been urging that quasi-finalistic types of explanations are called for in the theory of evolution as well as in that of development." Natural selection "in itself suffices to determine, to a certain degree, the nature of the end towards which evolution will proceed, it must result in an increase in the efficiency of the biosystem as a whole in finding ways of reproducing itself." He refers here to completely generalized processes, rather than to specific goals. It is rather easy to demonstrate how ludicrous the conclusions are which one reaches by over-extending the concept of goal-direction. For instance, one might say that it is the purpose of every individual to die because this is the end of every individual, or that it is the goal of every evolutionary line to become extinct because this is what has happened to 99.9 % of all evolutionary lines that have ever existed. Indeed, one would be forced to consider as teleological even the second law of thermodynamics.

One of Darwin's greatest contributions was to have made it clear that teleonomic processes involving only a single individual are of an entirely different nature from evolutionary changes. The latter are controlled by the interplay of the production of variants (new genotypes) and their sorting out by natural selection, a process which is quite decidedly not directed toward a specified distant end. A discussion of legitimately teleological phenomena would be futile unless evolutionary processes are eliminated from consideration.

2. *Seemingly or Genuinely Goal-Directed Processes*

Nature (organic and inanimate) abounds in processes and activities that lead to an end. Some authors seem to believe that all such terminating processes are of one kind and 'finalistic' in the same manner and to the same degree. Taylor (1950), for instance, if I understand him correctly, claims that all forms of active behavior are of the same kind and that there is no fundamental difference between one kind of movement or purposive action and any other. Waddington (1968) gives a definition of his term 'quasi-finalistic' as requiring "that the end state of the process is determined by its properties at the beginning."

Further study indicates, however, that the class of 'end-directed processes' is composed of two entirely different kinds of phenomena. These two types of phenomena may be characterized as follows:

A. *Teleomatic processes in inanimate nature.* Many movements of inanimate objects as well as physico-chemical processes are the simple consequences of natural laws. For instance, gravity provides the end state for a rock which I drop into a well. It will reach its end-state when it has come to rest on the bottom. A red-hot piece of iron reaches its 'end-state' when its temperature and that of its environment are equal. All objects of the physical world are endowed with the capacity to change their state and these changes follow natural laws. They are 'end-directed' only in a passive, automatic way, regulated by external forces or conditions. Since the end state of such inanimate objects is automatically achieved, such changes might be designated as *teleomatic*. All teleomatic processes come to an end when the potential is used up (as in the cooling of a heated piece of iron) or when the process is stopped by encountering an external impediment (as a falling stone hitting the ground). Teleomatic processes simply follow natural laws, i.e. lead to a result consequential to conco-mitant physical forces, and the reaching of their end state is not controlled by a built-in program. The law of gravity and the second law of thermo-dynamics are among the natural laws which most frequently govern teleomatic processes.

B. *Teleonomic processes in living nature.* Seemingly goal-directed behavior in organisms is of an entirely different nature from teleomatic processes. Goal-directed 'behavior' (in the widest sense of this word) is extremely widespread in the organic world; for instance, most activity connected with migration, food-getting, courtship, ontogeny, and all phases of re-production is characterized by such goal orientation. The occurrence of goal-directed processes is perhaps the most characteristic feature of the world of living organisms.

The definition of the term *teleonomic*. For the last 15 years or so the term teleonomic has been used increasingly often for goal-directed pro-cesses in organisms. I proposed in 1961 the following definition for this term: "It would seem useful to restrict the term teleonomic rigidly to sys-tems operating on the basis of a program, a code of information" (Mayr, 1961). Although I used the term 'system' in this definition, I have since become convinced that it permits a better operational definition to con-sider certain activities, processes (like growth) and active behaviors as the most characteristic illustrations of teleonomic phenomena. I therefore modify my definition, as follows: *A teleonomic process or behavior is one which owes its goal-directedness to the operation of a program.* The term teleonomic implies goal direction. This, in turn, implies a dynamic process rather than a static condition, as represented by a system. The combination of 'teleonomic' with the term system is, thus, rather incongruent (see below).

All teleonomic behavior is characterized by two components. It is guided by a 'program' and it depends on the existence of some end point, goal, or terminus which is foreseen in the program that regulates the behavior. This endpoint might be a structure, a physiological function, the attainment of a new geographical position, or a 'consummatory' (Craig, 1918) act in behavior. Each particular program is the result of natural selection, constantly adjusted by the selective value of the achieved endpoint.

My definition of 'teleonomic' has been labeled by Hull (1973) as a 'historical definition'. Such a designation is rather misleading. Although the genetic program (as well as its individually acquired components) originated in the past, this history is completely irrelevant for the functional analysis of a given teleonomic process. For this it is entirely sufficient to know that a 'program' exists which is causally responsible for the teleonomic nature of a goal-directed process. Whether this program had originated through a lucky macromutation (as Richard Goldschmidt had conceived possible) or through a slow process of gradual selection, or even through individual learning or conditioning as in open programs, is quite immaterial for the class of a process as 'teleonomic'. On the other hand, a process that does not have a programmed end, does not qualify to be designated as teleonomic (see below for a discussion of the concept 'program').

All teleonomic processes are facilitated by specifically selected executive structures. The fleeing of a deer from a predatory carnivore is facilitated by the existence of superlative sense organs and the proper development of muscles and other components of the locomotory apparatus. The proper performing of teleonomic processes at the molecular level is made possible by highly specific properties of complex macromolecules. It would stultify the definition of 'teleonomic' if the appropriateness of these facilitating executive structures were made part of it. On the other hand it is in the nature of a teleonomic program that it does not induce a simple unfolding of some completely preformed Gestalt, but that it always controls a more or less complex process which must allow for internal and external disturbances. Teleonomic processes during ontogenetic development, for instance, are constantly in danger of being derailed even if only temporarily. There exist innumerable feedback devices to prevent this or to correct it. Waddington (1957) has quite rightly called attention to the frequency and importance of such homeostatic devices which virtually guarantee the appropriate canalization of development.

We owe a great debt of gratitude to Rosenblueth *et al.* (1943) for their endeavor to find a new solution for the explanation of teleological phenomena in organisms. They correctly identified two aspects of such phenomena, (1) that they are seemingly purposeful, being directed toward a goal, and (2) that they consist of active behavior. The background of these authors was in the newly developing field of cybernetics and it is only natural that they should have stressed the fact that goal directed behavior is characterized by mechanisms which correct errors committed during the goal-seeking. They considered the negative feedback loops of such behavior as its most characteristic aspect and stated "teleological behavior thus becomes synonymous with behavior controlled by negative feedback." This statement emphasizes important aspects of teleological behavior, yet it misses the crucial point: *The truly characteristic aspect of goal-seeking behavior is not that mechanisms exist which improve the precision with which a goal is reached, but rather that mechanisms exist which initiate, i.e. 'cause' this goal-seeking behavior.* It is not the thermostat which determines the temperature of a house, but the person who sets the thermostat. It is not the torpedo which determines toward what ship it will be shot and at what time, but the naval officer who releases the torpedo. Negative feedbacks only improve the precision of goal-seeking, but do not determine it. Feedback devices are only executive mechanisms that operate during the translation of a program.

Therefore it places the emphasis on the wrong point to define teleonomic processes in terms of the presence of feedback devices. They are mediators of the program, but as far as the basic principle of goal achievement is concerned, they are of minor consequence.

3. *Recent Usages of the Term Teleonomic*

The term 'teleonomic' was introduced into the literature by Pittendrigh (1958:394) in the following paragraph:

Today the concept of adaptation is beginning to enjoy an improved respectability for several reasons: it is seen as less than perfect; natural selection is better understood; and the engineer-physicist in building end-seeking automata has sanctified the use of teleological jargon. It seems unfortunate that the term 'teleology' should be resurrected and, as I think, abused in this way. The biologists' long-standing confusion would be more fully removed if all end-directed systems were described by some other term, like 'teleonomic', in order to emphasize that the recognition and description of end-directedness does not carry a commitment to Aristotelian teleology as an efficient [sic] causal principle.

It is evident that Pittendrigh had the same phenomena in mind as I do,[1] even though his definition is rather vague and his placing the term 'teleonomic' in opposition to Aristotle's 'teleology' is unfortunate. As we shall see below, most of Aristotle's references to end-directed processes refer precisely to the same things which Pittendrigh and I would call teleonomic (see also Delbrück, 1971).

Other recent usages of the term that differ from my own definition are the following. B. Davis (1962), believing that the term denotes 'the development of valuable structures and mechanisms' as a result of natural selection, uses the term virtually as synonymous with adaptiveness. The same is largely true for Simpson (1958:520–521) who sees in 'teleonomic' the description for a system or structure which is the product of evolution and of selective advantage:

The words 'finalistic' and 'teleological' have, however, had an unfortunate history in philosophy which makes them totally unsuitable for use in modern biology. They have too often been used to mean that evolution as a whole has a predetermined goal, or that the utility of organization in general is with respect to man or to some supernatural scheme of things. Thus these terms may implicitly negate rather than express the biological conclusion that organization in organisms is with respect to utility to each separate species at the time when it occurs, and not with respect to any other species or any future time. In emphasis of this point of view, Pittendrigh [above] suggests that the new coinage 'teleonomy' be substituted for the debased currency of teleology.

Monod (1971) likewise deals with teleonomy as if the word simply meant adaptation. It is not surprising therefore that Monod considers teleonomy "to be a profoundly ambiguous concept." Furthermore, says Monod, all functional adaptations are "so many aspects or fragments of a unique primary project which is the preservation and multiplication of the species." He finally completes the confusion by choosing "to define the essential teleonomic project as consisting in the transmission from generation to generation of the invariance content characteristic of the species. All these structures, all the performances, all the activities contributing to the success of the essential project will hence be called teleonomic."

What Monod calls 'teleonomic' I would designate as of 'selective value'. Under these circumstances it is not surprising when Ayala (1970) claims that the term 'teleonomy' had been introduced into the philosophical literature in order 'to explain adaptation in nature as the result of natural selection'. If this were indeed true, and it is true of Simpson's and Davis's cited definitions, the term would be quite unnecessary. Actually, there is nothing in my 1961 account which would support this interpretation, and I know of no other term that would define a goal-directed activity or behavior that is controlled by a program. Even though Pittendrigh's discussion of 'teleonomic' rather confused the issue and has led to the subsequent misinterpretations, he evidently had in mind the same processes and phenomena which I denoted as *teleonomic*. It would seem well worthwhile to retain the term in the more rigorous definition, which I have now given.

3. THE MEANING OF THE WORD PROGRAM

The key word in my definition of 'teleonomic' is the term 'program'. Someone might claim that the difficulties of an acceptable definition for teleological language in biology had simply been transferred to the term 'program'. This is not a legitimate objection, because it neglects to recognize that, regardless of its particular definition, a program is (1) something material, and (2) it exists prior to the initiation of the teleonomic process. Hence, it is consistent with a causal explanation.

Nevertheless, it must be admitted that the concept 'program' is so new that the diversity of meanings of this term has not yet been fully explored. The term is taken from the language of information theory. A computer may act purposefully when given appropriate programmed instructions. Tentatively program might be defined as *coded or prearranged information that controls a process (or behavior) leading it toward a given end.* As Raven (1960) has remarked correctly the program contains not only the blueprint but also the instructions of how to use the information of the blueprint. In the case of a computer program or of the DNA of the cell nucleus the program is completely separated from the executive machinery. In the case of most man-made automata the program is part of the total machinery.

My definition of program is deliberately chosen in such a way as to avoid drawing a line between seemingly 'purposive' behavior in organisms and in man-made machines. The simplest program is perhaps the weight inserted into loaded dice or attached to a 'fixed' number wheel so that they are likely to come to rest at a given number. A clock is constructed and programmed in such a way as to strike at the full hour. Any machine which is programmed to carry out goal-directed activities is capable of doing this 'mechanically'.

The programs which control teleonomic processes in organisms are either entirely laid down in the DNA of the genotype ('closed programs') or are constituted in such a way that they can incorporate additional information ('open programs') (Mayr, 1964), acquired through learning, conditioning or through other experiences. Most behavior, particularly in higher organisms, is controlled by such open programs. Let me illustrate this with an example. A young cuckoo or cowbird, although raised in the nest of foster parents, associates after fledging with other members of its own species and pairs with them when reaching reproductive age. The information controlling all this is part of the inherited DNA program, a closed program. On the other hand, members of certain species of ducks and geese or estrildid finches lack such inborn information concerning the recognition of members of their own species. It is subsequently inserted into their open program by a special 'learning' process called imprinting. It is still unknown whether such an acquired program is coded chemically (e.g. RNA) or through a highly specific pattern of synapses or both. Once the open program is filled in, it is equivalent to an originally closed program in its control of teleonomic behavior.

Open programs are particularly suitable to demonstrate the fact that the mode of acquisition of a program is an entirely different matter from the teleonomic nature of the behavior controlled by the program. Nothing could be more purposive, more teleonomic than much of the escape behavior in many prey species (in birds and mammals). Yet, in many cases the knowledge of which animals are dangerous predators is learned by the young who have an open program for this type of information. In other words this particular information was not acquired through selection and yet it is clearly in part responsible for teleonomic behavior. Many of the teleonomic components of the reproductive behavior (including mate selection) of species which are imprinted for mate recognition is likewise only partially the result of selection. The history of the acquisition of a program, therefore, can not be made part of the definition of teleonomic.

The origin of a program is quite irrelevant for the definition. It can be the product of evolution, as are all genetic programs, or it can be the acquired information of an open program, or it can be a man-made device. Anything that does *not* lead to what is at least in principle a predictable goal does not qualify as a program. Even though the future evolution of a species is being set severe limits by its current gene pool, its course is largely controlled by the changing constellation of selection pressures and is therefore not predictable. It is not programmed inside the contemporary gene pool.

The entire concept of a program of information is so new that it has received little attention from philosophers and logicians. My tentative analysis may, therefore, require considerable revision when subjected to further scrutiny.

How does the program operate? The philosopher may be willing to accept the assertion of the biologist that a program directs a given teleonomic behavior, but he would also like to know how the program performs this function. Alas, all the biologist can tell him is that the study of the operation of programs is the most difficult area of biology. For instance, the translation of the genetic program into growth processes and into the differentiation of cells, tissues and organs is at the present time the most challenging problem of developmental biology. The number of qualitatively different cells in a higher organism almost surely exceeds one billion. Even though all (or most) have the same gene complement they differ from each other owing to differences in the repression and derepression of individual gene loci and owing to differences in their cellular environment. It hardly needs stressing how complex the genetic program must be, to be able to give the appropriate signals to each cell lineage in order to provide it with the mixture of molecules which it needs in order to carry out its assigned tasks.

Similar problems arise in the analysis of goal-directed behavior. The number of ways in which a program may control a goal-directed behavior activity is legion. It differs from species to species. Sometimes the program is largely acquired by experience, in other cases it may be almost completely genetically fixed. Sometimes the behavior consists of a series of steps, each of which serves as reenforcement for the ensuing steps, in other cases the behavior, once initiated, seems to run its full course without need for any further input. Feedback loops are sometimes important but their presence cannot be demonstrated in other kinds of behavior. Again, as in developmental biology, much of the contemporary research in behavioral biology is devoted to the nature and the operation of the programs which control behavior and more specifically teleonomic behavior sequences (Hinde and Stevenson, 1970). Almost any statement one might make is apt to be challenged by one or the other school of psychologists and geneticists. It is, however, safe to state that the translation of programs into teleonomic behavior is greatly affected both by sensory inputs and by internal physiological (largely hormonal) states.

4. TELEOLOGICAL SYSTEMS

The word 'teleological', in the philosophical literature, is particularly often combined with the term 'system'. Is it justified to speak of 'teleological systems'? Analysis shows that this combination leads to definitional difficulties.

The Greek word *telos* means end or goal. Teleological means end-directed. To apply the word teleological to a goal-directed behavior or process would seem quite legitimate. I am perhaps a purist, but it bothers me to apply the word teleological, that is *end-directed*, to a stationary system. Any phenomenon (discussed in Section 2), to which we can refer as teleomatic or teleonomic, represents a movement, a behavior, or a process, that is goal-directed by having a determinable end. This is the core concept of teleological, the presence of a 'telos' (an end) toward which an object or process moves. Rosenblueth *et al.* (1943) have correctly stressed the same point.

Extending the term teleological to cover also static systems leads to contradictions and illogicalities. A torpedo that has been shot off and moves toward its target is a machine showing teleonomic behavior. But what justifies calling a torpedo a teleological system when, with hundreds of others, it is stored in an ordnance depot? Why should the eye of sleeping person be called a teleological system? It is not goal-directed at anything. Part of the confusion is due to the fact that the term 'teleological system' has been applied to two only partially overlapping phenomena. One comprises systems that are potentially able to perform teleonomic actions, like a torpedo. The other comprises systems that are well adapted, like the eye. To refer to a phenomenon in this second class as 'teleological', in order to express its adaptive perfection, reflects just enough of the old idea of evolution leading to a steady progression in adaptation and perfection, to make me uneasy. What is the telos toward which the teleological system moves?

The source of the conflict seems to be that 'goal-directed', in a more or less straightforward literal sense, is not necessarily the same as purposive. Completely stationary systems can be functional or purposive, but they can not be goal-directed in any literal sense. A poison on the shelf has the potential of killing somebody, but this inherent property does not make it a goal-directed object. Perhaps this difficulty can be resolved by making a terminological distinction between functional properties of systems and strict-goal directedness, that is teleonomy of behavioral or other processes. However, since one will be using so-called teleological language in both cases, one might subsume both categories under teleology.

R. Munson (1971) has recently dealt with such adaptive systems. In particular, he studied all those explanations, that deal with aspects of adaptation, but are often called 'teleological'. He designates sentences 'adaptational sentences', when they contain the terms 'adaptation', 'adaptive', or 'adapted'. In agreement with the majority opinion of biologists he concludes that "adaptational sentences do not need involve reference to any purpose, final cause, or other non-empirical notion in order to be meaningful". Adaptational sentences simply express the conclusion that a given trait, whether structural, physiological, or behavioral, is the product of the process of natural selection and, thus favors the perpetuation of the genotype responsible for this trait. Furthermore, adaptation is a heuristic concept because it demands an answer to the question, in what way the trait adds to the probability of survival and does so more successfully than an alternate conceivable trait. To me, it is misleading to call adaptational statements teleological. 'Adapted' is an *a posteriori* statement and it is only the success (statistically speaking), of the owner of an adaptive trait, which proves whether the trait is truly adaptive (=contributes to survival) or is not. Munson summarizes the utility of adaptational language in the sentence: "To show that a trait is adaptive is to present a phenomenon requiring explanation, and to provide the explanation is to display the success of the trait as the outcome of selection" (p. 214). The biologist fully agrees with this conclusion. Adaptive means simply: being the result of natural selection.

Many adaptive systems, as for instance all components of the loco-motory and of the central nervous systems, are capable of taking part in teleonomic processes or teleonomic behavior. However, it only obscures the issue when one designates a system 'teleological' or 'teleonomic', because it provides executive structures of a teleonomic process. Is an inactive, not-programmed computer a teleological system? What 'goal' or 'end' is it displaying during this period of inactivity? To repeat, one runs into serious logical difficulties when one applies the term 'teleological' to static systems (regardless of their potential) instead of to processes. Nothing is lost and much to be gained by not using the term teleological too freely and for too many rather diverse phenomena.

It may be necessary to coin a new term for systems which have the potential of displaying teleonomic behavior. The problem is particularly acute for biological organs which are capable of carrying out useful functions, such as pumping by the heart or filtration by the kidney. To some extent this problem exists for any organic structure, all the way down to the macromolecules which are capable of carrying out auto-nomously certain highly specific functions owing to their uniquely specific structure. It is this which induced Monod (1971) to call them teleonomic systems. Similar considerations have induced some authors, erroneously in my opinion, to designate a hammer as a teleological system, because it is designed to hit a nail (a rock, not having been so designed, but serving the same function not qualifying!).

The philosophical complexity of the logical definition of 'teleological' in living systems is obvious. Let me consider a few of the proximate and ultimate causes (Mayr, 1961), to bring out some of the difficulties more clearly. The functioning of these systems is the subject matter of regulatory biology, which analyzes proximate causes. Biological systems are complicated steady state systems, replete with feedback devices. There is a high premium on homeostasis, on the maintenance of the *milieu interieur*. Since most of the processes performed by these systems are programmed, it is legitimate to call them teleonomic processes. They are 'end-directed' even though very often the 'end' is the maintenance of the status quo. There is nothing metaphysical in any of this because, so far as these pro-

cesses are accessible to analysis, they represent chains of causally inter-
related stimuli and reactions, of inputs and of outputs.

The ultimate causes for the efficiency and seeming purposefulness of
these living systems were explained by Darwin in 1859. The adaptiveness
of these systems is the result of millions of generations of natural selection.
This is the mechanistic explanation of adaptiveness, as was clearly stated
by Sigwart (1881).

Proximate and ultimate causes must be carefully separated in the discussion of teleological systems (Mayr, 1961). A system is capable of performing teleonomic processes because it was programmed to function in
this manner. The origin of the program that is responsible for the adaptiveness of the system is an entirely independent matter. It obscures definitions to combine current functioning and history of origin in a single
explanation.

5. The Heuristic Nature of Teleonomic Language

Teleological language has been employed in the past in many different
senses, some of them legitimate and some of them not. When the distinctions are made, that are outlined in my survey above, the teleological
Fragestellung is a most powerful tool in biological analysis. Its heuristic
value was appreciated already by Aristotle and Galen, but neither of them
fully understood why this approach is so important. Questions which begin
with 'What?' and 'How?' are sufficient for explanation in the physical
sciences. In the biological sciences no explanation is complete until a
third kind of question has been asked: 'Why?'. It is Darwin's evolutionary
theory which necessitates this question: No feature (or behavioral program) of an organism ordinarily evolves unless this is favored by natural
selection. It must play a role in the survival or in the reproductive success
of its bearer. Accepting this premise, it is necessary for the completion of
causal analysis to ask for any feature, why it exists, that is what its function and role in the life of the particular organism is.

The philosopher Sigwart (1881) recognized this clearly:

A teleological analysis implies the demand to follow up causations in all directions by
which the purpose [of a structure or behavior] is effected. It represents a heuristic
principle because when one assumes that each organism is well adapted it requires that
we ask about the operation of each individual part and that we determine the meaning of

its form, its structure, and its chemical characteristics. At the same time it leads to an explanation of correlated subsidiary consequences which are not necessarily part of the same purpose but which are inevitable by-products of the same goal-directed process.

The method of course, was used successfully long before Darwin. It was Harvey's question concerning the reason for the existence of valves in the veins that made a major, if not the most important, contribution to his model of the circulation of blood. The observation that during mitosis the chromatic material is arranged in a single linear thread led Roux (1883) to question why such an elaborate process had evolved rather than a simple division of the nucleus into two halves. He concluded that the elaborate process made sense only if the chromatin consisted of an enormous number of qualitatively different small particles and that their equal division could be guaranteed only by lining them up linearly. The genetic analyses of chromosomal inheritance during the next sixty years were, in a sense, only footnotes to Roux's brilliant hypothesis. These cases demonstrate most convincingly the enormous heuristic value of the teleonomic approach. It is no exaggeration to claim that most of the greatest advances in biology were made possible by asking 'Why?' questions. This demands asking for the selective significance of every aspect of the phenotype. The former idea that many if not most characters of organisms are 'neutral', that is that they evolved simply as accidents of evolution, has been refuted again and again by more detailed analysis. It is the question as to the 'why?' of such structures and behaviors which initiates such analysis. Students of behavior have used this approach in recent years with great success. It has for example led to questions concerning the information content of individual vocal and visual displays (Smith, 1970; Hinde, 1972).

As soon as one accepts the simple conclusion that the totality of the genotype is the result of past selection, and that the phenotype is a product of the genotype (except for the open portions of the program that are filled in during the lifetime of the individual), it becomes one's task to ask about any and every component of the phenotype what its particular functions and selective advantages are.

It is now quite evident why all past efforts to translate teleonomic statements into purely causal ones were such a failure: A crucial portion of the message of a teleological sentence is invariably lost by the translation. Let us take, for instance the sentence: 'The Wood Thrush migrates in the fall into warmer countries *in order to* escape the inclemency of the weather and the food shortages of the northern climates'. If we replace the words 'in order to' by 'and thereby', we leave the important question unanswered as to *why* the Wood Thrush migrates. The teleonomic form of the statement implies that the goal-directed migratory activity is governed by a program. By omitting this important message the translated sentence is greatly impoverished as far as information content is concerned, without gaining in causal strength. The majority of modern philosophers are fully aware of this and agree that 'cleaned-up' sentences are not equivalent to the teleological sentences from which they were derived (Ayala, 1970; Beckner, 1969).

One can go one step further. Teleonomic statements have often been maligned as stultifying and obscurantist. This is simply not true. Actually the non-teleological translation is invariably a meaningless platitude while it is the teleonomic statement which leads to biologically interesting inquiries.

6. ARISTOTLE AND TELEOLOGY

No other ancient philosopher has been as badly misunderstood and mishandled by posterity as Aristotle. His interests were primarily those of a biologist and his philosophy is bound to be misunderstood if this fact is ignored. Neither Aristotle nor most of the other ancient philosophers made a sharp distinction between the living world and the inanimate. They saw something like life or soul even in the inorganic world. If one can discern purposiveness and goal direction in the world of organisms, why not consider the order of the Kosmos-as-a-whole also as due to final causes, i.e. as due to a built-in teleology? As Ayala (1970) said quite rightly, Aristotle's "error was not that he used teleological explanations in biology, but that he extended the concept of teleology to the non-living world." Unfortunately, it was this latter teleology which was first encountered during the scientific revolution of the 16th and 17th centuries (and at that in the badly distorted interpretations of the scholastics). This is one of the reasons for the violent rejection of Aristotle by Bacon, Descartes and their followers.

Although the philosophers of the last forty years acknowledge quite generally the inspiration which Aristotle derived from the study of living nature, they still express his philosophy in words taken from the vocabulary of Greek dictionaries that are hundreds of years old. The time would seem to have come for the translators and interpreters of Aristotle to use a language appropriate to his thinking, that is the language of biology, and not that of the 16th century humanists. Delbrück (1971) is entirely right when insisting that it is quite legitimate to employ modern terms like 'genetic program' for *eidos* where this helps to elucidate Aristotle's thoughts. One of the reasons why Aristotle has been so consistently misunderstood is that he uses the term *eidos* for his form-giving principle, and everybody took it for granted that he had something in mind similar to Plato's concept of *eidos*. Yet, the context of Aristotle's discussions makes it abundantly clear that *his* eidos is something totally different from Plato's *eidos* (I myself did not understand this until recently). Aristotle saw with exraordinary clarity that it made no more sense to describe living organisms in terms of mere matter than to describe a house as a pile of bricks and mortar. Just as the blueprint used by the builder determines the form of a house, so does the *eidos* (in its Aristotelian definition) give the form to the developing organism and this *eidos* reflects the terminal *telos* of the full grown individual. There are numerous discussions in many of Aristotle's works reflecting the same ideas. They can be found in the *Analytika* and in the *Physics* (Book II), but particularly in the *Parts of Animals* and in the *Generation of Animals*. Much of Aristotle's discussion becomes remarkably modern if one inserts modern terms to replace obsolete 16th and 17th century vocabulary. There is, of course, one major difference between Aristotle's interpretation and the modern one. Aristotle could not actually *see* the form-giving principle (which, after all, was not fully understood until 1953) and assumed therefore that it had to be something immaterial. When he said "Now it may be that the Form (*eidos*) of any living creature is soul, or some part of soul, or something that involves soul" (P. A. 641a 18), it must be remembered that Aristotle's psyche (soul) was something quite different from the conception of soul later developed in Christianity. Indeed the properties of 'soul' were to Aristotle something subject to investigation. Since the modern scientist does not actually 'see' the genetic program of DNA either, it is for him just as invisible for all practical purposes as it was for Aristotle. Its existence is inferred, as it was by Aristotle.

As Delbrück (1971) points out correctly, Aristotle's principle of the

eidos being an 'unmoved mover' is one of the greatest conceptual innova-
tions. The physicists were particularly opposed to the existence of such a
principle by

having been blinded for 300 years by the Newtonian view of the world. So much so,
that anybody who held that the mover had to be in contact with the moved and talked
about an 'unmoved mover', collided head on with Newton's dictum: *action equals
reaction*. Any statement in conflict with this axiom of Newtonian dynamics could only
appear to be muddled nonsense, a leftover from a benighted prescientific past. And
yet, 'unmoved mover' perfectly describes DNA: it acts, creates form and development,
and is not changed in the process (Delbrück, 1971:55).

As I stated above, the existence of teleonomic programs – unmoved
movers – is one of the most profound differences between the living and
the inanimate world, and it is Aristotle who first postulated such a
causation.

7. KANT AND TELEOLOGY

The denial of conspicuous purposiveness in living organisms and the
ascription of their adaptive properties to the blind accidental interplay of
forces and matter, so popular in the 18th century, was not palatable to
discerning philosophers. No one felt this more keenly than Immanuel
Kant, who devoted the second half of his *Critique of Judgment* to the
problem of teleology. It is rather surprising how completely most students
of Kant ignore this work, as if they were embarrassed that the great Kant
had devoted so much attention to such a 'soft' subject. Yet, as in so many
other cases, Kant was more perceptive than his critics. He clearly saw two
points, first that no explanation of nature is complete that cannot account
for the seeming purposiveness of much of the development and behavior
of living organisms, and secondly that the purely mechanical explanations
available at his time were quite insufficient to explain teleological phenomena. Unfortunately, he subscribed to the prevailing dogma of his period
that the only legitimate explanations were purely mechanical ('Newtonian') ones, which left him without any explanation for all teleological
phenomena. He therefore concluded that the true explanation was
out of our reach and that the most practical approach to the study of
organisms was to deal with them 'as if they were designed'. Even though
he was unable to free himself from the design-designed analogy, he
stressed the heuristic value of such an approach: It permits us to make

products and processes of nature far more intelligible than trying to express them purely in terms of mechanical laws.

Kant's interest was clearly more in the explanation of 'design' (adaptation) than in teleonomic behavior, yet he thought that an explanation of design was beyond the reach of human intellect. Just 69 years before the publication of the *Origin of Species* Kant (1790) wrote as follows:

It is quite certain that we can never get a sufficient knowledge of organized beings and their inner possibility, much less explain them, according to mere mechanical principles of nature. So certain is it, that we may confidently assert that it is absurd for men to make any such attempt, or to hope that maybe another Newton will some day arrive to make intelligible to us even the production of a blade of grass according to natural laws which no design has ordered. Such insight we must absolutely deny to mankind. (quoted from McFarland, 1970).

Darwin removed the roadblock of design, and modern genetics introduced the concept of the genetic program. Between these two major advances the problem of teleology has now acquired an entirely new face. A comparison of Kant's discussion with our new concepts provides a most informative insight into the role of scientific advances in the formulation of philosophical problems. Equally informative is a comparison of three treatments of Kant's teleology, roughly separated by 50-year intervals (Stadler, 1874; Ungerer, 1922; and McFarland, 1970).

8. CONCLUSIONS

(1) The use of so-called teleological language by biologists is legitimate, it neither implies a rejection of physico-chemical explanation nor does it imply non-causal explanation.

(2) The terms teleology and teleological have been applied to highly diverse phenomena. An attempt is made by the author to group these into more or less homogeneous classes.

(3) It is illegitimate to describe evolutionary processes or trends as goal-directed (teleological). Selection rewards past phenomena (mutation, recombination, etc.), but does not plan for the future, at least not in any specific way.

(4) Processes (behavior) whose goal-directedness is controlled by a program may be referred to as *teleonomic*.

(5) Processes which reach an end state caused by natural laws (e.g.,

gravity, second law of thermodynamics) but not by a program may be designated as *teleomatic*.

(6) Programs are in part or entirely the product of natural selection.

(7) The question of the legitimacy of applying the term teleological to stationary functional or adaptive systems requires further analysis.

(8) Teleonomic (i.e. programmed) behavior occurs only in organisms (and man-made machines) and constitutes a clear-cut difference between the levels of complexity in living and in inanimate nature.

(9) Teleonomic explanations are strictly causal and mechanistic. They give no comfort to adherents of vitalistic concepts.

(10) The heuristic value of the teleological *Fragestellung* makes it a powerful tool in biological analysis, from the study of the structural configuration of macromolecules up to the study of cooperative behavior in social systems.

ACKNOWLEDGEMENTS

Earlier versions of the manuscript were read critically by Michael Ghiselin, Stephen Gould, R. A. Hinde, Peter Medawar, Ronald Munson, W. V. Quine, Perry Turner, and William C. Wimsatt, to all of whom I am deeply indebted for numerous criticisms and suggestions, by no means all of which I adopted. I am fully aware of many remaining shortcomings, but I hope that the eventual solution of the difficult problem of teleology has been helped by my analysis.

Postscript. A number of discussions of teleology have come to my attention since the manuscript was completed. R. Jakobson (1970) presents a perceptive analysis of the indispensable uses (as well as limitations) of teleological language. He stresses the remarkable similarity between language and the genetic program in their ability to combine simple elements ('letters') into 'words' and 'sentences' which convey information efficiently. William C. Wimsatt (1971) deals with functional systems which he analyzes in terms of selection processes. Although I entirely agree with his analysis, it would seem to me that he merely confirms that the term teleological is superfluous when applied to systems. However, his analysis (as does that of Munson) raises the problem of how to delimit functional against teleological language.

Harvard University

NOTE

[1] This is quite evident from the following explanatory comment I have received from Professor Pittendrigh by letter (dated February 26, 1970):

You ask about the word 'teleonomy'. You are correct that I did introduce the term into biology and, moreover, I invented it. In the course of thinking about that paper which I wrote for the Simpson and Roe book (in which the term is introduced) I was haunted by that famous old quip of Haldane's to the effect that 'Teleology is like a mistress to a biologist: he cannot live without her but he's unwilling to be seen with her in public'. The more I thought about that, it occurred to me that the whole thing was nonsense – that what it was the biologist couldn't live with was not the illegitimacy of the relationship, but the relationship itself. Teleology in its Aristotelian form has, of course, the end as immediate, 'efficient', cause. And that is precisely what the biologist (with the whole history of science since 1500 behind him) cannot accept: it is unacceptable in a world that is always mechanistic (and of course in this I include probabilistic as well as strictly deterministic). What it was the biologist could not escape was the plain fact – or rather the fundamental fact – which he must (as scientist) explain: that the objects of biological analysis are organizations (he calls them organisms) and, as such, are end-directed. Organization is more than mere order; order lacks end-directedness; organization *is* end-directed. [I recall a wonderful conversation with John von Neumann in which we explored the difference between 'mere order' and 'organization' and *his* insistence (I already believed it) that the concept of organization (as contextually defined in its everyday use) always involved 'purpose' or end-directedness.]

I wanted a word that would allow me (all of us biologists) to describe, stress or simply to allude to – without offense – this end-directedness of a perfectly respectable mechanistic system. Teleology would not do, carrying with it that implication that the end is causally effective in the current operation of the machine. Teleonomic, it is hoped, escapes that plain falsity which is anyhow unnecessary. Haldane was, in this sense wrong (surely a rare event): we can live without teleology.

The crux of the problem lies of course in unconfounding the mechanism of evolutionary change and the physiological mechanism of the organism abstracted from the evolutionary time scale. The most general of all biological 'ends', or 'purposes' is of course perpetuation by reproduction. *That* end [and all its subsidiary 'ends' of feeding, defense and survival generally] is in some sense effective in causing natural selection; in causing evolutionary change; but not in causing itself. In brief, we have failed in the past to unconfound causation in the historial origins of a system and causation in the contemporary working of the system … .

You ask in your letter whether or not one of the 'information' people didn't introduce it. They did not, unless you wish to call me an information bloke. It is, however, true that my own thinking about the whole thing was very significantly affected by a paper which was published by Wiener and Bigelow with the intriguing title 'Purposeful machines'. This pointed out that in the then newly-emerging computer period it was possible to design and build machines that had ends or purposes without implying that the purposes were the cause of the immediate operation of the machine.

BIBLIOGRAPHY

Ayala, F. J., 'Teleological Explanations in Evolutionary Biology', *Phil. Sci.* 37 (1970) 1–15.

Baer, K. E. von, 'Über den Zweck in den Vorgängen der Natur', *Studien, etc.* (1876) 49–105, 170–234, St. Petersburg.

Beckner, M., 'Function and Teleology', *J. Hist. Biol.* **2** (1969) 151–164.

Bergson, H., *Evolution Créative*, Alcan, Paris, 1907.

Braithwaite, R. D., *Scientific Explanation*, Cambridge Univ. Press, 1954 (also in Canfield, pp. 27–47).

Canfield, J. V. (ed.), *Purpose in Nature*, Prentice-Hall, Englewood Cliffs, N. J., 1966, p. 1–7.

Craig, W., 'Appetites and Aversions as Constituents of Instincts', *Biol. Bull.* **34** (1918) 91–107.

Davis, B. D., 'The Teleonomic Significance of Biosynthetic Control Mechanisms', *Cold Spring Harbor Symposia* **26** (1961), 1–10.

Delbrück, M., 'Aristotle-totle-totle', in *Of Microbes and Life* (ed. by J. Monod and E. Borek), Columbia University Press, New York, 1971.

Driesch, H., *Philosophie des Organischen*, Quelle und Meyer, Leipzig, 1909.

Hinde, R. A. (ed.), *Non-Verbal Communication*, Cambridge University Press, 1972.

Hinde, R. A. and Stevenson, J. G., 'Goals and Response Controls', in *Development and Evolution of Behavior* (ed. by L. R. Aronson *et al.*), Freeman, 1970.

Hull, D., *Philosophy of Biological Science*, Foundations of Philosophy Series, Prentice-Hall, Englewood Cliffs, N. J., 1973.

Jakobson, R., 'Linguistics', in *Main Trends of Research in the Social and Human Sciences*, Chapter 6, 1970.

Kant, I., *Kritik der Urteilskraft*, Zweiter Teil, 1790.

Lagerspetz, K., 'Teleological Explanations and Terms in Biology', *Ann. Zool. Soc. Vanamo* **19** (1959) 1–73.

Lehman, H., 'Functional Explanation in Biology', *Philos. Sci.* **32** (1965) 1–20.

Lovejoy, A. O., *The Great Chain of Being*, Harvard University Press, Cambridge, Mass., 1936.

MacLeod, R. B., 'Teleology and Theory of Human Behavior', *Science* **125** (1957) 477.

Mainx, F., *Foundations of Biology*, Foundations of the Unity of Science, I(9) (1955) 1–86.

Mayr, E., 'Cause and Effect in Biology', *Science* **134** (1961) 1501–1506.

Mayr, E., 'The Evolution of Living Systems', *Proc. Nat. Acad. Sci.* **51** (1964) 934–941.

McFarland, J. D., *Kant's Concept of Teleology*, Univ. of Edinburgh Press, 1970.

Monod, J., *Chance and Necessity*, Alfred A. Knopf., New York, 1971.

Munson, R., Biological Adaptation', *Phil. Sci.* **38** (1971) 200–215.

Nagel, E., 'The Structure of Teleological Explanations', in *The Structure of Science*, Harcourt, Brace, and World, 1961.

Pittendrigh, C. S., *Behavior and Evolution* (ed. by A. Roe and G. G. Simpson), Yale University Press, New Haven, 1958.

Raven, Chr. P., 'The Formalization of Finality', *Folia Biotheoretica* **V** (1960) 1–27.

Roe, A. and Simpson, G. G. (eds.), *Behavior and Evolution*, Yale University Press, New Haven, 1958.

Rosenblueth, H., Wiener, N., and Bigelow, J., 'Behavior, Purpose, and Teleology', *Philos. Sci.* **10** (1943) 18–24 (also in Canfield, pp. 9–16).

Roux, W., *Über die Bedeutung der Kerntheilungsfiguren. Eine hypothetische Erörterung*, Leipzig, 1883.

Sigwart, C., *Der Kampf gegen den Zweck. Kleine Schriften* **2**, Mohr, Freiburg, 1881, pp. 24–67.

Simpson, G. G., *The Meaning of Evolution*; a study of the history of life and of its significance for man. Yale University Press, New Haven, 1949.

Smith, W. John, 'Messages of Vertebrate Communication', *Science* **165** (1969) 145–150.

Sommerhoff, G., *Analytical Biology*, Oxford University Press, London, 1950.

Stadler, H., *Kant's Teleologie und ihre erkenntnistheoretische Bedeutung*, F. Dümmler, 1874.

Taylor, R., 'Comments on a Mechanistic Conception of Purposefulness, *Phil. Sci.* **17** (1950) 310–317 (also in Canfield, pp. 17–26).

Teilhard de Chardin, P., *Le Phénomène Humain*, Editions de Seuil, Paris, 1955.

Theiler, W., *Zur Geschichte der Teleologischen Naturbetrachtung bis Aristoteles*, Zürich und Leipzig, 1925.

Ungerer, E., *Die Teleologie Kants und ihre Bedeutung für die Logik der Biologie*, Bornträger, Berlin, 1922.

Waddington, C. H., *The Strategy of the Genes*, George Allen & Unwin Ltd., London, 1957, pp. 11–58.

Waddington, C. H., *Towards a Theoretical Biology* I, Edinburgh University Press, 1968, pp. 55–56.

Weismann, A., 'The Selection Theory', in *Darwin and Modern Science* (ed. by A. C. Seward), Cambridge University Press, 1909.

Wimsatt, W. C., 'Teleology and the Logical Structure of Function Statements', *Stud. Hist. Phil. Sci.* **3** (1972) 1–80.

J. E. MCGUIRE

FORCES, POWERS, AETHERS, AND FIELDS

Despite the large literature, no satisfactory account has yet been provided of the dynamics of conceptual change involved in the development of the physics of continuous media and fields.[1] The present paper will not consider that rather intractable problem directly, but will restrict itself to a more modest and preliminary consideration. What are the historical stages of conceptual development in the historical emergence of views concerning the propagation of physical action in electrical, magnetic or optical media? Little familiarity with this question is necessary for a realization of the differences between eighteenth-century natural philosophies which regard space as filled with intensive powers, and the elastic aethers of early nineteenth-century optics. Again, both these views are distinguishable from post-Maxwellian concepts of the continuous field. Yet these approaches came together to form part of an important nexus of ideas from which arose interpretations of the physical field as they developed until the end of the nineteenth century. Unless we are able to characterize these similarities and differences, little understanding will be possible of interactions over time between these domains of concepts. As we shall see, the thought of Faraday and Maxwell is somewhat differently related to eighteenth-century doctrines of substantive powers. And Lorentz's work bears a different relationship to the Maxwellian corpus than the ideas of Lord Kelvin. In general, unless we get clearer about descriptive problems of this sort, it will be difficult to give dynamic and historical explanations of the change over time in the refinement and legitimation of concepts.

Interpretations of nature which ground change on the action of a medium, whether construed in terms of intensive powers, elastic solids, or energy, fall into two overlapping periods. First, in an important way, certain features of continuous field explanations were conditioned by eighteenth-century British theories of matter conceived as substantive powers. If Faraday is an important figure in the history of field theory, this eighteenth-century background of ideas assumes important signifi-

Boston Studies in the Philosophy of Science, XIV. All Rights Reserved.

cance. During the course of that century, there occurred widespread repudiation of the epistemology and ontology of seventeenth-century science. Thinkers like Berkeley, Leibniz, Robert Greene, Joseph Priestley, James Hutton, and Kant attacked, largely for epistemological reasons, the conceptual foundations of theories of matter and force current since the scientific revolution.[2] Moreover, they and philosophers such as Reid and Stewart resolutely attacked the contention, which was emphasized in the writings of Hume, that the human mind cannot have an idea of substantive powers. As I hope to show, the shift from the seventeenth-century idea that matter is corpuscular or atomic in constitution to the articulation of a field view in the thought of Faraday, involved important staging posts in British thought: matter conceived as dispositional in character in the thought of Boyle, Hobbes and Locke; the primacy of force in the Newtonian tradition; and lastly, matter construed as a plenum of intensive, activating powers. Though Continental thinkers like the Leibnizians, the Wolffians formalists, Boscovich, Buffon, LaMettrie, d'Holbach and Kant all articulated views on activated matter or ex-pounded a dynamism of attractive and repulsive forces, there is no direct evidence that their doctrines can be linked significantly with the ideas of Faraday on the correlation of forces or the physical field.[3] Nor do Schelling and *Naturphilosophie* fare better.[4] Schelling's dialectic of nature is alien to the sensibility of Faraday's thought as it is to Grove's, despite the cultural influence of Wordsworth, Coleridge, Oersted and Davy.[5] However, the British tradition of thought which has been mentioned, and which has something in common with Continental ideas of self-acting matter, does seem to have influenced the genesis of some of Faraday's ideas on matter,[6] which in turn helped to shape his mature conceptions of the physical field.

Though eighteenth-century conceptions of activated matter can be shown to bear on the thought of Faraday and to a lesser extent on that of Maxwell and Kelvin, elastic solid optical aethers are significant in the early development of theories concerned with the propagation of action through space. Apart from the efforts of Descartes, Hooke, and Huygens to provide a medium for the action of light waves, conceived like sound waves to have longitudinal vibrations, the problem of polarization in-duced Fresnel and Young early in the nineteenth century to consider the properties of elastic and solid aethers which could form a basis for the propagation of transverse waves. Their aethers were not continuous, but were constituted of tiny, rigid and elastic spheres which filled all space thus characterized by differences in rigidity or density. Like later aether theorists, Fresnel wished to explain optical problems in mechanical terms, though he did not produce a truly dynamical theory of wave motion in elastic solids. This was the work of Poisson, Cauchy, and Green.[7] The last rejected the ontological necessity of Cauchy's Newtonian assumption of central forces between the aether particles and constructed a continuous mathematical aether on the Lagrangian version of d'Alembert's principle. Green's optical medium was conceived in terms of an elastic solid which was infinitely resistant to compression, and the resistance to distortion of which was small. By appealing to unknown internal properties of his aether, Green was able to construct highly sophisticated techniques based on the work of Lagrange, for integrating differential volume which represented the motion of optical waves from substance to substance. He was able thereby to eliminate, effectively, the longitudinal compressional wave which bedevilled elastic solid aethers.[8] From this mathematical aether, which yields the equations of motion and boundary conditions, Green was able to derive many of the laws established by Fresnel, Poisson, and Cauchy. James MacCullagh developed an alternative aether which he put on a unified dynamical basis in 1839. MacCullagh's aether was of constant density, ruled out any compressional wave, and was put in determinate mathematical form by use of the d'Alembert-Lagrange principles. Though criticized by Stokes on mathematical grounds, and by Lorentz and Rayleigh on experimental grounds,[9] MacCullagh's aether was used successfully by Fitzgerald and Larmor, to interpret Maxwell's electromagnetic theory. Moreover, Lord Kelvin and Glazebrook following him, was able, by analyzing various plausible models for the luminiferous aether, to generalize and strengthen Green's approach to optical phenomena. The attempt to use elastic solid aethers continued will into the 1890's. And these models were of interest to electromagnetic theorists like Maxwell, Larmor, and Lorentz, though in general there was some lack of contact conceptually between optical aether theories and electromagnetic theories of propagation both in Britain and on the Continent.

Maxwell, Kelvin, Heaviside, and Thomson, are exceptions to this observation. These theorists transformed into their work Faraday's

electrical concepts of 'lines of force', 'dielectric polarization' and 'electrotonic state'. These concepts were employed by Faraday in considering
the medium through which electric and magnetic action travels and which
did not presuppose aethereal stuff in which to undulate.[10] As Faraday
was influenced by a British movement of thought which interpreted nature
in terms of self-acting substantive powers, care must be taken in considering the use which Maxwell made of his ideas. Maxwell's early
training was in continuum mechanics and theories of elastic solids.[11]
Faraday's ideas were thus transformed by Maxwell into modes of consistent representation which include interpretations of elastic aethers. As
might be expected, the fusion of two such different ontologies is difficult
to characterize. In general, though Maxwell many times identifies the
optical and electromagnetic aether,[12] Faraday's notions are central to his
thought when he is considering the action of electromagnetic waves. There
is plausibility in Simpson's view, however, that Maxwell thought largely
in terms of optical aethers[13] inasmuch as he never set about to test
directly for the existence of his electromagnetic waves along the lines of
Hertz.

Since Faraday explicitly rejected aethereal media in 'Thoughts on Ray
Vibrations' (1846),[14] and thus combined his opposition to action-at-a-
distance with the affirmation of a medium involving only the action of
lines of force, he explicitly declared the field to be the basic reality. As
there are good grounds for assuming that the field view of reality is a
distinctively British conception, an important historical corollary can be
drawn. Interpreting reality in terms of action-at-a-distance was the hall-
mark of eighteenth-century Newtonianism. For epistemological reasons,
among others, this conception with its metaphysics of particles and
superadded forces was widely rejected in eighteenth-century Britain.[15]
These critics of action-at-a-distance did not oppose it simply because it
postulated a void rather than a plenum, or because it ground change,
in the action of forces rather than in fluid vortices. Thinkers like Priestley,
Hutton, William Nicholson, and Thomas Exley opposed insensible
particles with superadded forces on the epistemological grounds that
they, along with void spaces, were not possible objects of direct experience.[16] Like Kant, these writers wanted to articulate theories of
matter which were compatible with the nature and content of immediate
experience.[17] Accordingly, many natural philosophers argued that matter

did not act by the contact of solid particles, nor at a distance, but by the varied intensities of a universal stuff. By the end of the eighteenth century, therefore, the basic metaphysics of Newtonianism had been under attack for some time by a significant group of natural philosophers. The scene was ripe for Faraday to begin the transformation of these conceptions of physical action into the scientific concepts of the field.

On the Continent, theorists like Fechner, Weber, Helmholtz, Neumann, and Kirchhoff, using the Newtonian model of particles and intervening forces, developed in a mathematically sophisticated way the Ampère-Coulomb action-at-a-distance approach to electric propagation. Yet, the British were developing theories of electromagnetic action based on field views of reality, and this eventually drove another wedge into the Newtonian world-view. Accordingly, there is the historical irony of the Germans in electromagnetics and the Laplacian school in general stoutly defending the Newtonian tradition against the British, who by the 1860's had repudiated it on scientific, and mathematical, as well as on epistemological grounds.

In what follows, these themes will be surveyed in more detail. Attention will also be given to some of the ontological aspects of the existence of fields. In general, it will emerge that they were regarded as substantive entities throughout the nineteenth century; the ontological primacy of dispositional notions with respect to fields is a modern-day phenomenon. Though fields were regarded during the nineteenth century as self-subsisting in character, this in no way mitigated the importance of dispositional concepts in physical thought. Not surprisingly, they abound in the language of science during the period which gave birth to field theory. They were not, however, thought by contemporaries to be among the basic categories of that science.

Though fields were conceived in substantive terms, the sense in which this is so is complex. Obviously, at no time was there a fixed definition or interpretation of the term field. Nor, in general, should the intellectual historian impose such standards on the materials which he is investigating. From the historical point of view there are no pre-determined and unequivocal defining characteristics of the field, such as will satisfy the strict criteria which the modern philosopher demands. The intellectual historian aims in his investigations *to produce* a clearer understanding of his problems. It is thus with the development of field concepts. Deeper

understanding can only emerge at the end of any study; definitional
categories cannot be imposed at the beginning.

I

Since what will follow is somewhat complex in its ramifications, a
chronology of the main stages in the account will be helpful:

(1) *The latter part of the seventeenth century*: Following Boyle, Hobbes and Descartes,
Locke embarked on a programme of the causal reduction of matter to invisible powers.
These were exemplified in a micro-structure, which was characterized solely by primary
qualities. Thinkers like John Rowning, Benjamin Martin, Robert Smith, John Keill and
John Freind, continued to emphasize the primacy of force in mechanics, optics and
chemistry. The dynamic thrust of Newton's force-aether of 1717/18 was preserved in
various ways in the work of such diverse thinkers as Robert Shaw, Boerhaave, Benjamin Wilson, and Brian Robinson. This widespread concern with activity in nature,
centering on the idea of active substance, helped to orient interpretations of nature in
directions inimical to the mechanical philosophies based as they were on the passivity
of matter.

(2) *1710–1790* As the philosophy of powers was developed from Locke, powers
conceived in substantive terms were elevated to the realm of direct experience. This
view of powers has affinities with the Continental rejection of the primary/secondary
quality distinction represented principally in the work of Leibniz, Bayle, and in England, Berkeley and Hume. Thinkers like Robert Greene, Berkeley, Cadwallader
Colden, Hume, Priestley, James Hutton, and Robert Young rejected the seventeenth-
century relational conception of powers and dispositions along with the Newtonian
theory of invisible particles and forces. In their work, like Kant, these writers insisted
that the concepts of natural science be related to the contents of immediate experience.

(3) *1710–1800* The Philosophical debate on powers. There was a widespread rejection of the view of Berkeley and Hume on causation and powers by Priestley, Hutton,
Robert Young and the Scottish School of Philosophers and Scientists. These notions
were held to be neither meaningless nor inappropriate as categories of natural philosophy, despite Hume's polemics against the notion of efficacy in nature.

(4) *1740–1840* There was a widespread belief in the reality and essentiality of active
powers conceived as substantive entities which define the nature of material phenomena.
This view is represented in the work of Greene, Colden, Priestley, Hutton, Robert Young,
Reid, Robison, Stewart, Leslie, James Gregory, John Nicholson and Robert Exley.
These thinkers hold that either the notion of power is a legitimate category which applies descriptively to reality (Gregory, Stewart, Robison) or that powers exist substantively in *rerum natura*. Some, like Priestley, Hutton, Nicholson, and Exley, held that
activating powers are universally present and that they can be established by an analysis
of direct experience. Moreover, these thinkers generally argued that particles and
point centers were unnecessary as a basis for activity: powers form a continuous plenum
of intensifying activity. Priestley, Hutton and Nicholson argue explicitly for the ultimate
reduction of scientific thought to an immaterial and 'immechanical' basis. Cauchy,

Green and MacCullagh, developed discontinuous or mathematically continuous aethers to explain the action and propagation of transverse waves through media.

(5) *1844–1855* In all probability, under the influence of the British school of thought which considered substantive powers as the essence of matter and almost certainly having read or read about Priestley, Exley, and Nicholson, Faraday developed a theory of matter conceived as powers diffused throughout space, existing independently of point centers. The proximate stimulus for Faraday's rethinking his ideas on matter was Hare's 1840 criticism of his particulate theory of physical action. The field of force which became more prominent in Faraday's thought from 1844 onwards, was conceived as a substantive entity having embedded in it lines of force as directional structures.

(6) *1847–1873* Maxwell, and to a lesser extent Kelvin, transformed Faraday's models of lines of force residing *in* space, and the particulate polarization of particles in dielectrics, into a mathematical theory of luminiferous and electromagnetic media. As did Faraday, Maxwell considered these models as alternative modes of representation of the electromagnetic field. Both theories were thus conceived as 'field conceptions' of reality. In neither the thought of Faraday nor Maxwell do these theories represent earlier and later views. In the thought of both scientists, the field is a substantive and active entity. After 1850, in Faraday's thought, the field and matter are one: in Maxwell's thought, the aether-field is a state or form of matter. For neither is the field identified with space. Clifford sketched out such a view of physical propagation in 1873.[18] For Maxwell, energy was a substantive entity, though he never systematically considered the problems of interaction between the aether-field, energy, and ponderable matter.

(7) *1880–1900* Writing in the late 1870's and early 1880's G. F. Fitzgerald sought a mechanical basis for Maxwell's electromagnetic theory. Using Maxwell's equations for the transformation of energy of the medium, Fitzgerald's strategy was to relate the basic expressions of these equations to MacCullagh's optical aether. Like the latter, Fitzgerald generated his equations by using variational principles. J. H. Poynting, J. J. Thomson, and O. Heaviside developed the two modes of consistent representation explicit in Maxwell's work. Poynting and Thomson favored the approach afforded by the model of lines of force. In 1884, Poynting explicitly identified the field with omnipresent energy. He denied the need for an underlying material substratum which Maxwell, Fitzgerald and others thought necessary for the existence of the field. In 1889, Heaviside speculated that it was perhaps wrong to think of the aether as a kind of matter, rather it might be best to "explain matter in terms of the aether, going from the simple to the more complex".[19] As we shall see, this was not Heaviside's committed view. Such a position, however, was later developed by Larmor.

(8) *1892–1910* In Lorentz's work, there is a clear ontological *separation* of matter and the field. Developing his electron theory from the work of Weber in Germany, Lorentz clarified Maxwell's work enormously. By 1904, the problem was still that of connecting discrete particles and continuous fields. Theories of the aether developed in the later nineteenth century, different from the earlier elastic solid aethers, had recognized the importance of interaction between aether and matter. Using the electromagnetic approach of Wien, Abraham and Wiechert, Lorentz developed an electromagnetic view of physics which, in dealing thoroughly with problems of interaction between matter and the field, denied that classical concepts such as mass and momentum applied to the field or to matter passing through it. Also in the early 1890's Joseph

Larmor argued that matter was constructible out of the field. Larmor's aether-field
was not a species of ordinary matter as with the theories of MacCullagh and Fitzgerald; rather it was ontologically prior to matter. Thus, unlike Lorentz, Larmor did
not separate matter from the field. With Wiechert in Germany, he further argued
that the field was the sole and primitive reality out of which electrons and matter
were to be constructed. Until around 1910, many like Larmor and Gustav Mie saw
all physical action as being reducible in principle to the motion of massless electrical
centers in an omnipresent aether. The aim of this school of thought was the reduction
of mechanics and optics to electromagnetic theory. From an ontological point of view,
an extreme form of reduction was being urged: the dematerialization of matter. It
remained for Einstein to reduce space to the field in the General Theory of Relativity
and to show from symmetry conditions in the Lorentz transformations that the aether
no longer had a reason to exist. By 1910, physical theory began to take new paths,
which are not germane to this study.

II

The preconditions which made possible the emergence of the revolutionary concept of the physical field early in the nineteenth century go
back to the eighteenth century. An account of these preconditions affords
a fascinating illustration of the ways in which science interacts with
philosophical thought.

As P. M. Heimann and I have shown elsewhere,[20] the eighteenth
century witnessed a significant rejection of the mechanical philosophies.
This took place at three related levels: ontology, epistemology, and
explanatory strategy. The notorious difficulties of the problem of trans-
duction, that is, of the justification for inference from the visible to
the invisible realm, led thinkers to reject the insensible entities of seventeenth-century scientific thought.[21] Here, skepticism played a major role.
The skeptical attack on Descartes' thought by Huet and Foucher, found
reverberations in the philosophies of Bayle, Berkeley, and Hume. What
Berkeley saw, was that the ideal of simple, intuitive, and demonstrative
knowledge of nature, found in the work of Descartes, Locke and Newton,
could not be squared with the claim to know an invisible realm of
corpuscles and forces. This claim had to go. So, too, must the existence
of a realm of invisible entities, inferences to which the mechanical
philosophies could not justify with this conception of demonstrative
knowledge.[22] Nature could not be explained from the inside out. A
reinterpretation or a rejection of the traditional primary/secondary
quality distinction played a prominent part in this skeptical reappraisal
of scientific knowledge. Not only was this doctrine rejected by Bayle,

Berkeley, Leibniz, and Hume, but it was similarly rejected or modified by Greene, Jonathan Edwards, Maupertuis, Nicolas de Beguelin, Priestley, James Hutton, William Nicholson, and Thomas Exley.[23] In general, these thinkers argued that experience afforded no warrant for distinguishing two categories of qualities such that the primary alone were essential. Even were this so, the ascription of the primary qualities to invisible particles was not only unnecessary but unjustifiable. Matter and its properties were thus to be grounded in the contents of immediate experience, such that its action and the connections of its properties could not be held to emerge from the operations of invisible entities. This epistemological position led to the articulation of ontologies different from those of the mechanical philosophies. Matter was conceived as an active power, and a body *was* what it had the invariant *capacity* to do.[24]

Since Locke's treatment of powers in large measure influenced those in Britain who conceived nature in terms of active substance, it will be helpful to discuss his position in some detail. This will also help to illustrate the difficulties which were thought to be inherent in the explanatory strategies of the mechanical philosophies and to place in perspective the views on matter of Leibniz, Newton, and Kant.

The crux of the issue is how action is to be conceived in scientific explanation. Basing himself on the achievement of seventeenth-century mechanics, Locke reduced all forms of action to impulse. Locke thus sought to establish a unique explanatory nexus comprising insensible corpuscles characterized solely by primary qualities such as extension, motion, solidity, etc. All causal explanations are reducible to the action of a simple and minimal set of geometrical entities all of which fall under the same determinables, the primary qualities. It is these which have the 'power' to act on our sensory organs and which specify interactions among natural objects.

This mode of understanding the physical world is a classic example of explanation in terms of the micro-structure of a set of basic entities. Radically reductionist in character, it claims that in principle all forms of change arise from entities endowed with invariant non-dispositional properties.[25] For Locke, these were the primary qualities which provide the necessary grounds for the conditions governing the exercise of an object's capacities in varying contexts. Since the primary properties alone are held to be fully actual at all times, dispositions and capacities are to be explicable in terms of them. The secondary qualities are for Locke "nothing in the objects themselves but powers." They are dispositions to act upon the sensory organs since 'the power to produce any idea in our mind' is a 'quality of the subject wherein the power is'.[26] Secondary qualities or powers are correlated with the capacities of human sensation. Tertiary qualities, that is the powers of affecting things other than sensory organs, are, unlike the secondary, completely reducible to the causal nexus of primary qualities.[27]

Apart from the fact that Locke modeled his reductionism on the explanatory strategies of seventeenth-century mechanics with its inherent metaphysics of primary and secondary qualities, and its narrowly conceived conception of physical action, there are intrinsic difficulties of a philosophical nature in his programme. Along with the influence of skeptical thought, which raised epistemological difficulties about the invisible realm, these difficulties provide yet other preconditions relating to the ultimate emergence of fields. These difficulties surround the notion of power. As we have seen, Locke's aim was to reduce all physical modalities, capacities, and dependency on context, to a causal nexus which is geometrical, totally actual, and characterized fully by categorical statements. Quite obviously, powers are not fundamental entities, and they are dependent on observers. Tertiary qualities, though they are independent of observers, have the same ontological status as the secondary. Powers, for Locke, are thus relational in character. Following Boyle, he holds that the sun, for example, has the active power to melt wax only when wax is present; conversely, only when heat is present, has wax the passive power to be melted. Power is not, however, a mere potentiality for Locke. A disposition (power) is an aspect of a thing's nature; it is part of what the thing does as distinct from what merely happens to it. As part of a thing's nature, a power helps characterize the thing's capacity to act or behave in certain ways without ceasing to be itself. That there can be no reasonable anticipation of exceptions to the regularity of the sun's melting wax when the latter is present, is grounded in the micro-structure of the objects themselves.

Though Locke was able to give a coherent account of the activity of powers and dispositions within the general framework of the mechanical philosophies, that framework with its presuppositions, as we have seen, came under attack. Yet Lockean powers could be and were construed as endowing objects with active capacities which transcended passive matter in motion. Greene, Hutton, Robert Young and Stewart so interpreted Locke's view of the essence of matter as a set of intrinsic and active powers.[28] As these powers can manifest degrees of intensity throughout space, physical action was no longer thought to be reducible to impulse. These thinkers also wished to avoid notions, such as the void and invisible particles, which they held were not possible objects of immediate experience. Lockean powers, therefore, insofar as they were thought to be reducible to invisible primary qualities, were for the same reason rejected by these writers. Their opposition to a Lockean microstructure as a necessary ground for the operations of bodies, was strengthened by Locke's own denial that the mechanical properties of particles could be known, such as those which cause rhubarb to purge and hemlock to kill.[29] Not only because of the invisible nature of the constituents of bodies, but because Locke also asserted that there was no demonstrative way of associating particular secondary qualities with specific corpuscular configurations, these thinkers were probably further confirmed in their opposition to structural explanations of the mechanical philosophies.

III

If powers were not conceived by the eighteenth century as invisible yet actual dispositions according to Locke's interpretation of the mechanical philosophy, they were nevertheless thought to be warranted, if somewhat unclearly, by direct experience. The appeal to sensory experience with writers like Greene, Priestley, and Hutton is closely linked, as we have seen, with their modification or rejection of the primary and secondary qualities doctrine. Unlike Boyle and Locke, they were dubious about whether simple, intuitive *ideas* of primary qualities could be correlated directly with these qualities in the ways which the mechanical philosophies conceived them to inhere in bodies. This view was too simple, and it presupposed a privileged epistemology of primary ideas, and a privileged ontology of primary qualities. Moreover, seventeenth-century science had achieved little success with explanatory strategies based on these doctrines. After all, Boyle had done little more than *illustrate* the probable truth of the corpuscular hypothesis; Locke – though he gave empirical examples

like the color of porphyry and the texture of almond – failed to give
a significant instance of a scientific explanation based on reduction to
the micro-structure of matter; and Newton never succeeded in squaring
his theory of gravitation with the theory of primary qualities, a doctrine
which he accepted. Gravitation, an intensive power, thus had no intelligible grounding in the action of passive matter. This in itself provided
yet another precondition for the gradual transformation of the concept
of matter.

Faced with these epistemological, ontological and explanatory difficulties, which they perceived to be intrinsic to the metaphysics of
seventeenth-century science, thinkers as diverse as Greene, Priestley,
Hutton, Stewart, Nicholson, Robert Young, Thomas Exley, and John
Playfair developed a positive approach to the problem of matter. Why
must it be assumed, they probably asked themselves, that the action of
entities be reducible to the nature of their micro-structure? If secondary
qualities were held to be mind-dependent by the mechanical philosophies,
and to be powers by Locke, why not conceive primary qualities in the
same way? Tactile experience does not tell us that objects are non-
dispositionally solid and hard, but rather that they actively resist pressure.
This is a primitive experience; it is thus compatible with that experience
to conceive solidity as an active power, something fully actual at all
times. By analyzing experience in this way, and by holding that invisible
particles, forces, and powers cannot be its possible objects, many
eighteenth-century natural philosophers agreed with Berkeley, Leibniz
and Hume in their approach to primary qualities. During the course of
the eighteenth century, two related theses became more and more
prominent. That the essence of matter is resistance or action and that
matter is a set of active powers, fully realized without a mysterious or
invisible substratum.[30] Resistance qua active force is thus epistemologi-
cally *prior* to the ideas of solidity and impenetrability as modes of contact
between material bodies.

This view of matter had important implications for the notion of
contact action with thinkers like Maupertuis, Priestley, Hutton, Stewart,
Playfair, Robison, Leslie, Kant, Nicholson, Young and Exley.[31] Moreover, their rejection of contact action had an important bearing on the
mode of existence of powers and the whole notion of action by impulse
alone. The arguments against contact action concentrate on the concep-

tion of matter which it presupposes. Since the corpuscles of the mechanical philosophy were thought ultimately to be solid and non-
deformable, on coming into contact they would rebound instantaneously
thus violating the principle of continuity. Matter as power satisfies this
criterion since it admits of degrees of intensity. On the other hand, if
invisible repulsive forces are superimposed between particles as in the
Newtonian schema, particles could never come into contact without the
creation of an infinite force. Since resistance and solidity can be explained
by repulsive powers alone, solid particles become unnecessary.[32] While
the attack on contact action was a negative tack in the task of providing
a rationale for action-at-a-distance, ironically, it led to the rejection of
the latter conception insofar as attempts to show contact action were
unintelligible became yet another precondition for the emergence of the
field with the affirmation of finite propagation across space. Many who
opposed the intelligibility of contact action drew on Newton's *Opticks*
where, for example, he speaks of reflecting surfaces as being diffused with
repulsive powers.[33] Others, like the Scots, Stewart and Playfair, appealed
to the theories of matter of Priestley and Hutton,[34] though they did not
entirely subscribe to their views. With reference to the epistemology of
the latter, Stewart argued that there is no experience of solid bodies
coming into actual contact when matter is considered solely as an 'object
of perception'.

The effects – which are vulgarly ascribed to actual contact are all produced by repulsive
forces, occupying those parts of space where *bodies* are perceived by our senses, and
therefore, the correct idea that we ought to annex to matter considered as an object
of perception, is merely that of *power of resistance*, sufficient to counteract the com-
pressing power which our physical strength enables us to exert.[35]

Stewart still adhered, however, to a belief in the absolute existence of
matter independent of the contents of sensation. And like Reid he
maintains a modified version of the distinction between primary and
secondary qualities. Still, he was emphatically moving towards an experiential grounding of the action of matter. Similar arguments regarding
the grounding of the concept of matter are found in the writing of Robert
Young, Hutton, Leslie, Robison, Kant, Nicholson, and Exley.[36] Unlike
Boscovich and Kant, these thinkers did not argue explicitly that the
action of solid atoms moving in a void cannot satisfy the principle of
continuity favored by the Leibnizians.

That powers exemplify continuity of physical action in nature can, however, be recognized contextually in their writings. Powers, whether attractive or repulsive, are thus conceived to manifest a continuous series in degrees of intensity. All space is therefore a continuous plenum of intensified powers; and though filled, is filled in varying degrees. Moreover, matter conceived in this mode, satisfies a pervasive fact; the content of sensation manifests varying intensities in experience. With this view of sensation in mind, Robert Young clearly stated in 1788 the conception of matter which best correlated with such experience: "Fullness is an idea capable of intension and remission; the same extension may be filled with quantities of filling substance; it may be more or less full, in all possible degrees".[37] An explanation of differential density according to this conception of matter was based on the view that "our ideas of differences of densities in bodies, is that of different fullnesses."[38] It was no longer necessary to explain differences in the densities of bodies in terms of ratios between interstitial pores and invisible particles. Earlier, Greene expressed the same view of the action of matter, which:

we have also said is distinguished into the Expansive and Contractive Forces, which, and the Different Combinations of them, are the occasion of those Diversities of Matter we Feel and See to Exist in Like and Equal Portions of Space.[39]

This conception of matter, with its associated epistemology, was systematically articulated by Kant in his *Metaphysical Foundations of Natural Science* (1786). In this work, Kant gave a culminating rationale for a world-view substantially different from Newtonianism. However, Kant's work does not seem to have been widely known in Britain, where the field view of reality first appeared.[40]

If powers filled space with varying intensity, they were the essence of matter. As Priestley put it, "take away attraction and repulsion, and matter vanishes."[41] Later, in expounding Hutton, Playfair stated "that power is the essence of matter," adding that:

the supposition of extended particles as a *substratum* or residence for such power, is a mere hypothesis, without any countenance from the matter of fact. For if these solid particles are never in contact with one another, what part can they have in the production of natural appearances, or in what sense can they be called the residence of a force which never acts at the point where they are present?[42]

The writers under discussion unanimously agreed. Priestley, Hutton, Kant, Nicholson and Exley, among others, along with the rejection of the invisible realm, claimed that point centers to and from which powers can act are unnecessary. Powers were thus held to be universally and substantively present. And Hutton and Robert Young went so far as to deny that extension was essential to material substance: it was destitute of bodily form, non-solid, non-spatial and continuously active.[43] In this school of thought there is, therefore, the beginning of a truly dynamical theory of action, grounded in an active medium diffused with power. Such a conception of nature is incompatible with contact action between solid, passive bodies, and with the notion of invisible particles acting at a distance.

What was the ontological status of these powers? As they defined powers to be the essence of matter, eighteenth-century thinkers must have considered them in terms other than mere potentiality or the possession of dispositional properties. It was the *nature* of matter to be comprised of active powers; and in the possession of such a nature was the sole ground for conditions governing the capacities of physical objects. So long as the eighteenth century defined powers as essential to matter, as known through direct sensory experience, and as the sole grounds necessary for the action of physical phenomena, they conceived them under the category of substance. Things were thus identified with their operations: a thing of a certain kind is what performs the operations of that kind of thing.

It is scarcely surprising that the Lockean programme of reducing powers to a unique and essential micro-structure was rejected. And it is also important to notice that these natural philosophers, along with philosophers like Reid, Kant, and Stewart, did *not accept* the Humean view that powers are meaningless entities. On the contrary, they were seen as the substantial and essential basis for the existence of matter.

Though Locke's *Essay* was well-known to eighteenth-century thought, his work was not the only source in which difficulties in the programme of the mechanical philosophy could be found. Nor was his view of powers, though important, the only component in the background of thought from which arose interpretations of nature as a self-contained system of active powers. The Newtonian and Leibnizian conceptions of matter must also be considered, though the views of the latter seem to have had little direct influence in Britain except through the mediation of Boscovichean ideas in Scotland.

Newton's *Principia* and *Opticks* were held to be paradigms of rational and empirical thought during the Enlightenment. They were thus largely instrumental in forging that experimental rationalism so evident in the work of Maupertuis, d'Alembert, Euler, and propagandized by Voltaire. It was this attitude of mind, on the Continent, which tempered the excesses of skepticism, and which considerably weakened the Cartesian ideal of deductive knowledge. The Newtonian inductive programme with its ideological prohibition on hypotheses, and in harness with the Lockean notion of certainty through the chartered limitations of the human intellect, produced an interpretative framework which greatly enhanced its claims about the nature of reality.

It is now widely recognized that the early years of the eighteenth century witnessed in Britain the development of the Newtonian programme which sought to explain a wide range of phenomena in terms of homogeneous particles to which were superadded forces of repulsion and attraction.[44] This programme attempted to establish the force-laws of microscopic interactions and, in general, it attached considerable importance to the use of mathematics and techniques of quantification. Though cosmological and optical aethers began to gain ground in the 1740's in Britain, along with the development of imperceptible fluids in which inhered electrical, chemical, physiological, and magnetic properties, these movements of thought did not mitigate the importance of Newtonian forces for conceiving nature as intrinsically active. Apart from the fact that Berkeley, Boerhaave, and Peter Shaw saw Newton's 1717/18 'force-aether' as a self-contained source of activity,[45] Priestley, Hutton, Nicholson, Robert Young, and Exley expressly linked substantive powers with Newtonian forces. Priestley well expressed the attitude of these thinkers towards the primacy of force in nature when he explicitly connected a denial of solidity, and his theory that the essence of matter was extension and active powers, with the Newtonian doctrine of the paucity of matter and a nearly vacuous universe.[46] In this way, the notion of substantive powers and Newtonian forces were brought into the same interpretative framework. Moreover, insofar as the 1717/18 Newtonian force-aether was interpreted as an active substance, thinkers like James Hutton saw the distinction between forces and fluids as lying primarily with sorts and levels of interactions which could be supposed to exist among phenomena. Forces or powers, however, were the basic entities in which was grounded the self-contained activity of nature.

The conception of nature as a self-sufficient system of intrinsic activity, though variously interpreted, was widespread during the eighteenth century. For Leibniz, the intellect can only grasp the intelligibility of substance in the natural order insofar as it conceives the essence of substance as activity. Extension is an inadequate and incomplete idea; the doctrine of atoms and the void violates the principles of continuity and indiscernibility. Only a real 'power', 'entelechy' or 'force' will satisfy the need for permanence and continuity through change, since "extension is nothing but the continuous repetition of a presupposed striving and resisting substance and cannot possibly be substance itself."[47] As the principle of permanence and continuity, active power is a principle of intelligibility. As such, it is the ground from which is generated a series of determinate changes in place, all of which are quantifiable as are acceleration and impact.[48] Powers are thus laws of action having implicit in their nature all the manifestations of which they are capable. There is no arbitrary change or motion in nature; a series of actions is an expression of true unity, not a bundle of beads without a string. What is and what will be, emerges or will emerge, is an intelligible consequence of the intrinsic activity of simple substances.[49] Activity is the manifestation of the principle of continuity; intrinsic change through imperceptible degrees is grounded in this principle. Leibniz thus transformed the scholastic notion of potentiality into that of active power. A thing is as it acts and has the power to do so. Substance and causation are not two categories but two factors in the more concrete category of activity. And Leibniz argued further that the notion of physical activity must be grounded in a theory of simple, perduring, metaphysical substances.

Leibniz's notion of active power has much in common with British writers; it defines the essence of matter; and as the operations of powers are observable, uniform, and quantifiable, powers are different from hidden substantial forms. On the other hand, Greene, Priestley and Hutton construed the individuation of particular bodies in terms of combinations of attractive and repulsive powers grounded in a universal plenum. Leibniz opposed individuation as the instantiation of a universal principle of matter.[50] The individuating principle of any specific body must be unique to it; all change is intrinsic and there are no true interactions among objects. There is thus no sense in which things can 'interact' as 'something' being passed from one to another as in the ontology of the

Mechanical Philosophy. Unlike the ontologies of Priestley and Hutton, Leibnizian powers are therefore not a universal ground from which arise particular bodies and their operations. Kant was later to develop Leibnizianism in this direction.

Though the Clarke-Leibniz correspondence was published in England in 1718, Leibniz's theory of matter seems to have had little early influence. Certain affinities with their thought apart, there is no direct evidence that Leibniz was later known to Robert Young, Priestley, Hutton, Nicholson, or Exley. This is explained partly by the ascendancy of Newton and Locke, partly by the British emphasis on experiment and empiricism, but also by the unsystematic fashion in which Leibniz presented his seminal ideas. On the Continent, attempts by Wolff and Bilfinger at systematization produced a string of bloodless formalisms, which failed to convey the perceptive depth of Leibniz's ideas. Leibniz's philosophy came into its own in the writings of Johann Lambert and Johann Tetens, where his work became, along with Locke's the core of their opposition to Wolffian essentialism. Along with Christian Crusius, these writers criticized the Wolffians for confusing sufficient reason with physical causation and for supposing that the irreducible elements of experience can be got from conceptual analysis alone.[51]

The philosophical orientation of these thinkers had an important influence on Kant's late pre-critical writings. So, too, did the debates on problems of the actual infinite, which stimulated Euler, Boscovich, Béguelin, Kant, and others to mediate between Newton and Leibniz on the nature of space and matter. From the time of Bayle, the following difficulty occupied thinkers like Keill, Berkeley, Maupertuis, d'Alembert, Boscovich, Euler, and Lambert. As geometry demonstrates that space is infinitely divisible, matter, which occupies it, is similarly divisible.[52] Yet, matter was held to be composed of simple elements which, being simple, must therefore be indivisible.

In his *Monadologia Physica* (1756), Kant attempted to reconcile the notion of matter with the geometrical theory of space. The views of Leibniz and Newton on action are merged by Kant. Leibniz's simple and non-extended monads become centers of attractive and repulsive forces, which fill space with varying intensities. This anti-mechanical dynamism Kant continued to maintain in his post-critical thought, especially in his *Metaphysical Foundations of Natural Science*. Here, Kant argues in an anti-Newtonian vein that any given space can be thought of both as equally full yet as filled in varying degrees. Like Greene, Priestley, and Hutton, Kant argues that the varying intensities of sensation indicate that reality contains no vacuum, but is rather a universal plenum which can manifest degrees of intensity. Unlike Leibniz, Kant does not ground activity in the principle of continuity, but rather in the nature of sensation. Nor does he, like Leibniz, attempt to explain activity in substance on analogy to mental processes.

It is clear, then, that the field view of reality as it developed in the thought of the Leibnizians and Kant was closely related to the problem of space and matter. Both Leibniz and Kant rejected the view that atomicity fulfilled the metaphysical requirement of unity and simplicity in nature: like matter, atoms were divisible in principle and were thus not candidates for ultimate unity. In place of the composite aggregates of the mechanical philosophy, Leibniz and Kant posited the monadal simplicity of intensive powers as a ground for continuous activity in the physical world. In contrast to this, English thinkers came to reject the invisible realm of particles in consequence of their general criticism of seventeenth-century metaphysics, and their particular epistemological demand that reality correspond to the nature of sensation. They thus rejected the duality of primary and secondary qualities and the Lockean opposition between ideas and matter.

As we shall see, there is little evidence that Continental thought directly influenced nineteenth-century field theories. Nor is there clear evidence to show that there were significant interactions on problems of dynamism between Continental and British thought during the eighteenth century.[53] What seems clear enough, however, is that eighteenth-century anti-mechanistic thought greatly transformed the general interpretations of nature of that period. This, along with the particular preconditions which have been mentioned, made possible the later emergence of field concepts.

IV

Michael Faraday had been opposed to action-at-a-distance from 1821. In 1831, his discovery of electromagnetic induction led Faraday to explain the induction of an electric current between two coils of wire wound round an iron ring in terms of the creation of a 'peculiar state' or 'peculiar

condition' in the ring. This 'electrical condition of matter' he called the
'electrotonic state', a term later used by Maxwell.[54] In this interpretative
framework, an electric current was considered as the creation and dissolu-
tion of the electrotonic state. Though he associated it with particles of
matter, Faraday remained agnostic about the nature of this electrical condition. At this stage in his thought, Faraday was not so much proposing
a theory of matter, as a model for understanding the propagation of electric action.[55] By 1837, arising from his work on electrochemical phenomena, Faraday began to speak of a 'peculiar state of tension or polarity'.
He tells us: "with respect to the term *polarity*... I mean at present only a
disposition of force by which the same molecule acquires opposite powers
on different parts."[56] The 'electrotonic state' is now unambiguously
associated with the polarity of molecules. Matter is now conceived as being
subject to electrical tension, polarity being opposite electrical states on
different parts of the molecules.

In 1832, shortly after the formulation of the concept of electrotonic
state, Faraday temporarily abandoned the notion in favor of a theory of
'lines of force'. If a magnet is placed in a jar of mercury and the mercury
connected by wires to a galvanometer, rotating the magnet caused no
difference in the effect the magnet had on the needle of the galvanometer.
Considering this and a number of other related experiments, Faraday
concluded that there was a "*singular independence* of the magnetism and
the bar in which it resides."[57] He now began to consider magneticism in
terms of action along curved lines. And though they were associated with
particles of matter, nevertheless these 'lines of force' were in a sense
independent of them. The notion of the 'electrotonic state' began to
recede from the focus of Faraday's thought. He declared: "the reasons
which induce me to suppose a particular state [the electrotonic state] ...
have disappeared" and concluded that he was "not aware of any distinct
facts which authorize the conclusion that it [a wire] is in a particular
state."[58] These two modes of representation, lines of forces and molecular
polarizing, were to dominate Faraday's thought, and from 1832 were
largely conceived as alternatives. As he later emphasized, the "*electrotonic
state* ... would coincide and become identified with that which would then
constitute the physical lines of magnetic force."[59] Maxwell later expressed
this relationship by saying that the electrotonic state was measured by
the number of these lines passing through a point in space; conversely,

the electrotonic intensity at a point measures the number of lines of magnetic force.[60]

In his studies of electrostatic induction after 1838, however, Faraday continued to think of the polarization of particles in a dielectric medium as an existential condition. Lines of force curved in space were considered largely as imaginary: they were a means of 'expressing the direction of the power'.[61] By 1845, this view of lines of force was to change.

The work on electrostatic induction, however, was to raise difficulties in Faraday's particulate model of the action of the electrotonic state. These stemmed from the idea that a state of 'tension' polarized a line of 'contiguous' particles across space. Faraday assumed, further, that the difference between conduction and insulation in terms of particles in a state of polarization was not the mode of communication of force but its rate of communication from particle to particle. Induction occurred when there was ready communication between 'contiguous' particles; difficult communication gave rise to insulation. Faraday was not at all clear as to what he meant by 'contiguous'. Since he allowed, however, that contiguous particles did not touch, he was committed to holding that short range forces could act at a distance: "nothing in my present views forbids that [a] particle should act at the distance of half an inch."[62]

Robert Hare raised a fundamental difficulty. On what grounds could Faraday admit action-at-a-distance at the micro-level and deny it at the macro-level? Either you affirm or deny both, since a 'distance of half an inch' is perceptible. As Faraday pondered Hare's criticisms, he was finally led to abandon his particulate theory of matter.

In 'Speculation Touching Electric Conduction and the Nature of Matter'[63] (1844), Faraday rejected the possibility of forces acting across insensible distances. Since Faraday now rejected the notion of 'contiguous' particles as well, the difference between conducting and non-conducting bodies was due to the void alone. As space could be conceived as a non-conductor in non-conducting bodies and as a conductor in others, this meant ascribing to it causal and dispositional properties. But Faraday was quickly led to claim that "mere space cannot act as matter acts."[64] By 1850, he could claim unequivocally: "Any portion of space traversed by lines of magnetic power, may be taken as a field, and there is probably no space without them. The condition of the field may vary in intensity of power from place to place, either along the lines or across them...."[65]

By 1845, Faraday had reduced matter to active powers:

Recognizing or perceiving *matter* by its powers, and knowing nothing of any imaginary nucleus, abstracted from the idea of these powers, the phaenomena described in this paper much strengthen my inclination to trust in the views I have on a former occasion advanced....[66]

Knowledge of matter was thus limited to the idea of its powers, and there was no necessity to assume the existence of matter independent of the powers. Moreover, Faraday insisted that matter *qua* powers is known to the senses solely as a phenomenon of active resistance, and that we have no evidence of hard, impenetrable nuclei. Matter was now continuous, and there was no need to distinguish atoms from the intervening space. The difficulty of space being both an insulator and conductor was overcome since matter now continuously filled all space.

Heimann has shown that there are compelling grounds for assuming that Faraday drew on ideas of matter from the writings of Priestley, Exley, Charles Hutton, and William Nicholson.[67] Heimann points out that Faraday's rejection after 1845 of impenetrable particles, his insistence that we know matter only as a resisting power, and his rejection of point centers or nuclei, are similar in character and purpose to the discussion of these writers. Moreover, Heimann also shows that Faraday had reason to know their writings and that he had active access to them. We have, therefore, got good grounds for linking Faraday's speculation on matter and the field with eighteenth-century British thought on the philosophy of powers.

With his work on the action of magnets on crystals, Faraday came to doubt the existence of a state of polarity in the crystals. Action at poles was simply the maximum convergence of lines of force. By 1852, Faraday saw the spatial medium surrounding ponderable bodies "as *essential* to [them] ... it is that which relates the external polarities to each other by curved lines of power; and that these must be so related as a matter of necessity."[68] There was no need for an ether, and force resided in the medium, not in bodies themselves: "I cannot conceive curved lines of force without the conditions of a physical existence in that intermediate space."[69] In replacing the tension and polarization of particles of matter by the primacy of lines of force, Faraday denied the existence of polarity as a state of matter. And the lines now exist permanently, whether or not ponderable bodies are there to be acted upon, as directional entities in a field of activating powers.

If a case can be argued for the existence of continuity between Faraday's later views on the field and eighteenth-century British theories of matter, this is not so with Maxwell. In the first place, care must be taken in assessing the use which Maxwell made of Faraday's ideas. The Scot's basic orientation was in theories of elastic solids and continuum mechanics.[70] Even before his experience with Cambridge mathematical physics, Maxwell had published a paper on the subject of equilibrium in elastic solids.[71] His early work on Saturn's rings reveals a growing mastery of techniques of mathematical analysis which he later displayed in his papers on electrodynamics. In his first paper, 'On Faraday's Lines of Force (1855–56)',[72] Maxwell followed William Thomson's approach to electrical problems using mathematical analogies between heat flow, streamlines of frictionless incompressible fluids, current electricity, and lines of force. Moreover, Thomson had been the first to give a mathematical treatment of Faraday's conception of lines of force which could expand equatorially and shorten axially. This result, along with Thomson's 'Mathematical Theory of Magnetism' (1851)[73] and his theory of electrical images, influenced Maxwell's treatment of electro-magnetism in Part II of 'Faraday's Lines'. It is, therefore, nearer the mark to realize that Maxwell, like Thomson, was a Cambridge mathematical physicist who employed some of Faraday's concepts for explaining electromagnetic action by developing successive theoretical frameworks embodying various postulated entities and mathematical analogies. Thus, in no direct way did eighteenth-century concepts of matter enter into Maxwell's theorizing as they did into that of Faraday.

The relation of Maxwell's thought to Faraday's work can be made clearer by considering Maxwell's conception of constructing consistent representations of reality. Maxwell was influenced by Whewell's notion that knowledge of reality is ordered in a series of levels, each more or less complete in itself, for which the mind must find the appropriate ideas.[74] In the light of this conception, Maxwell was careful not to confuse a mathematical description of a given level with ontological commitment at another. The aim of physical theory must always be to present a consistent representation of the phenomena, a term which he translated from Gauss' 'construirbare Vorstellung'.[75] In this respect, Maxwell's main concern was to formulate a consistent representation of the propagation

142   J. E. MCGUIRE

of electrical action: "... what is its condition after it has left the one particle and before it has reached the other?"[76]

The question leads us "to the conception of a medium in which propagation takes place," and "we ought to endeavor to construct a mental representation of all the details of its action...."[77] Maxwell's relation to Faraday's ideas must be seen in this light. Maxwell transformed Faraday's concepts in the demand for consistent representation. Though Faraday's lines of force, for example, may be seen to indicate that electricity is a substantial entity existing in space, Maxwell insists that the use of analogy in the mode of representation prevents us from assuming too easily that electricity is "either a substance like water or a state of agitation like heat."[78] Following William Hamilton, Maxwell holds that knowledge is primarily an awareness of a "similarity between relations, not a similarity between the things related."[79]

Behind Maxwell's conception of modes of consistent representation, lies an important epistemological point of view. Maxwell's position has some affinities with the construction of explanatory frameworks which postulate theoretical entities, discussions of which are found in the writings of contemporaries like Whewell. The program of postulatory explanation posits the existence of entities along with consistent modes of interaction and then seeks to test the consequences of the model against experience. This procedure is similar to Maxwell's notion of consistent representation in that the main constraints which an explanatory model should satisfy are conceptual consistency and being testable against subsequent experience.

This explanatory strategy was first developed by LeSage in the eighteenth century to justify his aethereal hypothesis against the attacks of the inductivists.[80] Later it was advanced by Hershel and Whewell. And its epistemological demands allow for greater flexibility in explanation than is the case with seventeenth-century mechanical philosophies. As we saw, these philosophies demanded that explanations satisfy the epistemological requirement of clear and distinct ideas. All explanations were thus by way of reduction to microstructures, characterized by primary qualities. And to specify correspondence between the manifest properties of phenomena to be explained and the entities of their fine structure was the aim of these explanatory strategies. This clearly placed stringent epistemological constraints on explanation.

Following the eighteenth-century philosophies of powers, and reasoning from his wide experimental achievements, Faraday argued that the action of material phenomena should be conceived in ways which correspond to the nature of sensation. As we have seen, he conceived matter as resisting power of varying degrees. This view of matter allows for more flexibility in explanation. Although Faraday made ingenious use of the method of hypothesis, his main constraint on explanation was that phenomena be conceived in ways which were consistent with experience.

Maxwell's view of explanation was more flexible as it emphasized postulation and testibility. Consequently, his conception of consistent representation allowed the elaboration of models which embodies ideas from continuum mechanics and from Faraday's published corpus. And the main constraint that Maxwell imposed on explanation in electromagnetism was that possible models be consistent with the principle of energy conservation. This was never an issue for Faraday.

The employment of modes of consistent representation by Maxwell, and later by Tait, Thomson, and Larmor, throws light on British attitudes concerning the representation of physical fields. The British were certainly more pragmatic than the school of Weber and Neumann, wishing to carry analogies as far as they would go. They were not, however, afraid to abandon them when they began to break down or when too many different analogies seemed necessary. Maxwell's work exemplifies this. There is, of course, little doubt that the British hoped to find analogies which worked in various areas at once, and thus achieve a unified theory. Duhem misread this limited pragmaticism as full-blown pragmaticism.

Though Maxwell is in no way a disciple of Faraday, his debt to the latter is significant. The fundamental conceptual dichotomy which has been chartered in Faraday's thought is also present in Maxwell's work: the articulation of two different models for the representation of the field. As we have seen, in 'Faraday's lines', Maxwell employed the theory of the primacy of lines of force, which was characteristic of Faraday's later thought. He did not, however, discuss the nature of the physical state to which these lines of force correspond. In 'On Physical Lines of Force' (1861–62),[81] Maxwell attempted to specify the nature of this physical state. In 'Faraday's Lines', he had already formulated a new function which after Faraday he called the 'electrotonic function'. It provided equations connecting ordinary magnetic action, electromotive force, electromagnetic induction, and action between closed currents.[82] Later Maxwell was to identify this function as a generalization of Neumann's electrodynamic function. In 'Physical Lines', however, Maxwell wished to present a consistent mechanical representation of this function in terms of a medium which is "put into such a state that at every point the pressures are different in different directions. The direction of least pressure being that of the observed lines of force, and the difference of greatest and least pressures being proportional to the square of the intensity of the force at that point."[83] Maxwell conceived this medium in terms of Faraday's particulate model of polarization, opposite parts being in opposite electrical states. In particular, he wished to answer the question "what mechanical explanation can we give of this inequality of pressures in a fluid or mobile medium? The explanation which most readily occurs to mind is that the excess of pressure in the equatorial direction arises from the centrifugal force of vortices or eddies in the medium having their axes in directions parallel to the lines of force."[84] Normally in an incompressible fluid, the pressure is identical in all directions. But rotation causes centrifugal forces, which make each vortex contract longitudinally and exert radial pressure. Faraday proposed exactly this stress distribution for physical lines of force. And William Thomson had shown the rotatory character of magneticism in a medium, as opposed to the streamline disposition of electricity. Electricity in Maxwell's model became tentatively identified with a layer of particles interposed between each vortex and the next, "so that each vortex has a tendency to make the neighbouring vortices revolve in the same direction with itself."[85] Understanding the action of electric currents on the vortex medium in this manner, Maxwell viewed electricity as an entity disseminated through space.[86] It was no longer a fluid confined to conductors.

    In Part III of 'Physical Lines', Maxwell discussed the relationship between electric currents and the induction of charge through a dielectric. Already predisposed in terms of the analogy between streamlines and lines of electric force to view induction as nothing more than a special case of conduction, Maxwell, with the conception of electricity as disseminated in space, saw the possibility of representing this as *elastic displacements* in the vortex medium. Again, following the work of Faraday and Kelvin, Maxwell divided the magnetic medium into elastic cells within which resides stored energy. If the interposed particles of electricity are urged in any direction by their tangential action, they will distort each cell. The distortion of each cell can be pictured as the displacement of electricity within each molecule in a given direction. The effect over the whole dielectric is the general displacement of electricity in any given direction. Since electric particles surrounding a conductor are now capable of elastic displacement, an electric current is no longer confined like a fluid in a container.[87] This is the beginning of Maxwell's controversial concept of displacement current.

In 'A Dynamical Theory of the Electromagnetic Field' (1865),[88] Maxwell derived eight general equations of the electromagnetic field without any hypotheses regarding the mechanical form of omnipresent energy. Maxwell retained the idea of treating light and electromagnetism as processes in a common medium. His aim was to replace the derivation of equations for these two classes of phenomena from relatable mechanism, by an analysis of experiments based on the phenomena themselves.

At the beginning of this paper, Maxwell states that:

the theory I propose may... be called a theory of the *Electromagnetic field*, because, it has to do with the space in the neighbourhood of the electric or magnetic bodies and it may be called a *Dynamical* theory because it assumes that in that space there is matter in motion, by which the observed electromagnetic phenomena are produced. The electromagnetic field is that part of space which contains and surrounds bodies in electric or magnetic conditions.[89]

Here, Maxwell makes the field the ultimate reality, and he refers it to the space in which it resides. It must not be thought, however, that Maxwell associated the physical field with space. In 1874, he emphatically rejected Reimann's notion that "we must see the ground of ... [the] metric relation [of space] outside it in the binding forces which act upon it."[90] For Maxwell, the structure of space did not depend on the distribution of energy or matter. Lines of force and particulate polarization were *in* space; they were not *of* space itself.[91] Einstein's idea that "there is no space without a field" was contrary to Maxwell's views. Though Maxwell agrees with Faraday's conception of the field as lines of force depending on the distribution of matter, he did not identify materiality with forces diffused through space as did Faraday in *Thoughts on Ray-Vibrations*.[92]

In 'Dynamical Theory', Maxwell stressed the importance of the 'intrinsic energy' of the electromagnetic field. He unambiguously declared that he wished "to be understood literally" regarding the nature of this energy, and that it

resides in the electromagnetic field, as well as in... bodies themselves, and is in two different forms, which may be described without hypothesis as magnetic polarization and electric polarization, or, according to a very probable hypothesis, as the motion and strain of one and the same medium.[93]

Notice that Maxwell still hankers after a substratum to which energy is essentially connected. As he says at the close of the *Treatise on Electricity and Magnetism,*

...whenever energy is transmitted from one body to another in time, there must be a medium or substance in which the energy exists after it leaves one body and before it reaches the other, for energy, as Torricelli remarked, 'is a quintessence of so subtle a nature that it cannot be contained in any vessel except the inmost substance of material things'.[94]

Notice, too, that Maxwell takes the criterion of finite velocity of propagation to presuppose the existence of an omnipresent medium in more than a dispositional sense. Unless the field exists substantively, energy conservation would be violated.

Though 'Dynamical Theory' did not base itself primarily on either particulate polarization or lines of force, in order to avoid difficulties associated with the notion of displacement, Maxwell stressed the primacy of lines of force in the 'Notes on the Electromagnetic Theory of Light' (1868) and in the *Elementary Treatise* (1881). In this work, however, Maxwell transformed the theory of the primacy of the lines of force by incorporating features first used in the theory of particulate polarization. The categories of particulate polarization and the primacy of lines of force do not conform to a distinction between earlier and later ideas in Maxwell's thought. Both, moreover, were 'field' theories for Maxwell.

What role did the aether play in Maxwell's thought? This question demands caution. Unlike Faraday, Maxwell was not antithetical to the aether for the conveyance of radiations. Moreover, he was well acquainted with the undulatory theories of Fresnel, Green, MacCullagh, and Fitzgerald which supposed energy to reside in the medium. This approach was acceptable to Maxwell and was further strengthened by his opposition to "the dogma of Cotes, that action-at-a-distance is one of the primary properties of matter, and that no explanation can be more intelligible than this fact."[95] Again, Maxwell was struck with the way the velocity of propagation of electromagnetic disturbances is the same as that of light. This provided strong reason for believing that light is an electromagnetic phenomenon, and that

the combination of the optical with the electrical evidence will produce a conviction of the reality of the medium similar to that which we obtain, in the case of other kinds of matter, from the combined evidence of the senses.[96]

Maxwell thus concluded that his "theory agrees with the undulatory theory in assuming the existence of a medium which is capable of becoming a receptacle of two forms of energy." [97] Though Maxwell held that knowledge of reality was ordered in a series of levels, knowledge of energy distributions at the level of the field, demanded the presence of an underlying medium. Though we know directly only the relations of the electrotonic function, imposed upon the field, Maxwell never ruled out knowledge of the medium itself in which the forms of energy reside.

Maxwell's systematic drive for modes of consistent representation – whether particulate polarization or lines of force – leaves no doubt about what he held to exist: matter, the aether, conceived as a state of matter, and energy. The field is thus a substantive entity conceived in terms of ordinary matter, fully actual at all times, and related to an underlying substratum. But the electromagnetic field is also a dynamical theory since there is "matter in motion, by which the observed electromagnetic phenomena are produced." [98] Consistent with this relational view of knowledge Maxwell held that electromagnetic science was based on measured operations of the dispositions of matter which manifest varied characteristics of the omni-present field. And the field and matter exist at the same ontology level.

v

The dichotomy between lines of force and molecular polarization can be seen in the work of the British Maxwellians J. H. Poynting and J. J. Thomson, who adopted the lines of force approach relating their work through Maxwell to Faraday. Poynting, in adopting the idea of tubes of force and unit cells, also related his ideas to Maxwell's *Elementary Treatise* and represented the energy of the field as being stored within the unit cells. "A space containing electric currents may be regarded as a field where energy is transformed at certain points into electric and magnetic kinds." [99] Poynting is the first person to identify explicitly field and energy concepts. Faraday's concept of forces diffused through space is transformed into a theory of the diffusion of energy in the field. As Poynting said: "When a body is acted on by a force, and moves so as to acquire energy, there is no reason to suppose that the energy is given to it immediately, with reference to the intervening space, than there is to suppose that the forces act at a distance – energy enters from the surrounding space, if so, the energy may well be in the surrounding space, *whether the body is there to be acted upon or not*." [100] Space is thus occupied by a field of energy which is the fundamental entity in Poynting's ontology. A field certainly real but scarcely material. Hence, his theory was non-mechanical in its rejection of the need for a material substratum. Electricity was simply a mode of motion of energy.

Both Poynting and Thomson avoided displacement. Thomson replacing it by a quantity which he defined in terms of the number of tubes of force passing through a plane surface. Both adopted the viewpoint of the primacy of lines of force, and though they were regarded as fundamental entities, neither scientist attempted any theory of their constitution. In their theories the field functioned as a non-mechanical but electromagnetic reality. As Poynting put it:

...progress is more likely to be made if we are content with an electromagnetic explanation – if we merely carry down to the molecules and their interspaces the electric and magnetic relations which we find between large masses and around large circuits, and leave the ether out of account. [101]

Heaviside was opposed to the use of lines of force. His adoption of Maxwell's theory was based on symmetries existing in the proportionalities between displacement and electric force. For Heaviside, though he was unable to specify their nature, the fluxes – displacement and magnetic induction – were fundamental quantities. In general, Heaviside developed his categories from Maxwell's *Treatise*. In that work, displacement, electrotonic state, magnetic induction, and electric and magnetic forces, are all vector quantities expressing states of polarization, and having opposite properties in opposite directions. In a letter to Hertz in 1889, Heaviside stated his view of the aether: "It often occurs to me that we may be all wrong in thinking of the ether as a kind of matter (elastic solid for instance) accounting for its properties by those of the matter in bulk with which we are acquainted; and that the true way could we only see how to do it, is to explain matter in terms of the ether, going from the simplest to the more complex."[102] Unlike Maxwell, who saw the ether as a form of matter, Heaviside saw the possibility of reducing matter to a primitive aether. It was only a possibility, however, for in his *Electromagnetic Theory* Heaviside argued that energy fluxes arise from the internal structure of the aether. Therefore the "ether must be regarded as a form of matter, because it is the recipient of energy, and that is the characteristic of ordinary matter."[103]

G. F. Fitzgerald took another tack. As Schaffner has shown,[104] Fitzgerald well exemplifies the belief of the late nineteenth century that the establishment of a mechanical basis for Maxwell's theory would constitute progress in its advancement. Fitzgerald was in the 'Dublin tradition' which approached optics and aether theory through the least action principles of Hamilton. And he set out to generalize MacCullagh's aether so as to include Maxwell's theory with an aim to explain reflection and refraction on the basis of electromagnetics. Larmor later stated the program succinctly: "... the conception of an elastic aethereal medium that had been originally evolved from consideration of purely optical phenomena, is capable of direct natural development so as to pass into line with the much wider and more recent electrodynamic theory which was constructed by Maxwell on the basis of purely electrical phenomena, – in fact largely as the dynamical representation and development of Faraday's idea of a varying electrotonic state in space, determined by the changing lines of force."[105] Fitzgerald thus sought to reduce the Faraday-Maxwell electromagnetic theory to MacCullagh's optical aether thereby giving the former a mechanical basis. This is another instance of optical and electromagnetic aethers being brought together in generalized form: thus in terms of formal symmetry of their mathematical structures the lines of force approach was fused with the elastic media models of nineteenth century optics.

Joseph Larmor brought the work of Fitzgerald and Maxwell to its culmination. Not only did Larmor bring the electromagnetic aether to its most developed state, but he was able to incorporate the Lorentz-Fitzgerald contraction hypothesis within a general explanation of aberration phenomena. In a series of memoirs on 'A Dynamical Theory of the Electric and Luminiferous Mediums' (1894–97), and in his *Aether and Matter* (1900) Larmor's expressed intention was "to develop a method of evolving the dynamical properties of the ether from a single analytical basis."[106] Larmor admitted a stagnant aether of the Fresnel type, and further supposed that the motion of matter does not affect it. Matter, moreover, was composed of positive and negative electrons embedded in the universal aether. Larmor's position was radical: "Matter may be and likely is a structure in the aether, but certainly aether is not a structure made of matter. This introduction of a suprasensual aethered medium, which is not the same as matter, may of course be described as leaving reality behind us: and so in fact may every result of thought be described which is more than a record of comparison of sensations."[107] There is little doubt that Larmor considered the mechanics of matter as arising from the action of the electromagnetic aether. Unlike Fitzgerald and Heaviside, he did not embark on a program of reducing electromagnetic and optical action to a mechanical basis. The aether was conceived as an ultra-primitive reality. "An aether of the present type can hardly on any scheme be other than a medium, or method of construction if that term is preferred, prior to matter and therefore not expressible in terms of matter."[108] Though matter was conceived as being "constituted of isolated portions each of which is of necessity a permanent nucleus or singularity in and belonging to the aether."[109] electrons are mobile singularities which move through the aether "much in the way that a knot slips along a rope." [110]

For Larmor the electromagnetic aether was the fundamental and sole reality existing at all times in a fully actual way as a ground of physical action. He asserted that "all that is known (or perhaps need be known) of the aether itself may be formulated as a scheme of differential equations defining the properties of a *continuum* in space, which it would be gratuitous to further explain by any complication of structure."[111] Within this framework, Larmor was able to absorb the electron theory of Lorentz construing electrons as excitations in the aether itself. Using the least action approach of the 'Dublin School' Larmor thus claimed that his electric aether was different from that of MacCullagh and Green which "virtually identifies aether with a species of matter."[112] Moreover, since the differential operations exactly describe at any time or place the action of this aether, Larmor concluded that its reality could not be in serious doubt. In this respect Larmor's thought was moving in the same direction as that of Wien, Wiechert and Abraham, all of whom saw the program for physics as reducing mechanics and optics to electromagnetic theory.

As with the other theorists discussed in this section, space does not permit a full discussion of the evolution of Lorentz's thought. He was early influenced by the electromagnetical tradition of Weber and Helmholtz in which the sole constituents of nature were electrical in character and moving in accordance with one dynamical law. These thinkers had stipulated, as least in a mathematical sense, that mass was velocity-dependent, that electrical propagation was finite, and that mechanical action and reaction did not apply to electromagnetical processes. In his doctoral thesis of 1875, however, Lorentz applied Maxwell's electromagnetic theory to the problem of the reflection and refraction of light. From this background and the sorts of problems in which he was interested such as radiation, Lorentz set out to fuse continental electron theory and Maxwell's dynamics of the electromagnetic field.

Prior to 1889, Lorentz held that mass contraction applied only to the positions and shapes of electrons as they passed through the field.[113] By 1899 arising from the null result of the Michelson-Morley experiment, and a unification of the principles of his transformation theory: the velocity-dependent electric force; the motion-dependent molecular force of his 1892 theory; and the 'local time' requirement for the transformation from a moving to a resting system with respect to the aether, Lorentz was led to an extraordinary conclusion. The invariance of the field equations conceived in terms of this framework meant that all matter – electrical and ponderable – contracts. This meant that mass is velocity-dependent. And Lorentz's interpretation of the Michelson-Morley experiment asserted that there is no matter to which Newtonian mechanics uniquely applies. Unlike Larmor, Lorentz did not believe that electrons and ponderable bodies could be regarded as structures or singularities in the aether.[114] Rather, he proposed two separate entities: movable electrons (a term he first used in 1898); and an immobile aether. Electrons had a finite radius and a uniform charge density, and they were deformable in the direction of motion. In any event, the deformable electron was at odds with the rigid sphere electron theory proposed by Abraham. Lorentz could not accept the latter theory as it was at variance with his theorem of corresponding states.[115] In any event, Lorentz had established the field as an independent entity, ontologically separate from matter.

By 1900, Poincaré, having considered the fact that Lorentz's theory does not preserve Newton's third law, postulated the existence of electromagnetic momentum. This concept was taken up by Max Abraham, who like Wien, Langevin and Wiechert, was an uncompromising advocate of the electromagnetic view of reality.[116] Their sole aim was to replace kinetic and potential energy with magnetic and electrical energy and thus replace the mechanical point of view with the electromagnetic one.[117]

Though Lorentz did not agree entirely with their uncompromising reductionism in broad outline his view of reality, as expressed in his 1906 'The Theory of Electrons', has affinities with the theories of Abraham, Wien, Langevin, Wiechert and Mie.[118] The entities which his equations described – the immobile aether and the electric charge – were fundamentally non-mechanical: electromagnetic mass and momentum. Thus the non-dispositional concept of Newtonian mass as applicable to electrons and bodies was replaced by the concept of velocity-dependent mass, related to the electromagnetic concept of self-reflection.

What ontological status did Lorentz accord his aether-field? Was it material or immaterial? Since Lorentz unambiguously separated matter and electrons from the field, the latter was scarcely to be conceived as a species of the former. In the first place, Lorentz's aether is positional, as it serves as an absolute frame of reference for the equation of electrodynamics.[119] The aether interpreted in this way becomes vacuous, of course, as soon as there is established complete reciprocity between all reference frames with regard to the theory of transformation. This was precisely the consequence of Einstein's interpretation. In 1901, Lorentz seemed to accept the merits of this approach. Referring to the aether theories of MacCullagh and Fitzgerald *inter al.* he claimed that "they become more and more artificial the more cases are required to be explained in detail. Of late the mechanical explanations of what is going on in the aether were, in fact, driven more and more to the background. For many physicists the essential part of a theory consists in an exact, quantitative description of phenomena ..."[120] Later, however, in discussing Einstein's theory of relativity Lorentz was more definite in his commitment to the aether. "I cannot but regard the ether, which can be the seat of an electromagnetic field with its energy and its vibrations, as endowed with a certain degree of substantiality, however different it may be from all ordinary matter."[121] Here Lorentz is expressing a point of view which has affinities with Maxwell's position: that the field is associated with an underlying aethereal reality as the unique ground of its action. Unlike Maxwell, Lorentz makes it clear that though the aether is in a sense substantial, it need not be considered as a form of ordinary matter. Whether it was a dematerialized entity for Lorentz, is an open question. In any event, as Hirosige has pointed out, Lorentz's notion of a charged particle so necessary to his electron theory was derived from the Continental theories of electromagnetics. This notion in turn played an important role in Lorentz's classification of the electromagnetic field as an independent physical reality.[122] Moreover, Lorentz's theory resolved macroscopic bodies into a system of charged particles moving in the electromagnetic field.

Apart from speculation regarding the materiality or immateriality of the aether, there was also the general question of the way in which the aether was known to exist. Like most physicists of his time, Lorentz was impressed with Hertz's experimental detection of electromagnetic waves. These experiments were generally thought to reveal the existence of the electromagnetic aether in which these waves undulated. MacCormmach, however, points to another aspect of the problem: the fact that the properties of the aether were described exactly by elegant equations.[123] By contrast, to Lorentz and his contemporaries, the mechanics of ponderable bodies seemed little more than an inelegant system of approximation. In comparison to the simple elegance of field propagation, therefore, complete knowledge of the complexities of ordinary matter seemed more problematic. Accordingly, the aether, though suprasensible, was exactly known through the mathematical determination of its action. Its existence was thus not in doubt.

It should be clear, by now, that nineteenth-century electromagnetics produced no unique conception of the physical field. There were as many field theories as there were thinkers concerned with the propagation of action across space. To characterize action in dielectrics, Faraday first used the term 'magnetic field' in 1845, in a study of the action of magnets on glass. [124] In his 1851 paper 'On the Theory of Magnetic Induction in Crystalline and Non-Crystalline Substances', William Thomson stated: "Any space at every point of which there is a finite magnetic force is called 'a field of magnetic force'; or, *magnetic* being understood, simply 'a field force'; or, sometimes, 'a magnetic field'."[125] Later in 1865 in his 'A Dynamic Theory of the Electromagnetic Field' Maxwell stated that "the theory I propose may therefore be called a theory of the electromagnetic Field, because it has to do with the space in the neighborhood

of the electric or magnetic bodies."[126] Thomson and Maxwell doubtless
had in mind Faraday's conception of physical action in space when they
stated their definitions: and like Faraday they did not discuss possible
mechanisms as a ground for the field.

Again, there is nothing in the work of the thinkers discussed to suggest
that they took the notion of the field to be inherently dispositional, a
power which must be analyzed counterfactually.[127] Certainly when considering real and possible actions of magnetic phenomena, nineteenth-
century scientists used the counterfactual mode of analysis. This did not
mean that they regarded fields as basically dispositional. On the contrary,
the evidence strongly suggests that they regarded fields as having a mode
of existence as real as that of substance. The elastic solid theorists and
Maxwell thought of aethers as analogous to ordinary matter; Kelvin and
Fitzgerald saw the action of media as being subsumed under the mechanics of ponderable bodies; Larmor and Wiechert thought of ordinary
matter as being constructable out of, or reducible to the field, an abstract,
dematerialized entity; and Lorentz and Abraham saw the field as an
omnipresent reality, ontologically distinct from ordinary matter. All of
these theorists thus construed the field as a self-subsisting and real entity.
But the difference between Fitzgerald, Maxwell and Larmor is instructive.
Insofar as the former conceive the aether as a form or state of matter, they
imply that it is subsumable under the laws of mechanics, and that the
field is locatable through the stable behavior in it of objects like magnets.
From the fact that the field is spatio-temporally locatable through the
stability of expectations regarding the behavior of test bodies in it, and
that the laws of mechanics were applicable to it, thinkers like Maxwell
and Fitzgerald concluded that the aether existed like ponderable bodies
as a substance. The aether was the stuff of the fields. Poynting and Lorentz
separated the field ontologically from matter; while Larmor conceived it
as the sole reality. The field thus existed because of the stability of its
behavior mapped out by exact equations; but it did not exist in the category of material substance. It was substantial enough, however, for Larmor and Wiechert to conclude that matter existed merely as a structure in
the field, and for Lorentz to hold that the field was an independent reality.
It then made little sense to ask what the field itself was made of.

This paper has confined attention to isolating some of the turning points
in the history of the emergence of field concepts as a prolegomenon to

understanding the dynamics of conceptual change involved. Though eighteenth-century theories of matter provided an important precondition for the articulation of field theories, they cannot be linked up to Faraday and his intellectual background in any straightforward way. Substantive powers are not the same as electromagnetic fields; on the other hand powers and fields have conceptual affinities which neither share with the traditions of mechanical atomism. Certainly there are many technical problems in the thought of Maxwell and Kelvin, such as the controversy on the violation of energy principles between the British and the Continentals, which *precipitated* their immediate interest in considering electromagnetic models based on the action of the medium. And there was also the continuing background of thinking in terms of optical media and continuum mechanics. Furthermore, with Lorentz and Abraham there were problems in the theory of electrons which conditioned their view of the field as an underlying reality. Yet, Faraday's ideas were there early in the nineteenth century to be translated by Maxwell and others into various models of consistent representation. Once the field approach became established in England, and later in Germany and France, the problem became that of explaining the refinement and legitimation of field-theoretical concepts.

University of Pittsburgh

NOTES

[1] For a good general account of Field theory see Mary B. Hesse: *Forces and Fields*, London 1961. Hesse has an analysis of Euler's development of the notion of a mathematical field in his hydrodynamics. See also Kenneth F. Schaffner's important *Nineteenth-Century Aether Theories*, Pergamon Press, Oxford 1972; R. A. R. Tricker: *The Contributions of Faraday and Maxwell to Electrical Science*, Pergamon Press, 1966; L. Pearce Williams: *The Origins of Field Theory*, New York 1966; and E. T. Whittaker: *History of the Theories of Aether and Electricity*, London 1951.
[2] This development is fully documented in P. M. Heimann's and J. E. McGuire's 'Newtonian Forces and Lockean Powers: Concepts of Matter in Eighteenth-Century Thought', *Historical Studies in the Physical Sciences* 3 (1971), 233–306.
[3] Trevor H. Levere: *Affinity and Matter*, Oxford, 1971, Chapter 4.
[4] *Ibid.*
[5] *Ibid.*, Chapters 3 and 4.
[6] P. M. Heimann, 'Faraday's Theories of Matter and Electricity', *British Journal for the History of Science* 5 (1971), 235–257.
[7] Schaffner, *op. cit.*, 40–58.
[8] *Ibid.*, 47.

[9] *Ibid.*, 59–75.
[10] Michael Faraday, *Experimental Researches in Electricity*, Vol. 3, New York 1965, Dover Reprint, 450–451.
[11] L. Campbell and W. Garnett, *The Life of James Clerk Maxwell*, London 1882.
[12] W. D. Niven (ed.), *The Scientific Papers of James Clerk Maxwell*, Dover Reprint, 1965, 500, 577–588; *A Treatise on Electricity and Magneticism*, Dover Reprint, 1954, 431–450.
[13] T. K. Simpson, 'Maxwell and the Direct Experimental Test of his Electromagnetic Theory', *Isis* **57** (1966), 411–432.
[14] Faraday, *op. cit.*, Note 10.
[15] Heimann and McGuire, *op. cit.*, Note 2.
[16] *Ibid.*
[17] *Ibid.*
[18] W. K. Clifford, *Mathematical Papers* (ed. by R. Tricker), London 1882, 21 f.
[19] Schaffner, *op. cit.*, Note 1, 90.
[20] *Op. cit.*, Note 2.
[21] J. E. McGuire, 'Atoms and the "Analogy of Nature": Newton's Third Rule of Philosophizing', *Studies in the History and Philosophy of Science* **1** (1970), 3–58.
[22] *Ibid.*
[23] John Herman Randall, Jr., *The Career of Philosophy* **2**, Columbia University Press, New York.
1970, Chapters 1–6; Heimann and McGuire, *op. cit.*, Note 2.
[24] Heimann and McGuire, *op. cit.*, Note 2, 261–281.
[25] John Locke: *Essay Concerning Human Understanding*, London 1828, Book II.
[26] *Ibid.*, Book II, Chapter VIII, Section 8.
[27] *Ibid.* For an interesting discussion of capacities and natures in connection with Locke and structural explanation see Milton Fisk, 'Capacities and Natures', *Boston Studies in the Philosophy of Science*, Vol. VIII (1971), pp. 49–62, and Ernan McMullin, 'Capacities and Natures: An Exercise in Ontology', *Boston Studies in the Philosophy of Science*, Vol. VIII (1971), pp. 63–81.
[28] Heimann and McGuire, *op. cit.*, Note 2, 255–295.
[29] *Op. cit.*, Note 25, Book IV, Chapter 3.
[30] Heimann and McGuire, *op. cit.*, Note 2.
[31] Richard Olson, 'The Reception of Boscovich's Ideas in Scotland', *Isis* **60** (1969), 91–103; Heimann and McGuire, *ibid.* I am grateful to Olson for reminding me in a private communication about the complexities in the conception of matter in the thought of men like Stewart and Playfair.
[32] Olson, *ibid.*
[33] Isaac Newton: *Opticks*, Dover, New York, 1952, 266.
[34] Heimann and McGuire, *op. cit.*, Note 2, 293–294.
[35] Dugald Stewart: *Philosophical Essays*, 3rd ed., Edinburgh 1818, p. 123. Similar views are expressed by John Playfair – "if it be true that in the material world every phenomenon can be explained by the existence of power, the supposition of extended particles as a *substratum* or residence for such a power, is a mere hypothesis". *The Works of John Playfair*, 4 vols., Edinburgh 1822, Vol. 4, p. 85.
[36] Olson, *op. cit.*, Note 31.
[37] Robert Young: *An Essay on the Powers and Mechanisms of Nature*, London 1788, p. 34.
[38] *Ibid.*, p. 344.

[39] Robert Greene: *The Principles of the Philosophy of the Expansive and Contractive Forces*, Cambridge 1727, 409.
[40] Trevor Levere, *op. cit.*, Note 3, 29–34.
[41] Joseph Priestley, *The Theological and Miscellaneous Works of Joseph Priestley* (ed. by J. T. Rutt), 25 vols., London 1817–1831, Vol. 3, p. 238.
[42] *The Works of John Playfair*, 4 vols., Edinburgh 1822, Vol. 4, p. 85.
[43] Heimann and McGuire, *op. cit.*, Note 2, 288.
[44] Arnold Thackray: *Atoms and Powers*, Cambridge 1970.
[45] P. M. Heimann, 'Nature is a Perpetual Worker: Newton's Aether and Eighteenth-Century Natural Philosophy', *Ambix* 20 (1973), 1ff.
[46] Priestley, *Works* 3, 230.
[47] *Die Mathematischen Schriften von G. W. Leibniz* (ed. by Buchenan and Cassirer), p. 257.
[48] Philip P. Wiener (ed.), *Leibniz Selections*, New York 1951, pp. 119–137.
[49] *Ibid.*, pp. 137–162.
[50] *Ibid.*, pp. 96–98.
[51] Randall, *op. cit.*, Note 23, 62–64.
[52] Herman J. de Vleeschawer; *The Development of Kantian Thought*, London 1962.
[35] Trevor, Levere, *op. cit.*, Note 3.
[54] Michael Faraday, *op. cit.*, Note 10, 60. For my interpretation of Faraday I am indebted to P. M. Heimann's 'Faraday's Theories of Matter and Electricity', *British Journal for History of Science* 5 (1971), 235–257.
[55] Heimann, *ibid.*, 241.
[56] Faraday, *op. cit.*, Note 10, Vol. 1, para. 1304.
[57] *Ibid.*, Vol. 1, 220.
[58] *Ibid.*, Vol. 1, 242.
[59] *Ibid.*, Vol. 3, 3269.
[60] Maxwell, *op. cit.*, Note 12, 552–553.
[61] Faraday, *op. cit.*, Note 10, **1**, 1231.
[62] *Ibid.*, **1**, 1616.
[63] *Ibid.*, **2**, 284.
[64] *Ibid.*, **3**, 2787.
[65] *Ibid.*, **3**, 2806.
[66] *Ibid.*, **3**, 2225.
[67] Heimann, *op. cit.*, Note 54.
[68] Faraday, *op. cit.*, Note 10, 3, 3277.
[69] *Ibid.*, 3, 3258.
[70] Maxwell, *op. cit.*, Note 11.
[71] Maxwell, *op. cit.*, Note 12, 30–73.
[72] *Ibid.*, 155–229.
[73] *Ibid.*, 188–299.
[74] L. Campbell and W. Garnett: *The Life of James Clerk Maxwell*, London, 1882, 215, 235–244.
[75] P. M. Heimann, 'Maxwell and the Modes of Consistent Representatia', *Archives for History of Exact Sciences* 6 (1970), 171–213. I am indebted to Heimann's important study.
[76] James Clerk Maxwell, *A Treatise on Electricity and Magnetism*, New York 1954, Vol. 2, p. 493.
[77] *Ibid.*, p. 493.

[78] *Ibid.*, Vol. 1, p. 79.
[79] *An Elementary Treatise on Electricity*, Oxford, 1881, p. 52.
[80] Laurens Laudan, 'G. L. LeSage: A Case Study in the Interaction Between Physics and Philosophy', *Proceedings II International Leibniz Congress*, forthcoming.
[81] *Scientific Papers, op. cit.*, Note 12, pp. 451–513.
[82] *Ibid.*, pp. 188–209.
[83] *Ibid.*, pp. 467.
[84] *Ibid.*, p. 455.
[85] *Ibid.*, p. 468.
[86] *Ibid.*, pp. 477–488.
[87] *Ibid.*, p. 488.
[88] *Ibid.*, pp. 526–597.
[89] *Ibid.*, p. 527.
[90] P. M. Heimann, *op. cit.*, Note 75, 182.
[91] *Ibid.*, p. 182.
[92] *Ibid.*, p. 182.
[93] *Scientific Papers, op. cit.*, Note 12, p. 564.
[94] *Ibid.*, p. 493.
[95] *Ibid.*, p. 492.
[96] *Ibid.*, p. 431.
[97] *Ibid.*, p. 432.
[98] *Scientific Papers, op. cit.*, Note 93, p. 527.
[99] J. H. Poynting: *Collected Scientific Papers*, Cambridge 1920, 175.
[100] J. H. Poynting, *ibid.*
[101] J. H. Poynting, *op. cit.*, Note 12, p. 267.
[102] This letter is located in the Deutsches Museum, Munich.
[103] Schaffner, *op. cit.*, Note 1, p. 90.
[104] *Ibid.*, pp. 84–111.
[105] Joseph Larmor, *Aether and Matter*, Cambridge 1900, Preface, VIII.
[106] *Ibid.*
[107] *Ibid.*, VI.
[108] *Ibid.*
[109] *Ibid.*, VII.
[110] *Ibid.*, p. 86.
[111] *Ibid.*, p. 78.
[112] Schaffner, *op. cit.*, Note 1, p. 94.
[113] For a fine discussion of the electromagnetic world-view, see Russell McCormmach, 'H. A. Lorentz and the Electromagnetic View of Nature', *Isis* **61** (1970), 459–497.
[114] *Ibid.*, 495.
[115] H. A. Lorentz, *The Theory of Electrons*, Dover Reprint, 1952, pp. 214–215.
[116] McCormmach, *op. cit.*, Note 113.
[117] Stanley Goldberg, 'The Abraham Theory of the Electron: The Symbiosis of Experiment and Theory', forthcoming.
[118] *Op. cit.*, Note 115.
[119] *Ibid.*, pp. 186–218.
[120] Schaffner, *op. cit.*, Note 1, p. 106.
[121] Lorentz, *op. cit.*, Note 115, p. 230.
[122] Tetu Hirosige, 'Origins of Lorentz's Theory of Electrons and the Concept of the Electromagnetic Field', *Historical Studies in the Physical Sciences* **1**, 151–209. Hirosige's

study is an important clarification of the process in physical thought which led to
the separation of ponderable matter and the electromagnetic field.

[123] McCormmach, *op. cit.*, Note 113, 494. Larmor put the point thus: "It may fairly
be claimed that the theoretical investigations of Maxwell, in combination with the
experimental verifications of Hertz and his successors in that field, have imparted
to this analytical formulation of the dynamical relations of free aether an exactiness
and precision which is not surpassed in any other department of physics, even in the
theory of gravitation". *Op. cit.*, Note 105, pp. 163–164.

[124] Faraday, *op. cit.*, Note 10, Vol. 3, p. 2252.

[125] William Thomson, *Reprint on Papers on Electrostatics and Magnetism,* London 1884,
pp. 472–473.

[126] Maxwell, *op. cit.*, Note 89, 527.

[127] For a recent discussion of the notion of the field as a dispositional entity see
Howard Stein, 'On the Notion of the Field in Newton, Maxwell and Beyond', *Minnesota Studies in the Philosophy of Science*, Vol. 5, 1970.

DUDLEY SHAPERE

NATURAL SCIENCE AND
THE FUTURE OF METAPHYSICS

Does physical science, in its employment of such terms as 'electron', 'photon', and 'space-time', make claims about what exists? The immediate reply of common sense would undoubtedly be affirmative; and yet powerful philosophical forces have aligned themselves against this view, however obvious it may appear at first glance. For, it has been urged, such terms as the above are 'theoretical'; and whatever meaning they may have either is, and must be, wholly exhausted by some set of 'observation terms', or else, if there is any additional component of meaning of such terms – if, that is, they are only 'partially interpreted' in observational terms, – that surplus meaning does not, and could not, consist in reference to any 'unobservable' entities behind the scenes of, and causing, the scientist's experiential data. The use of such terms, it is concluded, does not involve any claim that something referred to by them (other than a set of observed or observable data) literally exists. Still more generally and more positively, the allegation is that any scientific terms which appear to refer to unobservable entities – and 'electron', 'photon', and 'space-time' are said to be such terms – serve a function other than making claims about existence, a function which I will summarize by saying that they are alleged to be mere *conceptual devices*. Other writers have used more specific names for such terms or the ideas they represent, among them 'logical constructs', 'convenient fictions', 'shorthand summaries', 'intervening variables', 'models', 'abstractions', 'idealizations', 'instruments' – all, often, prefixed by a rather pejorative 'mere'.

This general type of solution to the so-called 'problem of the ontological status of theoretical entities' has been characteristic of most of the positivistic tradition in the philosophy of science. However, it might at first glance seem to offer solace even to the bitterest opponents of positivism; for the solution seems to suggest that, since science itself does not provide a key to the understanding of reality, perhaps there is some other subject that does. The positivistic retort has been the verifiability theory of meaning, or some other analysis of meaning which is alleged to render questions about ultimate reality, whether raised in science or in other circumstances, senseless (or, more cautiously, 'scientifically meaningless').

Recent years have seen growing doubts regarding the positivistic ideas which form the basis of its doctrine that 'theoretical entities' are mere conceptual devices. Yet the criticism of the foundations of that doctrine leave the status of the doctrine itself uncertain : given the destructive criticisms of the sharp, clean distinction between 'theoretical' and 'observational', and of the verifiability theory and associated views, what are we to say now about the 'ontological status' of electrons, photons, space-time, and other such scientific entities? Are they 'entities'? Or were the positivists, despite their erroneous arguments, right in their conclusion that they are mere conceptual devices to facilitate the organization and prediction of experience?

With increasing disaffection from positivism, increasing numbers of philosophers have begun to advocate realistic interpretations of theoretical entities. Nevertheless, I do not believe that complete success has yet been achieved in these efforts. In particular, the precise way in which the question of ontological status is relevant to the concerns of the practicing scientist himself needs to be clarified. The view is still prevalent that, as far as his scientific work goes, it is a matter of complete indifference to the scientist whether electrons 'really exist' or not. Many scientists have reinforced this view by expressing a shoulder-shrugging 'So what?' attitude toward what they regard as essentially an irrelevant philosophers' debate – though, as always, we must ask whether those scientists, in such remarks, are really reporting their own reasoning practice, or only reflecting what they have been told by a generation of positivism. In any case, it must be confessed that it is difficult to see what could even be meant by saying that 'realism' is implied by or presupposed by or somehow involved in science without at the same time showing that that 'realism' is somehow reflected in scientific reasoning and practice. Without such clarification, realistic interpretations of science can leave us with little more than did the positivistic view which they are meant to combat: for whereas positivism held that the 'realism' claim was scientifically meaningless, the newer views, where they fail to show how the alleged realistic aspects of scientific concepts play an actual functioning role in scientific reasoning, say only that "realism" is scientifically irrelevant; or, if it is alleged to be relevant, the exact nature of that relevance is not clear.

It seems to me that if a realistic interpretation of 'theoretical entities' is to be defended, that defense must lean heavily on an analysis of exactly how the 'realistic' aspect of such concepts functions in scientific reasoning itself. The present paper will attempt to sketch the broad outlines of such a defense. I wish to argue not only in favor of a realistic interpretation of such terms as 'electron' in science, but to do so by arguing that claims about the existence of such entities play a definite and important role in the rational processes of science, specifically through their working contrast with what I called, above, conceptual devices. Thus I hope to counter not only the positivistic view that existence-claims about theoretical entities are meaningless, but also the view that, even under a realistic view, this realism is a philosophical overlay on science, a conclusion superadded to scientific work, which could go on in complete indifference to any claims about the existence or non-existence of, for instance, electrons.

Before proceeding to the argument, however, a methodological remark seems appropriate. In trying to understand science, we must, of course, look closely at cases of scientific reasoning. But the sorts of cases to be examined must be selected judiciously. The reasoning involved in cases from the early stages of scientific history is apt to be too confused, particularly with regard to what was being alleged to exist. On the other hand, cases taken from immediately contemporary science are also liable not to be completely satisfactory: the dust has often not settled sufficiently. Questions – or at least some questions, including ours – about what is going on in early science and in immediately contemporary science are more likely to receive illumination from cases where the reasoning is clear and explicit than to give it.

According to the Lorentz theory of the electron, that particle cannot be a geometrical point, having zero radius. This results fundamentally from the fact that the electrostatic energy of a charged sphere of radius r and charge e is (except for a numerical factor) equal to e^2/r; this formula implies that a charged sphere of zero radius would have infinite energy, or, if we apply the Einstein relation $E=mc^2$ between energy E and mass m, infinite (rest) mass. However, the electron does not have infinite energy or mass. Nevertheless, for certain purposes – for the solution of certain problems – and under certain circumstances, it is convenient and possible to treat the electron *as if* it were a point-particle.

We have written the solution of the potential problem as a sum of boundary contributions and a volume integral extending over the source charges. These volume integrals will not lead to singular values of the potentials (or of the fields) if the charge density is finite. If, on the other hand, the charges are considered to be surface, line, or point charges, then singularities will result.... Although these singularities do not actually exist in nature, the fields that do occur are often indistinguishable, over much of the region concerned, from those of simple geometrical configurations. The idealizations of real charges as points, lines, and surfaces not only permit great mathematical simplicity, they also give rise to convenient physical concepts for the description and representation of actual fields.[1]

We see, in this passage from Panofsky and Phillips, that it is on *scientific* grounds that treatment of the charged particle as a dimensionless point is considered an 'idealization': that is, the conclusion that the electron *cannot really be* a dimensionless point is not, in this case, a logical or epistemological overlay superadded to the science concerned – a conclusion drawn solely from a more general and sweeping philosophical thesis, to the effect, for example, that '*All* scientific concepts are idealizations', or that 'All bodies are (or must be) extended', or that 'our ordinary concept (usage)' of the expression 'material object' and related terms implies that talk of dimensionless material objects is absurd.

Furthermore, not only is the *impossibility* of considering electrons really to be dimensionless points based on purely scientific considerations; the *rationale* for considering them *as if they were* – the possibility of so treating them, and the reasons why it is convenient to do so – are also scientific in character. As Panofsky and Phillips note, the fields that occur when we consider the source charges to be localized in a point are 'often indistinguishable, over much of the region concerned, from those of simple geometrical configurations'. It is thus possible to treat the source charges (at least in many problems) as if they were concentrated at a point, *even though we know, on purely scientific (i.e., not philosophical or linguistic) grounds that electrons cannot really be that sort of thing.* Furthermore, it is *convenient* to treat them in that way: for "The idealizations of real charges as points, lines, and surfaces not only permit great mathematical simplicity, they also give rise to convenient physical concepts for the description and representation of actual fields."

The electron as it really is cannot, therefore, have the zero-radius characteristic which is attributed to it for the sake of dealing with certain problems; and this distinction – between the electron as it really is and the electron as idealized because it is possible and convenient to treat it in

NATURAL SCIENCE AND THE FUTURE OF METAPHYSICS 165

a certain way – is one which, in this case at least, is made on purely scientific grounds.

A further example of such idealization is given by the status of the classical concept of a rigid body according to the special theory of relativity. Classically, a 'rigid body' is one in which the distances between any two of its constituent parts (particles) remains invariant. If a force is applied to the body at any point, then in order for the body to remain rigid in this sense, i.e., in order for the distances between any two points to remain the same, that force must be transmitted instantaneously to all other parts of the body. In other words, the force must be transmitted with infinite velocity. But according to the special theory of relativity, energy and momentum (and hence forces) cannot be transmitted with a velocity greater than that of light. Therefore, with the application of a force at one of its points, all the parts of a body cannot begin moving simultaneously; the body must be deformed. It is thus impossible, according to the special theory of relativity, that there should exist any such things as rigid bodies in the classical sense. Nevertheless, it is often convenient and possible to use that conception working with the theory. (Note, incidentally, that if the state of motion of elementary particles like electrons is conceived as being completely specifiable in terms of their position, velocity, and rotation as wholes, without reference to any internal structure, then the impossibility of rigid bodies according to special theory of relativity *precludes* elementary particles from having any extension. Thus, whereas the Lorentz theory precluded the electron from having a zero radius, the theory of relativity is generally taken as *requiring* it to have this characteristic. Theories may specify how an entity *must be* as well as how it *must not be*.)

In the cases examined, the general features of the use of idealizations may be summarized as follows: (i) there exist certain problems to be solved; (ii) mathematical techniques exist for dealing with those problems if the entities dealt with (or their properties) are considered in a certain way, even though it is known, on the basis of the theory, that those entities cannot really be that way; and (iii) in many cases, it is provable that the difference between a realistic and an idealized treatment will be insignificant (e.g., below the limits of accuracy required by the problem at hand; or below the limits of experimental accuracy) relative to the problem to be solved.

Yet another type of 'conceptual device' is exemplified in Bohr's ignoring, in his original 1913 presentation of his theory of the atom, of the motion of the nucleus. In this case, too, the supposition that this treatment cannot be realistic is based on scientific information. However, the present case differs from the previous ones in a crucial respect: the information which leads to the supposition that this is not a 'realistic' treatment is based on background information that is independent of the postulates of the new theory being proposed. This difference (and other differences which I will not go into here) make it advisable to distinguish this sort of conceptual device from those treated above and called 'idealizations'; I propose to call the present type 'simplifications', though that word, like the word 'idealization', is usually used in a vague and ambiguous way. In any case, what serves as the basis for considering the nucleus not to move while the orbital electrons do, is information taken from the fields of classical electricity and Newtonian mechanics. For the nucleus, in the Bohr theory, is assumed to have a charge opposite (viz., positive) that of the orbital electrons; but opposite charges attract one another; and Newtonian mechanics tells us that if a smaller body is moving in an orbit around a larger one, the latter also will move in response to attraction by the smaller, and both bodies will describe orbits about a common center of the attractive force. It was therefore reasonable to assume that the nucleus should move in response to the motions of the orbital electrons, and hence to suppose that ignoring such motion constituted a simplified (i.e., non-realistic) treatment of the situation. Such a supposition, of course, is a hypothesis which *can* turn out to be incorrect; the danger in the present example is a very real one, inasmuch as certain aspects of the classical theories of mechanics and electromagnetism were *rejected* by the very Bohr theory to which they were now being applied to determine the distinction between a realistic and simplified treatment.

What have been considered thus far are ways in which the distinction between realistic treatments and conceptual devices is set up in the scientific reasoning process. The discussion has thus centered on what can be described as the statics of the distinction. For the main purposes of this paper, it is perhaps enough to have shown how the distinction arises within science, on the basis of scientific considerations. But it may not be completely out of place to suggest one way in which the distinction functions 'dynamically', so to speak. The example I will give shows how the distinction serves to guide the scientist, on some occasions, in selecting a reasonable line of investigation to undertake in the attempt to answer certain scientific problems that have arisen. The example is this: certain mysterious lines in the spectrum of the star Zeta Puppis were ascribed by Bohr, on the basis of his theory, to helium. However, it was found that there was a small but significant discrepancy between the observed positions of those lines and the positions calculated on the basis of Bohr's theory. The difference was large enough to suggest that Bohr's theory might be incorrect. However, by a more exact calculation in which he took into account the motion of the nucleus which he had in his original presentation ignored, Bohr attained a spectacular agreement between the predictions of his theory and the observed positions. The situation may be described in a generalized way as follows: a problem arose for the theory; and a reasonable line of research to try in attempting to answer the problem (and one which, in this case, happened to work) was to look at areas in which simplifications had been made. One might even speak here of a general principle of non-rejection of theories: that when a discrepancy is found between the predictions of the theory and the results of observation or experiment, do not reject the theory as fundamentally incorrect before examining areas of the theory in which simplifications have been made which might be responsible for the discrepancy. (Hanson might have spoken here of a principle in the logic of discovery; however, such a name is misleading, inasmuch as there is no guarantee that following this principle – or even *any* such principle – will necessarily eventuate in a solution of the problem. It is thus less misleading to speak of plausible or reasonable lines of research than to speak of a logic of discovery here.)

We have seen that there are, in physics, cases in which a distinction is made, on scientific grounds, between the way (or ways) in which an entity can or cannot exist, and the way (or ways) in which, for the sake of dealing with certain scientific problems, it is possible and convenient to treat that entity (as idealized or simplified, e.g.). This distinction may now be put in a more general way as follows. On the one hand, we have assertions that certain entities do, or do not, or might, or might not exist; or, putting the point linguistically, we have terms which can occur in such contexts as '... exist(s)', '... do (does) not exist', '... might exist', '... might not exist'. (It should be emphasized that what are of relevance are not the *terms* involved, but rather their uses: thus in a problem in which we treat the electron as a point-charge, we may refer to the point-charge *itself* by the term 'electron'. The context of usage, however, will indicate that the term is in such a case concerned with the idealization and not directly with electrons as they really are.) There are *clear* cases of such terms, or of such uses of terms. Sometimes they have to do with entities which are presumed to exist ('electron'), though they might not (or might not have). Sometimes they have to do with (purported) entities which, though they have, at some time, been claimed to exist, do not ('ether', 'phlogiston'). And finally, still others refer to (purported) entities whose existence is claimed (on presumably good grounds) by some good theorists, but whose existence or non-existence has not yet been established ('quark'). It should be noted that this class of terms includes also many terms (uses) which have, in the positivistic tradition, been classified as 'observational' – 'table', 'planet' – as well as terms (uses) usually classified in that tradition as 'theoretical'.

On the other hand, we have expressions like 'point-particle (in the Lorentz theory)', 'classical rigid body (from the viewpoint of special relativity)', which do not designate (purported) entities, though they are related, in the ways discussed above, to terms which do. Thus their reference to (e.g.) existing things is indirect.

The first type of terms (or uses) may be called 'existence–terms'; or, inasmuch as, in the cases considered above, what are alleged to exist are certain sorts of entities, they may be called 'entity-terms'. Not all terms in science which refer to something 'non-idealized', of course, are naturally classed as entity-terms. Many of them are more naturally referred to as having to do with 'properties', or with 'processes', or with 'behavior of entities', for example. And, furthermore, there are many borderline cases which are not easily brought under any of these headings. Finally, there are, as has been suggested, differences between the uses so classified which might lead us to draw finer-grained distinctions, or even, perhaps, for some purposes, to classify them differently.

Nevertheless, it seems to me that the analysis given above is sufficient to undermine the positivistic argument that the term 'electron' in physics is a mere conceptual device. Not only does science make existence-claims about electrons and other 'theoretical entities', to use the positivistic jargon, but those claims are made as an integral part of the scientific reasoning process, playing a definite and important role in the rational development of science. Nor, for this very reason, is the question of whether science makes existence claims one which is irrelevant to the working of science; the scientist himself may not be fully cognizant of the reasoning he himself employs; but that is no basis for denying that the reasoning takes place. One might as well deny that a tennis player has a backhand because he does not talk about it, or is unable to say how he manages to use it. Finally, the conceptual devices of science – or at least those types of conceptual devices illustrated here (and which I have called 'idealizations' and 'simplifications') – are parasitic, in the logic of their employment, on existence-claims.

But there is one further line of defense to which the 'conceptual device' interpretation of theoretical terms can retreat. For, it might be replied, the above arguments show only that there are *degrees* of being a conceptual device: instead of saying that the distinction is between, say, 'idealizations' and 'entity-terms' involved in existence-claims, what we should say is that the distinction is between 'more idealized' and 'less idealized' treatments. And even though the word 'exists' and its cognates are used in a natural way in connection with words like 'electron', such usage must not be taken literally. Such might be a positivistic reply to the above arguments; but even the opponent of positivism might claim that the sense of the term 'existence' used in these arguments – in contrast to conceptual device terms – is irrelevant to some more literal or philosophical sense of 'exists'. In my attempt to counter the 'So what?' attitude of some scientists toward the philosophical problem of 'ontological status', have I perhaps only opened the door to a *converse* 'So what?' problem: that my arguments are not relevant to any question about the real existence of electrons.

However, there are considerations which militate against this converse 'So what?' position. For there are some very important features of the term 'exists' as used in the above-analyzed contexts which are shared with usual uses of that term; and if this is true, then a still heavier burden of proof is placed on those who wish to allege that the above arguments are irrelevant to questions of 'real' existence. For to say that '*A* exists' implies, in usual contexts, at least – among other things, surely – the following:

(i) *A* can interact with other things that exist. Particles that exist can interact with other particles that exist, and, derivatively, can have effects on macroscopic objects and be affected by them. 'Convenient fictions' or 'constructs' or 'abstractions' or 'idealizations' cannot do this, at least in any ordinary sense.

(ii) To say that 'A exists' implies that A may have properties which are not manifesting themselves, and which have not yet been discovered; and, contrariwise, it is to say that some properties currently so ascribed may be incorrectly so ascribed. We may be wrong in saying that a certain property of an entity has a certain quantitative value. Or we may be wrong in thinking that that property is fundamental – it may be, to use Leibniz's colorful phrase, a 'well-founded phenomenon', being a manifestation of some deeper reality (e.g., as Wheeler claims that many properties of particles may be explainable as mere manifestations of an underlying 'geometrodynamic field'). Or, again – though these kinds of cases are rarer, especially in more sophisticated stages of scientific development, – we may be wrong in thinking that the entity has the property at all. Finally, we may be wrong in thinking we have exhausted all the properties of the entity, and may discover wholly new ones: spin, strangeness. These features are all hard to understand if electrons, for example, are mere 'convenient fictions'. (Note that what counts as a 'property' is also specified on scientific grounds.)

(iii) To say that 'A exists' is to say that A is something about which we can have different and competing theories. From the theoretical work of Ampere and Weber to that of Lorentz, from the experimental work of Faraday on electrolysis to Millikan's oil-drop experiment, there was an accumulation of reasons for holding that electricity comes in discrete units. The notion of the electron thus acquired what amounts to a theory-transcendent status: it was an entity about which theories – theories *of* the electron – were constructed. It is indeed ironic that the term 'electron' – often taken as a paradigm case, in the philosophical literature of the positivistic tradition, of a 'theoretical' term, should have this status; for the comparability of different, competing theories is now seen to be, not (at least not solely) their sharing of a common 'observational vocabulary', but rather their being about the same sort of entity. The erstwhile 'theoretical term' is now seen to be the source of what is perhaps the most important aspect of the 'comparability' of competing theories: for electrons are what those theories are in competition about.

There thus appear to be powerful considerations suggesting, first, that scientific theoretical terms like 'electron' are involved in claims about what exists in the universe; that these claims are often intimately involved in the detailed reasoning that goes on within science; and, finally, that such existence-claims have important features in common with the way 'existence' is discussed in ordinary and in philosophical contexts.

There is no guarantee that the reasoning-processes of science themselves will not change in the future, as they have evolved in the past, and that the above-argued features of science will no longer be such. But I see no clear evidence that these features have been or are about to be abandoned. And if these contentions are valid, then an analysis of the reasoning behind existence-claims in science, and of the implications of such claims, would have much to say that is of relevance to a number of the traditional problems of metaphysics.

University of Illinois (Urbana)

NOTE

[1] Panofsky, W. K. H. and Phillips, M., *Classical Electricity and Magnetism*, Reading, Mass., Addison-Wesley, 1962, p. 13.

STANISA NOVAKOVIC

IS THE TRANSITION FROM AN OLD THEORY TO A NEW ONE OF A SUDDEN AND UNEXPECTED CHARACTER?

It seems that the title of this paper is not a particularly clear one, despite its extraordinary length. So, I would like first of all to clear up all uncertainties as regards the real content of my paper.

If I may now make use of the opposite method, following the view which (with many reasons) associates clearness with conciseness, I can summarize the essence of my paper in just three words: I am going to discuss the problem of the *structure of scientific revolutions*. If the terms 'sudden' and 'unexpected' in the title of my paper may also be to some extent misleading, then I want to declare that I had in mind primarily some aspects of that conception of scientific revolutions of Thomas Kuhn, which will be discussed later on in considerable detail.

Contrary to theory of political revolutions, theory of scientific revolutions is of a recent character. The investigations of the structure of scientific revolutions have just begun. But, while some of the philosophers of science make serious efforts in this direction (I have in mind Popper and Kuhn, as well as – in a particular sense – Lakatos and Feyerabend), there are already several of them who deny the possibility of any real scientific revolution. (I am thinking of Toulmin and Kneale.)

Energetic breaking with old scientific tradition, and the transition to new scientific conceptions is not, naturally, of a recent character; but, scientific methodology did not say anything serious about this problem until quite recently. In fact, bearing in mind the history of scientific methodology, it seems that one can assert that the prevailing view about the development of scientific knowledge was the model of development-by-accumulation, which is closely connected with inductivist methods.[1]

I shall turn first to Popper's conception of the structure of scientific revolutions.

In order to get a clear understanding of this conception, it has to be put into the framework of the more general, and in fact the central problem of Popper's methodology – into the framework of his conception of growth of scientific knowledge. The most general method of growth of scientific knowledge for Popper is the method of trial and error.

But, we have also to be aware of a few other characteristics of the view of Popper. So, among other things, he maintains that we are never making any observation without a theory in mind. For this theory, Popper *does not say* that it is *arbitrary*, since, as a matter of fact, it is deeply rooted in our background knowledge, which is comprising the whole system of our preceding assumptions and theories, metaphysical as well as scientific. But, since he is denying any role of induction in our scientific reasonings, *he says* that it is the *product of our guessings*. We can read:

Scientific theories are not the digest of observations, but they are inventions – conjectures boldly put forward for trial, to be eliminated if they clashed with observations; with observations which were rarely accidental, but as a rule undertaken with the definite intention of testing a theory by obtaining, if possible, a decisive refutation. ([10], p. 46)

Now, in approaching our observations with a theory that is just a bold conjecture, we do not have any warranty that we shall even notice, let alone take seriously, those facts that do not agree with our theory. So, it is necessary to introduce yet another element – the element of critical thinking, the only element which can secure the rationality of scientific knowledge; scientific knowledge being perhaps the only human activity where the mistakes are systematically criticized, and so, in the process of time, often corrected. Accordingly, critical thinking is, for Popper, the main instrument in the attempt to refute any scientific theory. "Observations are used," these are Popper's words, "only if they fit into our critical discussion." ([10], p. 197)

It is only natural that the growth of scientific knowledge conceived in this manner, which is at the same time rational and empirical, is in perfect agreement, or – if you prefer stronger claim – demands the existence of rival hypotheses or theories. And, as a matter of fact, Popper holds that this is the case in greater part, or in the greatest majority of important scientific issues, since, in order to be able to find out all the shortcomings of the old theory, we need a new one. ([9], p. 87 and [10], p. 246) So, the most frequent case of refutation of a theory is the refutation by the crucial experiment, which is designed to decide between two (or even more) rival hypotheses.

This is, as it were, the general framework of Popper's conception of the structure of scientific revolution. We have seen that Popper does not maintain anything like the inductivist development-by-accumulation of scientific knowledge, but rather development by "repeated overthrow of scientific theories and their replacement by better or more satisfactory ones." ([10], p. 216) These new theories, though they have to explain old facts too, may explain them *in a different way* than the old theory; anyway, the new theories are requested to be as bold and unexpected as possible. This means that for Popper the scientific revolution is the basic and essential component of the growth of scientific knowledge. Popper says:

... Our critical examinations of our theories lead us to attempts to test and to overthrow them; and these lead us further to experiments and observations of a kind which nobody would ever have dreamt of without the stimulus and guidance both of our theories and of our criticism of them. For indeed, the most interesting experiments and observations were carefully designed by us in order to test our theories, especially our new theories. ([10], p. 216)

But I would like to warn you that we do not have here a theory of revolution in permanence; I am issuing this warning since both Lakatos and Watkins are bringing out the later interpretation. Namely, Popper is pointing out that once a hypothesis has been proposed and tested and has proved its mettle, it may not be allowed to drop out without good reason. ([9], p. 53) And in *Conjectures and Refutations* we can read:

The dogmatic attitude of sticking to a theory as long as possible is of considerable significance. Without it we could never find out what is in a theory – we should give the theory up before we had real opportunity of finding out its strength; and in consequence no theory would ever be able to play its role of bringing order into the world, of preparing us for future events, of drawing our attention to events we should otherwise never observe. ([10], p. 312)

At the same time, I have to admit that the revolutionary periods are for Popper the most valuable and most cherishable, since they represent the genuine creative nature of science. Yet, on the other hand, as Larry Laudan pointed out to me, in spite of the generally accepted interpretation, the question may be raised whether it is appropriate at all to speak of Popper's theory as a theory of scientific *revolution*. The point is that it seems as if Popper has no way of distinguishing between global scientific revolutions and minor changes of scientific theories. I am inclined to take this objection as quite correct. I may even add that the absence of any means of distinguishing between scientific revolutions and small replacement of theories is responsible for that interpretation of Popper which sees him as maintaining the theory of revolution in permanence, which I already rejected. Still, I have some reservations in the sense that I hold that the primary intention of Popper was to explain exactly those radical changes of our views on nature which are usually called scientific revolutions. My reasons are the following:

(1) Popper is exclusively interested in the most universal theories and the structure of their change;

(2) he is not at all interested in tinkering with theories, in any kind of theory – modification, rehabilitation, or reconditioning;

(3) his basic requirement for a new theory is that it has to be as bold and 'improbable' as possible, and so, consequently, as different as possible in comparison with the old theory;

(4) theories that replace one another during the scientific revolution need not be incompatible theories;

(5) if all Popper's conditions for theory change in science have been fulfilled, the change has to be a radical one;

(6) all the examples he uses may come under the heading of scientific revolutions.

The difficulty remains that in Popper there are no criteria for establishing whether we are dealing with scientific revolutions, or with minor changes of scientific theories.

As for the very structure of scientific revolutions in Popper, if my interpretation is correct, we can differentiate two cases:

(1) We are starting from a single theory, which we submit to empirical tests, or, which we try to criticize and then eventually reject. We will have enough grounds to reject the theory if and only if our observations that contradict this theory at the same time corroborate the falsifying hypothesis, i.e. low-level empirical hypothesis which describes those effects which are presented by our basic observational statements. Such experimental refutation of a theory compels scientists to search for a better theory. (Here, of course, one should bear in mind that Popper is very well aware of the fact that it is impossible to achieve conclusive disproof of any theory.) ([9], pp. 108, 222, 501)

(2) We are starting with two (or more) rival hypotheses which came into existence at the same time, i.e. within a very short period of time. In the course of the critical discussion that follows immediately, such an experiment is conceived that will refute (at least) one of the proposed hypotheses. In this second case there is greater possibility to recognize

not only the weak but also the strong points of a theory, so that theoreticians may – if both of the theories get refuted – come out with such a new theory in which the best elements of both theories will be preserved. ([9], p. 87, and [10], p. 315)

So, the structure of scientific revolutions has got here quite clear outlines. As one may expect, Popper does not consider the transition from an old to a new theory as of sudden character. As a matter of fact, the process of transition is for him rather complex. Nevertheless, in some formulations of Popper, as well as in many interpretations of his views, the process of falsification – as the essential element of the transition – looks very simple, even naive and unconvincing. To start a different interpretation, there is no such thing as conclusive falsification. Popper is quite aware of this, he even says that explicitly ([9], p. 50), though generally, I concede, one may get just the opposite impression since he does not seem to think that this admission may force him to qualify his rigid falsificationist position. In my opinion, this position needs qualification, but I will here restrict myself to the following question: what is the function of this inconclusive falsification in the process of theory change in science. Now, first of all, we are not confronted here with any kind of simple comparison of predictions, deduced from a theory, with experimental observations. The point is that these predictions are deduced together with auxiliary hypotheses and initial conditions. So, in case our observations contradict our predictions, it is necessary not only to check our observations, but also to show that attempts to neutralize such a refutation by modifying one or another of the auxiliary hypotheses are unsatisfactory, either because they themselves could be refuted, or are *ad hoc*, and that is not so simple to achieve. If all this has been done, it still does not necessarily entail the rejection of a given theory, especially not in the sense that scientists would completely give up using the theory, even when they have at their disposal the better one. What does this entail then? It entails only that we have positive indications that the theory had encountered serious problems, that it had – as Popper would say – created new and significant problems for further investigation, the solution of which requires a theoretical innovation. This interpretation of Popper's falsification principle, which partly has been suggested to me in my conversations with Alan Musgrave in 1968, may at the same time help us to understand better the suggestion of Popper that science should be visualized not so much as progressing from theory to theory, but as *progressing from problems to problems* – to problems of ever increasing depth. ([10], p. 222)

As for 'unexpectedness', one may say that, for Popper, the new theory really has to be as unexpected as possible, but in a special sense – in the sense of including as much new empirical content as possible, of giving as much new information as possible, of being as bold as possible in guessing about the world around us, in being as 'improbable' as possible, in order to be falsifiable to the greatest possible degree.

I will now turn to Kuhn to discuss his conception of the structure of scientific revolutions.[2]

I am assuming that you are more or less familiar with Kuhn's technical terms 'normal science', and 'paradigm', which have been debated exceptionally extensively during the last ten years. So, I will immediately start with Kuhn's definition of scientific revolution. Kuhn says:

When [the normal science goes astray] – when, that is, the profession can no longer evade anomalies that subvert the existing tradition of scientific practice – then begin the extraordinary investigations that lead the profession at least to a new set of commitments, a new basis for the practice of science. The extraordinary episodes in which that shift of professional commitments occurs are the ones known in this essay as scientific revolutions... – ... Each of them necessitated the community's rejection of one time-honored scientific theory in favor of another incompatible with it. ([4], p. 6)

Or in the same book:

Scientific revolutions are here taken to be those noncumulative developmental episodes in which an older paradigm is replaced in whole or in part by an incompatible new one. ([4], p. 91)

On the basis of these definitions, it is already possible to notice some essential new characteristics of the views of Kuhn:

(1) Revolutions are extraordinary, and rare episodes in science, while the long periods of 'normal science' would represent that regular state of science which is generally most convenient for scientists too; moreover, in Kuhn's opinion, it is only during the periods of normal science that progress seems both obvious and assured:

(2) the new prevailing scientific theory or paradigm is incompatible or incommensurable with the old one. ([4], pp. 162ff.)

But let us see now how scientific revolutions come into existence. During the period of 'normal science', which begins immediately after substitution of one paradigm for another, the new paradigm is fleshed out, articulated, and extended; if scientists in this period encounter the facts incompatible with the paradigm, these facts are ignored. This period may be characterized as 'the period of accumulation'. Yet, when all the possibilities for development within a certain paradigm get exhausted, when the problem solving activities within a paradigm get unsuccessful – scientists then become more and more aware of the existence of new anomalies, which is the pre-condition for arising of any acceptable change of theories. ([5], p. 250 and [4], pp. 75, 67) Kuhn says: "The awareness of anomaly [may last] so long and penetrate so deep that one can appropriately describe the fields affected by it as in a state of growing crisis." ([4], p. 67) And further on: "... All crises close with the emergence of a new candidate for paradigm and with the subsequent battle over its acceptance..." ([4], p. 84)

The transition from a paradigm in crisis to the new paradigm is of sudden character; in fact, according to Kuhn, this transition cannot be made a step at a time, since it does not take place 'forced by logic and neutral experience', i.e. by rational arguments. It happens in the form of conversion, so that it has to be sudden and all at once, like the gestalt switch. ([4], pp. 121, 149)

This is Kuhn's thesis of a sudden character of scientific revolutions, of the momentary character of the substitution of paradigms – the thesis that provoked many comments and severe criticism. But, as we shall see, many of the critical remarks have been too strong in the sense that they usually overlooked several places in Kuhn's book which have introduced important qualifications into this thesis of his.

A few of the critics justifiably pointed out that Kuhn's new paradigm cannot have any real pre-history. And indeed, by Paradigm-monopoly and Incompatibility theses, scientists should think quite differently until the last moment before the conversion takes place. This strain of thought made John Watkins assert that Kuhn has to maintain the unmanageable thesis that the *switch* to the new paradigm must be regarded as the very same thing as the *invention* of the new paradigm. My Yugoslav colleague, Djuro Susnjic, also objects that Kuhn does not make a difference between *acceptance* of a paradigm which may be a relatively short-lasting process, and *creation* of a paradigm, which as a rule has to be long-lasting and a very complex process that cannot be performed during the state of crisis, let alone during the short moment of the very transition from an old to a new paradigm. ([14], p. 35, and [12], p. 68)

Trying to support the above mentioned interpretation of his, Watkins quotes the following passage from the famous book of Kuhn:

The new paradigm, or a sufficient hint to permit later articulation, emerges all at once, sometimes in the middle of the night, in the mind of a man deeply immersed in crisis. ([4], p. 89)

Now, in my opinion, this passage cannot be taken as confirmation of Watkins' above mentioned identification of *switching* from one to another paradigm with *inventing* a new paradigm. This passage, as a matter of fact, has quite different import. I think that it simply expresses another belief of Kuhn, which though criticizable too, must not be associated with the interpretation of Watkins; the belief being that the theories are invented or created 'in one piece'.

I think all this is rather obvious, but maybe I shall offer a few passages from Kuhn's book which can show that Kuhn is not asserting anything Watkins would like him to assert.

The novel theory, [says Kuhn] seems a direct response to crisis [i.e. to the period which is not so momentary and which precedes revolution, or the very switch from old to new paradigm, *S.N.*]... Finally, ... the solution [to any particular case of crisis, *S.N.*] had been at least partially anticipated during a period when there was no crisis in the corresponding science; and in the absence of crisis those anticipations had been ignored. ([4], p. 75)

Or: "Often a new paradigm emerges, at least in embryo, before a crisis has developed far or been explicitly recognized..." ([4], p. 86). Or, the last quotation:

Philosophers of science have repeatedly demonstrated that more than one theoretical construction can always be placed upon a given collection of data. History of science indicates that, particularly in the early developmental stages of a new paradigm, it is not even very difficult to invent such alternates. But that invention of alternates is just what scientists seldom undertake, except during the pre-paradigm stage of their science's development and at very special occasions during its subsequent evolution. ([4], p. 76)

Naturally, in Kuhn's opinion, there are cases when considerable time elapses between the first recognition of crisis of a theory and the emergence of a new one, but that is all he wants to admit in connection with otherwise sudden, unexpected and radical process of change of paradigms. So, Watkins is quite right in making another objection of his:

That it [paradigm] might emerge before a crisis has developed *at all*, and might itself *generate* a crisis, is excluded by Kuhn's idea of paradigm-dominance within normal science. ([14], p. 31)

But, in connection with Kuhn's conception of change of paradigms, it is also interesting to notice that some of his critics (notably Lakatos and Watkins) were inclined to understand it as 'irrational leap-of-faith'. Kuhn denies this interpretation, and in the 'Postscript' to his book, he explains more fully his view, i.e. the difference between *persuasion* and *conversion*. In order to be persuaded to adopt a new theory, a scientist usually has to translate that theory into his own language. However:

To translate a theory or worldview into one's own language is not to make it one's own. For that one must go native, discover that one is thinking and working in, not simply translating out of, a language that was previously foreign. ([6], p. 204)

Since 'going native' happens all at once, and one cannot choose to do it, it has the features of a 'conversion experience'. Kuhn also suggests that these conversion experiences come more easily to the young, and that without such an experience a scientist

may use the new theory nonetheless, but he will do so as a foreigner in a foreign environment, an alternative available to him only because there are natives already there.... The conversion experience that I have likened to a gestalt switch remains, therefore, at the heart of the revolutionary process. ([6], p. 204)

But scientific revolution may be characterized not only as *sudden* but also as *unexpected*, though the sense of *unexpectedness* which Kuhn may use is quite different from that which we ascribe to Popper. The transition from a paradigm in crisis to a new one is unexpected for Kuhn in the sense that the new paradigm is 'incompatible' or 'incommensurable' with the old one. The new paradigm is a reconstruction of the field from new fundamentals, so that the profession will change its view of the field, its methods, and its goals; in other words, after a revolution, scientists work in a different world. ([4], pp. 84, 134, 149) It has to be added that this idea of 'incommensurability', which at first seemed as if it may be of logical nature, ceased to be a logical affair and turned out to be a psychological matter, leaving for good our logico-methodological field of inquiry.[3]

It is a fact that on page six of his *The Structure of Scientific Revolutions*, Kuhn has indicated the possibility of extending the concept of revolution to the so-called 'micro-revolutions'. But, obviously enough, what he discusses in his book, as well as his examples, are concerned with macro-revolutions. In his 'Postscript', Kuhn is even more explicit in asserting the possibility of small (limited to a very small group of scientists) changes of revolutionary character, but since Kuhn is leaving it at that, we do not have enough stimulus to discuss this problem. ([4], p. 181)

Anyway, my impression is that scientific revolutions, as Kuhn understood them, are conceived far too profoundly (leading even to incommensurability), and in that sense, unrealistically. It seems to me that even in political revolutions, the changes that take place in the sphere of economics and politics are never so profound as the revolutionary leaders claim them to be. In the beginning, they usually claim to achieve profound changes, but later on they just try to describe events as if radical changes really took place. In fact, what happens both in political and in scientific revolutions – though many concerned parties try to avoid admitting that – is the following: when the new paradigm develops enough, it is not difficult to see many points it has in common with the old one, i.e. that it is not so profoundly new as the majority would like to think at first. The reason why this is so in science, lies primarily in the fact that theorists, in order to create a new theory, have to lean heavily on the vast amount of background knowledge which they are, for the time being, taking for granted.

Methodologically, however, the greatest shortcoming of Kuhn's conception is his Paradigm-monopoly thesis, which forbids the simultaneous existence even of two, let alone three or more, paradigms or rival theories; or, in other words, forbids any (even limited) proliferation of scientific theories. In my opinion, this thesis means neglecting some of the essential and not so difficult to notice facts in the history of science; namely, 'normal science' may evolve with scientists working within the framework of two (or more) paradigms simultaneously. Examples are not lacking.

So, we have the simultaneous development of rival theories of matter all the way from Antiquity to the present time. ([11], p. 54–55) Or, we can take as example ([12], pp. 59–60) the case of contemporary sociology which is evolving through the discussion of two paradigms of equal importance. These paradigms are: the functionalist and the marxist paradigm. Let us see this case in some more detail.

What is the picture of reality offered by functionalism? Roughly speaking, the picture may be sketched in the following way:

   Functionalist's position:

   (1) Society is a *stable* system;

(2) Parts of this system are well *integrated*;

(3) Every part of the system has its well defined *function*, primarily oriented towards the continuance of the system;

(4) Smooth functioning of the system is based on *agreement* about the fundamental values of the system;

(5) Social *order* is the basic presupposition for successful functioning of the system;

(6) Any deviation from this *order* is considered as disfunctional, divergent, and *pathological*.

This is, in main lines, the conception of social reality of the functionalist sociology. All questions that this sociology asks, and all answers it gives are *determined* and *limited* by its basic presuppositions, by its conception of social reality, by its paradigm, to use Kuhn's language.

Let us see now the picture of reality offered by marxism:

(1) society is an *unstable* system;

(2) some important parts of the existing system are *not integrated*, and are even incapable of being integrated into it;

(3) some parts of the system, by their very location in this system, are directed towards *destruction* of the system;

(4) in a class society, there *cannot be any agreement* on fundamental values of the system;

(5) *class struggle* is the basic characteristic of the system;

(6) the parts which are disturbing the system represent emergence of something new, progressive, and *revolutionary*.

This basic comparison – which is more illustrative than exhaustive – of conceptions of social reality (paradigms) that constitute the foundations of contemporary sociology, clearly shows that marxist sociology is founded on the conception of reality which is *directly opposite* to that of the functionalist sociology. [Where functionalists speak of stability, marxists speak of instability. Where functionalists emphasize integration, marxists lay stress on conflict. Where functionalists insist upon balance, marxists point to disturbance of balance. Where functionalists assert agreement, marxists put forward the struggle for power. Where functionalists request order, marxists affirm criticism of all that exists. Where functionalists suggest homogeneity, marxists propose revolution.]

This situation, which is inconceivable in Kuhn's methodology, generates the following problem: which conception of social reality is theoretically and methodologically more fruitful, more correct, or more true, functionalist or marxist? It seems that we will not make any big mistake if we say – though this conclusion should come after careful critical investigation – that both conceptions may be fruitful, that both of them correctly describe one aspect of social reality, that their adequacy depends on the given socio-historical period, and that both are partially true.[4]

Apart from these basic objections, I would like to give some brief comments on Kuhn's concept of crisis, as well as on his thesis that in 'normal science' there is no correction of the dominant paradigm.

The main reasons that provoke the emergence of crisis within the framework of a paradigm, in Kuhn's opinion, are the following: in course of time, every paradigm exhausts its own possibilities, since the number of relevant and significant questions that may be put on the basis of the given paradigm is limited; or, in other words, the accumulation of questions that cannot be answered by the given paradigm becomes more and more unbearable. This means that the main cause of crisis is the growing pressure from the outside. This view obviously has some truth in it, but at the same time, it signifies in fact important impoverishment in realizing the possible reasons for crisis. One can point to a few other reasons which Kuhn does not mention:

(1) changes in the reality itself: namely, while the reality of the astronomical and physical sciences remains relatively the same through quite a long period of time, and we just change our perception of that reality, subject to the development of our theoretical standpoints or value systems, the reality of the social sciences behaves differently – it is open to relatively frequent and qualitative changes;

(2) theoretical pressure; namely, independently of existence or non-existence of any other kind of reasons, a paradigm may encounter a crisis if it is theoretically questioned either from a different metaphysical standpoint or simply by discovering inconsistencies or incoherences in it.

Kuhn's view that 'normal science' is firmly tied to the dominant paradigm, and that it confines itself exclusively to the articulation of those empirical or theoretical phenomena which are given in the paradigm, implies also his opinion that within the 'normal science' no correction of paradigm can take place. ([4], pp. 24, 121) It seems to me that this rather narrow-minded view of Kuhn can be successfully criticized. In my opinion, in the articulation or specification of a paradigm we are deter-

mining the set of facts that confirm as well as the set of facts that contradict that paradigm. In that situation, does the scientist really just neglect
the contradictory facts until there are no possibilities for development
of a paradigm, or does he nevertheless very often make corrections or
modifications of the paradigm in order to dispose of a number of contradictory facts? Is it really possible to dismiss criticism completely from the
domain of 'normal science'? No, I think it is not.

Turning to Lakatos, I shall try immediately to present what, in my
opinion, are the new elements in his conception of the structure of
scientific revolutions. There are two such elements:

(1) *extreme* relativizing of the possibility of falsifying a scientific theory,
and;

(2) the thesis that there are really no transitions from the *old* to the
new theory, since we are always deciding among a maze of simultaneously
existing incompatible theories.

I will discuss these elements in turn. Let us first have a closer look at the
first element of relativizing the possibility of falsification of a scientific
theory. It will be best to let Lakatos himself speak:

> One can easily see, [says Lakatos] that when we devise an experiment in order to test,
> to criticize a theory, we always have to use some 'observational theories', or 'touch-
> stone theories' (or 'interpretative theories'), uncritically if we want to make its 'falsifica-
> tion' possible.... It depends on our methodological decision which theory we regard as
> the touchstone theory and which one as being under test; but this decision will deter-
> mine in *which* deductive model we shall direct the *modus tollens*. Thus a 'potential
> falsifier', B, of T_1, given some touchstone theory T_2, may refute T_1; but the same state-
> ment B, if regarded as a potential falsifier of T_2 given T_1, as touchstone theory, refute
> T_2... – ... The problem becomes still more difficult if we realize that the theories we
> criticize always contain, in addition, a '*ceteris paribus*'. This '*hidden background knowl-
> edge*' makes mono-theoretical refutation utterly irrelevant. ([7], pp. 156–158)

This conception is, of course, closely connected with the second element,
i.e. with Lakatos' claim that there is really no transition from *the old* to
the new theory; yet, for a moment I will neglect this connection. Where I
see the extreme of his position is when he asserts that "mono-theoretical
refutation is utterly irrelevant." In such a strong formulation, I assume,
this thesis is unacceptable. Let us suppose that it is always possible to
turn around the relationship explanatory theory – interpretative theory,
though it seems to me that it is not even very often possible; namely, in any
particular period in the history of science, scientists are pretty sure which

theories they take as explanatory and which as interpretative, i.e. which theories they are taking as true even if they are at the same time willing to admit that it may someday turn out that these theories are not true, so that they may be forced to reconsider the position of their interpretative theories – to pull them back to the level of explanatory theories, or even to reject them completely. But, if we suppose that the relationship explanatory theory–interpretative theory can always be turned around, I do not see why the 'hidden background knowledge' – which Popper recognizes too – should make mono-theoretical refutation utterly irrelevant. On the contrary, I take it as a strong and reliable indicator of which theories to take as interpretative and which not to take as interpretative.[5] So this formulation of Lakatos' – if we are not confronted here with an uncontrolled exaggeration, which may easily arise during the critical debate – contradicts another standpoint of the author.

Namely, Lakatos rightly criticizes Popper for making no difference between 'refutation', and 'rejection' (or 'elimination'). His remark that simple refutation does not provide us with sufficient reason for elimination of the 'falsified' theory is quite correct. At the same time, and contrary to the above mentioned strong formulation, Lakatos, in fact, grants quite a significant role to a refutation – he says that refutation enhances the problematical tension, or problem fever, of the body of science, indicating in that way the urgent need of revising it in some yet unexpected way. ([7], pp. 155, 161)

A second new element of Lakatos' methodology, so far as the structure of scientific revolutions is concerned, is his thesis that there is no transition from *the old* to *the new* theory. I will quote again from Lakatos, now a very brief, but condensed statement:

No theory forbids some state of affairs specifiable in advance; it is not that we propose a theory and Nature may shout NO. Rather, we propose a maze of theories, and Nature may shout INCONSISTENT. ([7], p. 161)

Consequently, there is no establishing of a new theory which might be preceded by experimental refutation of the old theory, but only deciding between a bunch of incompatible theories. If anybody may be curious about the reasons why Lakatos does not want to make the difference between the old and the new theory, it is given in another methodological standpoint of his: we should not say for any theory that it is old, since theories may just temporarily retreat from the battle (under the assault of opponent's arguments), and need not consider themselves defeated; they may come back several times, always in a new, content-increasing version, and with a verification of some of their novel content. ([7], p. 176)

No advantage for one side [says Lakatos], can ever be regarded as absolutely conclusive. There is never anything inevitable about the triumph of a programme. Also, there is never anything inevitable about its defeat. Thus pigheadedness, like modesty, has more 'rational' scope. The scores of the rival sides, however, must be recorded and publicly displayed at all times. ([15])

This view, however, means in fact radical denial of any genuine scientific revolution. Science is reconstructed as a permanent instability of quite a number of rival theories, with temporary overwhelming now of the first, now of the second, now of the nth theory, which often may change in the leading position. In any case, we have always a number of theories present, out of which one is for the time being the dominant theory, one (or maybe a few) are temporarily in the phase of retreat, one (or maybe again a few) are appearing as the new ideas, one (or maybe quite a number) have retreated for good.[6] In such a colorful restlessness of theories – where the only essential thing is to keep the scores of the rival sides recorded and publicly displayed at all times – there is really no place for the pattern of classical scientific revolution, where the old theory, after a shorter or longer period of absolute supremacy, and after a shorter or longer war, surrenders its place to the new one. But, in this scheme, there is also no room for that remarkable continuity in science, which Kuhn pointed out, to which Popper did not pay any serious attention, and which Lakatos pretends to include in his methodological reconstruction (in the sense that scientists show strong tendency to ignore counter-examples for a very long time, by looking at them as recalcitrant and residual examples, and to hope that in the process of development of their theories, they will be able to get rid of these kinds of examples). So, it seems to me that Lakatos is in favour of this standpoint more declaratively, since it does not really match with his above mentioned views on the growth of scientific knowledge.

But, I will return to these problems again in the next part of this paper, where I discuss the views of Paul Feyerabend on the problem of revolutionary changes of scientific theories.

Feyerabend also starts both from Popper and from Kuhn. His *prin-*

ciple of tenacity, he admits, is taken over from Kuhn. Yet, of course, he elaborates it in his own special way. Feyerabend is sure that the principle of tenacity is reasonable because theories are capable of development, because they can be improved, and because they may eventually be able to accommodate the very same difficulties which in their original form they were quite incapable of explaining; but, at the same time, Feyerabend is introducing his *principle of proliferation*, since he is at variance with Kuhn's suggestion that scientists should stick to their paradigms to the bitter end, until disgust, frustration, and boredom make it quite impossible for them to go on. ([2], p. 203, 205)

Feyerabend is strongly against Kuhn's model: normal science – revolution – normal science – revolution, etc., "in which professional stupidity is periodically replaced by philosophical outbursts only to return again at a 'higher level'." ([2], p. 208)

Compared with the model of Kuhn, great advantage is on the side of the view which allows the possibility for theory-improvement, and at the same time accepts the principle of proliferation, because it permits us to be open-minded all the time and not only in the middle of a catastrophe.

Is it not also the case, [says Feyerabend], that anomalous facts are often *introduced* by the critics of a paradigm, rather than *used by them* as a starting point for criticism? And if that is true, does it not follow that it is proliferation rather than the pattern normalcy – proliferation – normalcy that characterizes science? ([2], p. 208)

On the basis of the suggestion I just quoted, that critics *often introduced* rather than *used* anomalous facts as a starting point for criticism of a paradigm, as well as the fact that it is often the case that quite a number of scientists just in the middle of revolution engaged themselves in attempts to resolve old, little puzzles and anomalies, Feyerabend has come to the conclusion (to be fair, with qualificatory phrase 'it seems') that it is not the puzzle-solving activity that is responsible for the growth of our knowledge, but the active interplay of various tenaciously held views. Moreover, we have to secure a proper place for the new ideas that emerge continuously, though we are inclined to pay attention to them only during revolutions. But since, for Feyerabend, this change of attention does not reflect any profound structural change (such as for example a transition from puzzle-solving to philosophical speculation and testing of foundations), it is nothing but a change of interest and of publicity. ([2], pp. 208–209)

With a little imagination and a little more historical research, Feyer-abend believes it is possible to come to the conclusion that proliferation not only *immediately precedes* revolutions, but that it is there *all the time*.

Science as we know it [says Feyerabend], is not a temporal succession of normal periods and of periods of proliferation, it is their *juxtaposition*. – ... I shall therefore speak of the normal *component* and the philosophical *component* of science, and not of the normal *period* and the *period* of revolution. ([2], p. 212)

So, for Feyerabend, scientific revolution is dialectically overcome, i.e. negated and at the same time preserved, – negated as a temporally ob-servable phenomenon, or as adequate pattern for explaining certain events in the history of science, but preserved as one of those two opposing components of science, which make the basis of dialectical development of science in any period of its history.

This proposed solution of the problem of the relationship between the principle of tenacity and the principle of proliferation is obviously one possible solution, but it requires a particular conception of knowledge. This conception of knowledge we find in Feyerabend's paper 'Reply to Criticism':

Knowledge so conceived is not a process that converges towards an ideal view; it is an ever increasing ocean of alternatives, each of them forcing the others into greater articulation, all of them contributing via this process of competition, to the develop-ment of our mental faculties. All theories, even those which for the time being have receded into the background, may be said to possess a 'utopian' component in the sense that they provide lasting, and steadily improving, measuring sticks of adequacy for the ideas which happen to be in the center of attention. ([1], p. 225)

Criticism of the conception of scientific revolution or the principle of proliferation of Feyerabend, I leave for the concluding remarks; since it represents an indication of my own view on this problem.

Now, I would like to point to an additional difficulty in the views of Feyerabend. I think of his thesis that the theories which eventually replace one another are incommensurable, i.e. that the meanings of the terms (words) that appear in describing even empirical facts depend on the theory within which they are used, which is then another reason to deny any function at all of refutation in the process of theory change in science. The most frequent objection against the incommensurability thesis boils down to the following: if a theory T^1 is incommensurable with the given theory T, then it cannot contradict T, and, consequently, cannot be used to criticize T. Feyerabend admits the relevancy of this objection, but, at the same time, points out that there are methods of criticism which do not depend on the existence of similarity of meanings. Making a difference between *external questions* (referring to the relation between different theories), and *internal questions* (referring to the features of the world as seen from the point of view of a single theory), Feyerabend believes that external questions can be decided without bringing in meanings at all. He says:

> We simply compare two infinite sets of elements with respect to certain structural properties and inquire whether or not an isomorphism can be established. Of course, there is still the question as to which theory is to be preferred. Usually, one tries to solve the question on the basis of *crucial experiments*. ([1], p. 232)

Naturally, one would not think of 'crucial experiment' in the sense of choosing, on observational basis, between two incompatible theories, since in that case we again need to establish a certain minimum of similarity of meanings. Feyerabend is convinced that there may be empirical evidence against one and for another theory without any need for similarity of meanings. *First*, we may introduce incompatibility of theories as above (lack of isomorphism with respect to certain basic relations, plus consequences following therefrom) and then confirm one of two incompatible theories. *Secondly*, both theories may be able to reproduce the 'local grammar' of sentences which are directly connected with observational procedures so that we may say that T^1 has been confirmed by the very same evidence that refutes T if there is a local statement S whose associated (=theoretical) statement in T^1 confirms T^1 while its associated (=theoretical) statement in T refutes T. *Thirdly*, we may construct a model of T within T^1 and consider its fate. ([1], p. 233)

Feyerabend is right in pointing out that one can consider a relation between different or rival theories without bringing in meanings at all. But, the question remains: what about other kinds of relations, if we want for example, to bring in internal questions too? Furthermore, it seems to me that all his arguments, designed to prove that there are methods of criticism which do not depend on the existence of similarity of meanings, are not convincing enough. As a matter of fact, my intuition is that no scientific theory T^1 can be put as a real alternative in explaining reality to any other scientific theory, if T^1 determines experience E^1 which is different than experience E'', which is determined by T''. These theories (or scientists who maintain them) are simply concerned with different facts, and they just do not enter any relation of rival views. In fact, it seems to me that by introducing the possibility that "both theories may be able to reproduce the 'local grammar' of sentences which are directly connected with observational procedures", Feyerabend is indirectly introducing again some kind of comparison of some kind of basic statements and, therefore, some kind of similarity of meanings.[7]

If I may say a few words about the position of Lakatos and Feyerabend in respect to those characteristics of scientific revolution that appear in the title of may paper ('sudden' and 'unexpected'), I would like to suggest – taking the risk of a not quite adequate interpretation – that the reconstruction given by Lakatos does not offer a sufficient possibility to speak either of the sudden or of the unexpected character of the transition from an old to a new theory; the main reason being that Lakatos prefers to speak of simultaneous recordings of the scores of the rival theories. The principle of proliferation of Feyerabend also excludes the sudden character of the transition, but his principle of incommensurability (which is not of psychological character, as in Kuhn), includes such kind of unexpectedness which is very close to maintaining irrelevancy of any real comparison of the old and the new theory (but I already made my comments on that point).

In discussing the problems of scientific revolution, and in criticizing some of the theses of Popper and Kuhn, Toulmin and Kneale – each in his own way, and with different arguments – claim that there is really no scientific revolution at all.

Toulmin, for example, maintains that the 'absoluteness' of the transition involved in Kuhn's conception of scientific revolution provided in fact the only criterion for recognizing that one had occurred at all. And, in Toulmin's opinion, once we acknowledge that *no* conceptual change in science is ever absolute, we are left only with a sequence of greater and lesser conceptual modifications differing from one another in degree. ([13], p. 41)

This inference from the non-existence of the absolute absoluteness to the existence of absolute relativity does not have much strength, I presume.

Kneale sees this problem in a different perspective. All philosophers of science who are trying to develop some theory of scientific revolutions are, as Kneale takes it, expressing their disagreement with the possibility that science may achieve stability by acceptance of a single theoretical framework. In Kneale's opinion, there seems to be no good sense in which nature can be said to be so complex as to require perpetual revolution in science. For, if by 'the infinite complexity of nature' is meant only the infinite multiplicity of the particulars it contains, that is no bar for scientists in their endeavour to succeed in covering theoretically all the particulars. If, however, what is meant is that there cannot be a finally satisfactory theory, there is no point in trying to approach such a theory, either by scientific reforms or by scientific revolution. For if there is no truth, there cannot be any approximation to truth, and the whole enterprise of theory making is futile. So, in Kneale's opinion, a scientist may someday formulate a comprehensive theory which is wholly true and, therefore, incapable of refutation. But, at the same time, Kneale does not wish to say that this will happen. The human race may be too stupid for the task, or it may be overtaken by some natural disaster while it is still far from its scientific goal. Still less does he want to say that if we produce the theory we want, we can also be sure that we have done so. For in order to attain such certainty, we should have to know everything about nature, including the fact that we knew everything about nature, which is impossible in principle. But, Kneale believes that the ideal of comprehensive true theory remains unaffected by these reservations. ([3], pp. 36, 38)

As a short comment on Kneale, I would say that there is not much point in making such optimistic prognostications about the future, where we have learned the comprehensive truth about the reality, though the reason may not be that of the reservations as regards the intellectual capacities of the human race or the possibility of natural disaster. My reasons would rather be the following: first, it seems to me that it was not so long ago that science became aware that it has wide-open new horizons for further investigation and enlarging knowledge in front of itself so that it must be still far away from a comprehensive theory which will be wholly true; secondly, nature is not infinite only in the sense of infinite number of particulars existing in it, i.e. infinite variety of these particulars, but also in the sense of infinite possibilities for change and transformation that no comprehensive scientific theory could ever predict and so incorporate in itself.

It seems appropriate, as a kind of conclusion, to give now just a few more comments on some of the main problems that have been raised in this paper.

(1) First of all, despite various direct or indirect arguments against the existence of scientific revolutions, I think that the critical comments I put forward in this paper against that kind of view provide the basis for us to conclude that it would not be unjustifiable, both historically and methodologically, if we continue with further investigations of different aspects of the complex problem of scientific revolutions.

(2) The principle of tenacity, I consider as a very important one. Of course, it must not imply that 'normal science', which is going on within the framework of a theory, cannot bring about various modifications and adaptations of the given theory, and become the source of crisis and revolutionary changes of the theory.

Or, to put it in other words: though the principle of tenacity involves certain disregardings (neglecting, evading) of contrary facts or arguments, it does not imply that this disregarding may result out of our blindness or theoretical decision not to take into consideration any empirical or theoretical refutation. On the contrary, the disregarding in question is usually meant to be of a temporary character, in order to give the theory the real opportunity to show its strength, rightly assuming at the same time that in the course of its development some of the difficulties will be eliminated.

(3) As for the sudden character of the revolutionary transition, it is necessary to differentiate between: (a) *creation* of a theory, and (b) *acceptance* of this theory. As regards the creation, it cannot be done 'in one piece', and consequently cannot be of sudden character. This would be really expecting too much, even of a genius, since scientific *theories* are complex products which assume considerable time and peaceful work to achieve their clear and satisfactory formulations. This is not to deny that just an *initial idea* may come and usually comes quickly to the mind of a scientist. In this sense, I would like to maintain that the periods of revolutionary transition, or even the periods of crises, are in fact very unfavourable as far as the *creation* of scientific theories is concerned. What really happens during a revolution in science? A group of scientists is trying with various arguments to defeat the old theory and to establish the supremacy of a new one; but a new one that has not been *created* just

on the spot during the very revolution, since we cannot really start a revolution without having a new theory. This does not mean, of course, that revolutionary periods are unfavorable for the further *development*, let alone *application* of the new theory. On the other hand, the *acceptance* of a scientific theory by a community of scientists may be, and as a rule is, of a sudden character, especially if it is understood in the sense of conversion, or, to put it differently, if it does not include that period of time that is necessary for convincing (persuading) scientists that they should reject the old and accept the new theory (in the sense that they actually believe in it, and not in the sense that they just start using it for any purpose they may have in mind).

(4) As for the unexpected character of the revolutionary transition, I reject it in the sense of incommensurability. But, it can be preserved in the sense that any new theory just has to bring something unexpected in our apprehension of the reality in order to be considered as new at all.

(5) Finally, a suggestion for a different interpretation of the principle of proliferation: namely, in the sense in which the great majority of philosophers of science nowadays consider the mere accumulation of observations to be nonsensical if not based on a theory, however fallible or temporary it may be; in a similar sense, I presume, the principle of proliferation of theories may be criticized. So, I claim that theoretically speaking, the principle of proliferation, without an explicitly specified device for limiting the theoretically infinite number of possible hypotheses means just nonsensical piling of various theories. It seems to me that, in fact, theories have to emerge out of observations conflicting with some existing theories (of course, if they could not have been accommodated within these theories), or else by recognizing theoretical shortcomings of the existing theories.[8] From the practical standpoint, however, any principle of proliferation would be more than desirable (since in practice we so seldom encounter the proliferation of theories).

Apart from this, as regards the simultaneous existence of several theories dealing with a particular problem situation, it may be useful to make some distinctions. First of all, one should differentiate between the new-emerging theories and the potential set of theoretical possibilities that in fact represent modifications or adaptations of some of the old (or previously existing) theories. Within this latter set of the 'alternative candidates', one can further differentiate between the theories which have been already – quite recently or long ago, it does not matter – in use, and those theories which needed to wait 'for better times' because of the undeveloped character of a particular scientific discipline.[9]

It is only natural that in any problem situation, we may have to, but need not, deal with all the above mentioned kinds of hypotheses. Accordingly, there may be such cases, when we will have to deal with single hypothesis or with two or more theories in turn until we decide which one to reject, or etc....

University of Pittsburgh and University of Belgrade

NOTES

[1] It is very interesting to find out why logical positivists did not pay any attention to this problem, too, but that does not belong directly to the topic of this paper. I am in agreement with many points that have been made in connection with this problem in W. D. Siemens' article 'A Logical Empiricist Theory of Scientific Change' in *PSA 1970, Boston Studies in the Philosophy of Science*, Vol. VIII (ed. by R. C. Buck and R. S. Cohen), 1971, pp. 524–535.

[2] It is one of Kuhn's fundamental claims that his new image of science is emerging out of his studies of the history of science. Nevertheless, it seems to me that the greater part of his book may be taken as a kind of prescriptive methodology of science. That is my excuse for approaching his views from a purely methodological standpoint without trying any historical case-study criticism. As a matter of fact, all those philosophers of science who are trying to form their conception of science through extensive historical studies cannot give us but their own reconstruction, which may be disputable both from a historical and a methodological standpoint.

[3] So, I will have to say more about incommensurability in connection with the views of Paul Feyerabend.

[4] Compare the attempt of proving that marxist paradigm is superior to orthodox economy paradigm in the paper of Paul M. Sweezy, 'Towards a Critique of Economics', *Monthly Review* 8 (1970).

[5] Here we have, in my opinion, a wrong interpretation of some new views of Popper (of Popper$_2$, as Lakatos would put it). So, Lakatos says that according to Popper science may grow without refutations leading the way. And then, misquoting the section 30 of *The Logic of Scientific Discovery* (according to him, Popper holds the view that *refutation cannot compel* the theorist to search for a better theory; though in that section Popper merely remarks that *refutation need not always be that which compels* the theorist to search for a better theory), Lakatos comes up with the thesis that refutation cannot precede the better theory. ([7], p. 166; [8], pp. 121–122.) This interpretation is wrong since Popper, even in the newest papers published, takes refutation as the main source of new problems in science; and new problems, as a matter of fact, lead the way in science. As for the practical import of the thesis that refutation cannot precede the better theory, it remains completely unclear and unproved.

[6] Closely connected with this is Lakatos' new account of 'crucial experiment'. He wants to claim that if it happens that a theory, after sustained effort, cannot fight its

comeback, then the experiment is proved – but only *with hindsight* – 'crucial' for its rejection. ('Criticism and Methodology ...', p. 176)

[7] In his paper 'Ontological and Terminological Commitment and the Methodological Commensurability of Theories', in *PSA 1970, Boston Studies in the Philosophy of Science*, Vol. VIII (ed. by R. C. Buck and R. S. Cohen), 1971, pp. 507–518, S. A. Kleiner insists on the possibility of changing some aspects of the meaning of non-logical expressions in a theory without generating incompatibility and *a fortiori* without generating methodological incommensurability between the old and new theories.

[8] This, then, represents also my main critical point against the principle of proliferation of Feyerabend, which has been explicitly identified with 'an ever increasing ocean of alternatives'.

[9] Here is one more argument emphasizing the significance which history of science has for men of science (especially if we are making difference between scientists-creators, and scientists-technologists. [See my paper, 'Two Conceptions of Science and Humanism', *Praxis* 1–2 (1969).])

BIBLIOGRAPHY

[1] Feyerabend, P. K., 'Reply to Criticism', *Boston Studies in the Philosophy of Science*, Vol. II (ed. by R. Cohen and M. Wartofsky), D. Reidel Publishing Company, Dordrecht-Holland, 1965.

[2] Feyerabend, P. K., 'Consolations for the Specialist', in *Criticism and the Growth of Knowledge* (ed. by I. Lakatos and A. Musgrave), Cambridge University Press, 1970.

[3] Kneale, W., 'Scientific Revolution for ever?', *The British Journal for the Philosophy of Science* 9 (1967).

[4] Kuhn, T. S., *The Structure of Scientific Revolutions*, Univ. of Chicago Press, 1962.

[5] Kuhn, T. S., 'Reflections of My Critics', in *Criticism and the Growth of Knowledge*, (ed. by I. Lakatos and A. Musgrave), 1970.

[6] Kuhn, T. S., 'Postscript' to the new edition of *The Structure of Scientific Revolutions*, 1970.

[7] Lakatos, I., 'Criticism and the Methodology of Scientific Research Programmes', *Proceedings of the Aristotelian Society* 69 (1968–1969).

[8] Lakatos, I., 'Falsification and the Methodology of Scientific Research Programmes', in *Criticism and the Growth of Knowledge* (ed. by I. Lakatos and A. Musgrave), 1970.

[9] Popper, K. R., *The Logic of Scientific Discovery*, Harper and Row, New York, 1965 [1934].

[10] Popper, K. R., *Conjectures and Refutations*, Basic Books, New York, 1962.

[11] Popper, K. R., 'Normal Science and Its Dangers', in *Criticism and the Growth of Knowledge* (ed. by I. Lakatos and A. Musgrave), 1970.

[12] Susnjic, D., *Resistances to Critical Thought*, Zodijak, Beograd, 1971 (in Serbo-croat).

[13] Toulmin, S., 'Distinction Between Normal and Revolutionary Science', in *Criticism and the Growth of Knowledge* (ed. by I. Lakatos and A. Musgrave), 1970.

[14] Watkins, J. W., 'Against "Normal Science"', in *Criticism and the Growth of Knowledge* (ed. by I. Lakatos and A. Musgrave), 1970.

[15] Lakatos, I., 'History and Its Rational Reconstruction', in *PSA 1970, Boston Studies in the Philosophy of Science*, Vol. VIII, pp. 91–136.

WILLIAM BERKSON

SOME PRACTICAL ISSUES IN THE RECENT
CONTROVERSY ON THE NATURE OF
SCIENTIFIC REVOLUTIONS

Do questions like Novakovic's have any practical importance? Does it matter whether the transition from one theory to another is slow or quick, continuous or discontinuous? One might think that the only question of practical importance is whether the new theory is *better* than the old. Yet thinkers with most intense concern for practical questions have discussed the issue. And the discussions have been heated. Why is this?

The reason is, I think, that their greatest concern has been with the problem: What methods promote the growth of science, and which hinder its advance? They have had different and conflicting answers to this question. All sides have turned to the history of science to find confirmation of their own views and refutation of their opponents. That is, they argued that important advances in science were in fact made using the methods they advocate. Thus Novakovic's question has come to be debated.

It is true that the bulk of arguments have been over the interpretation of history. But I think at bottom the greatest concern has been over the practical question of what are the best methods. This shows in the frequent change from description to recommendation in the debate. In any case, it seems to me that the practical question is most important. For methodology is in essence a practical subject: The reasons for accepting or rejecting a method must ultimately be whether or not the method promotes the aim being pursued. I do not deny that the historical arguments have relevance. But it seems to me that the discussion may be clarified by putting the practical question first. When historical arguments are brought in, I think their relevance should be explained.

Before commenting on Novakovic's paper directly, I would like to briefly survey the debate, with the practical question foremost. Then, I think, the reasons for my comments will be made clear. There are many interesting practical questions regarding the earlier theories of method – the theories of Bacon, Duhem, etc. I will restrict myself to the issues in

Boston Studies in the Philosophy of Science, XIV. All Rights Reserved.

the Popper-Kuhn controversy, since this is the subject of Novakovic's paper.

Popper and Kuhn personally both began with the same problem, a practical one.[1] What characteristics are responsible for the progressive character of physics, and the relative stagnation in such sciences as psychology and sociology? They gave quite different answers.

Popper thought that the progressive character of physics lay mainly in its use of three techniques: First, physicists continually search for new explanations, in a bold and imaginative way. Secondly, they seek particular theories which potentially can be refuted by experiments, especially experiments against the existing theories. Third, they subject these theories to rigorous criticism, especially criticism in the form of experimental testing. Popper thought that the refutability of a theory depended on the researcher's attitude toward the theory, and not merely the formal characteristics of the theory.[2]

To make a theory refutable, certain rules of research are followed. The main rules Popper mentioned are: (1) When a theory seems to come in conflict with experiment, the conflict should never be removed by a surreptitious change in meaning of terms of the theory. (2) When theories are modified, or new theories introduced, the most testable ones should be put forward first. (Popper developed a theory of degrees of testibility, or refutability.) (3) When a theory contradicts an experimental result which has earlier been tentatively accepted as correct, the theory should be rejected rather than the experimental fact.

Such, according to Popper, are the most progressive methods of research.

Kuhn, on the contrary, has accounted for the progressiveness of physics mainly by the circumstance that scientists are trained in a common tradition or 'paradigm', and apply that tradition to solve problems or 'puzzles' facing the paradigm. Kuhn divides scientific growth into two stages, which should use quite distinct methods – the 'normal' and 'extraordinary' stage; It is the normal stage which accounts for the relatively rapid growth of physics. Here, Kuhn claims, it is the application of uncritically accepted ideas which makes for progress. Thus Kuhn recommends that normally the existing tradition should not be criticized, and alternatives to its main ideas should not be sought.[3] Along these lines, Kuhn recommends for science education that students be indoctrinated in the existing ideas for a long time before they should be allowed to criticize the tradition, or suggest alternatives. In fact, these latter activities should not even be engaged in by fully trained scientists, except in times of crisis, of 'extraordinary science'. The times of crisis occur when application of the paradigm repeatedly fails to solve many of the puzzles facing it. At such times progress waits on a genius to radically alter the tradition, and anything goes.

Such are the main practical recommendations of Kuhn, directly opposed to Popper's, at least in the 'normal' stage.

Now let us turn to the arguments. I shall first consider the arguments not dependent on history of science.

Popper's basic argument for the designing of empirically refutable theories is this: When we make the theories experimentally refutable, we open the possibility of using a very powerful technique to discover any errors in our theory: namely observation and experiment. When the theory lacks refutability, it becomes much more difficult to discover whether, and where, it is mistaken. Popper's argument for the consistent application of criticism, both empirical and non-empirical, is similar: Without criticism we can never discover where our ideas are mistaken, and so can never advance beyond them. We can only apply, but not supercede them. The argument for boldness in the creation of theories, and for the invention of many alternatives, is that these new ideas may capture some aspect of nature which the previous ones have failed to do. And, one may add, they can guide us to new discoveries by pointing the way to crucial experiments with the older ideas. As for the subsidiary rules mentioned, Popper thinks that without these we could not have, in practice, refutable theories. And thus we would lose the great advantages of refutability.

Such are Popper's non-historical arguments for the methods he advocates. Kuhn, to my knowledge, has given no non-historical arguments. Later I will consider what arguments he seems to have in mind.

Now let us consider the arguments against Popper and Kuhn, taking again the non-historical arguments.

Kuhn has pointed out the main practical conflict between his and Popper's theory: "My claim has been – and it is my single genuine disagreement with Sir Karl about normal science – that with such a theory [basis of a paradigm] in hand the time for steady criticism and theory proliferation has passed." [4] The only argument by Kuhn I can find against Popper's recommendation of maximizing criticism and search for alternatives is this: "Some scientists [Kuhn grants] must, by virtue of a value system differing in its applicability from the average, choose it early, or it will not be developed to the point of general persuasiveness. The choices dictated by these atypical value systems are, however, generally wrong. If all members of the community applied values in the same high risk way, the group's enterprise would cease." [5]

Would the progress of science be retarded by an unrestrained application of criticism? I see no reason to believe that it would. Kuhn seems to fear that if scientists disbelieve the existing dominant theories, they would never work on problems of applying the theories. But this consequence does not necessarily follow. Application and development of a theory is necessary for decisive criticism, especially by experiment. And we may think that the theory is worth applying to explain known phenomena, though we disagree with it: We may adopt it as a guide to research, while believing it inadequate, and while searching for alternatives. Kuhn seems to make the mistake of identifying *choice of a problem of application* with *suspension of criticism and search for alternatives*. But it is neither necessary nor advisable to make the identification, I think. (This point I will return to later.) In sum, Kuhn's argument rests on a mistake, and does not threaten Popper's recommendations.

Agassi, however, has given arguments with more bite. Agassi has denied that two of Popper's subsidiary rules are necessary to support his main practical suggestions. In fact Agassi argues that they work against the main suggestions. The first rule is the requirement that when theory and previously agreed upon fact conflict, we drop the theory and accept the fact. This, says Agassi, does not take into account that the supposed fact may be wrong, a possibility which Popper himself emphasizes. [6] Agassi suggests that we regard experimental reports as part of the world to be explained by science. The reports may be explained as either true, *or false*. The important thing is that we try to make our explanation independently testable. In other words, we may propose to reject a previously supposed fact as an incorrect interpretation of experiment – this on the basis of our new theory. As long as we try to make our explanation independently testable, we do not go contrary to the main progressive suggestions of Popper. In fact, by denying that we need reject the theory and not the fact, we open up further possibilities of discovering our errors: namely our errors of interpretation of the experiments. Thus Agassi's suggestion is more in line with Popper's main suggestions, than the rule of Popper himself.

The second rule which Agassi rejects is Popper's suggestion that we always introduce and criticize the most testable theories first.[7] Agassi points out that it may be difficult at first to make a theory testable, but it may nevertheless be very suggestive, fruitful, and eventually testable. A metaphysics, or general world view, for instance, may be used as a research program. Scientists may search for testable scientific theories using the things which the world view supposes exists. These untestable world views thus can be valuable research programs, and should not be barred from science. Instead of regarding highest testability as an entrance requirement for scientific discussion we should regard it as a goal. By regarding testability as a goal we keep the progressive suggestion of *seeking* refutable theories. But we do not bar the less refutable ones, as this would limit the imagination in a way which could be harmful to science. This view, Agassi thinks, is more in harmony with Popper's emphasis on the value of bold speculation, than is Popper's own rule.

Agassi's suggestions do seem to me to strengthen and add to Popper's main recommendations, while rejecting the subsidiary rules.

Now let us turn to the criticism of Kuhn's recommendations. The heart of Kuhn's recommendation is the temporary suspension of critical judgement of the prevailing theories, and the cessation of the search for alternatives. This recommendation is so outrageous to Popperians that they regarded it as a decisive criticism to point out that it was in fact Kuhn's view. Here I am referring to the recent book *Criticism and the Growth of Knowledge*, concerning the controversy. This did not faze Kuhn, who emphasized that it was indeed his view. Popper, for instance, said: Yes, normal science, that is uncritical science, does exist: Isn't it a tragedy! Then he went on to try to explain how Kuhn could be misled into such a rotten view. He does not state what is to him the obvious, namely the arguments he gives elsewhere against an uncritical science. I have already stated these, but let me recall them: If we are not critical we cannot discover where our theories are mistaken, and if we do not create new alternatives we can never advance beyond our existing theories.

Does Kuhn give any reply to these criticisms? I don't think so. There are some mitigating qualifications Kuhn makes. First, he says that unity of paradigm should not be imposed in pre-paradigm science. Thus he restricts his advice to physics and other 'mature' sciences. Further, he allows that some individual diversity and novelty is helpful at times to lead to a new paradigm. But these are qualifications which only soften, but do not change Kuhn's basic position: Kuhn does not say that criticism and theory proliferation should be *eliminated*, but he does say it should be *minimized* and discouraged. While Kuhn's position is rendered more plausible, its practical import is almost the same; In all cases criticism and search for alternatives should be minimized. And so the criticisms of Popper apply. For Kuhn's theory to have a different practical import, he would have to carefully distinguish between those cases where criticism is to be encouraged, and those where it is to be discouraged. But such a distinction is lacking.

Let me now consider some of the historical arguments, pro and con. First Popper. To show that criticism has been a necessary ingredient in scientific progress, Popper has pointed to the fact that Newton's theory, strictly speaking, contradicts both Galileo's theory of falling bodies, and Kepler's theory of planetary orbits. And he has noted that Einstein's theory also formally contradicts Newton's. The contradiction between successive improvements of theory indicates that criticism has played an important part. Kuhn does not deny this, but he does say that this argument ignores normal science, and only applies to extraordinary science. It is normal science where criticism should be suspended. But why and where should it be suspended? Here Kuhn gives no historical counter-example to Popper's claim.

Kuhn argues for his idea of normal science by pointing out that long research efforts on, for instance, Newton's theory exist, and that they have led the way to such revolutions as Einstein's. To explain such long continuous research efforts was one of the central and interesting problems Kuhn tackled, and which Popper did indeed ignore. But for these efforts to be examples of Kuhn's 'normal science' and 'paradigms', they must have the features that Kuhn claims to have. Do they? Here we come to Kuhn's last stand.

In examining extended research efforts it is useful to distinguish between two main types: First, there are metaphysical research programs, like atomism. Second, there is applied science, which applies existing testable theories to particular cases. These particular cases are investigated to test the general theory, to explain the particular phenomenon, or for purposes of technology. Failure to distinguish these two can lead to confusion, as one program is concerned mainly with application of existing testable theories, and the other with search for alternatives to the existing theories. For either research effort to count as 'normal science' guided by a 'paradigm', it must have these characteristics: (1) There must be only one theory or world view aiding the research effort. (2) The scientists must be uncritically committed to carrying out the program: They must not criticize the basic idea, nor offer alternatives.

In considering whether Kuhn is historically correct about research programs, I would like to consider three questions in turn: First, do any examples of such paradigms *exist*? Second is, are these characteristics *prerequisites* to progress? The third is, are these characteristics *ever needed* as an aid to progress? To the first question I immediately answer yes, Kuhn is right, such examples exist. The main one, perhaps, is Newton's theory as applied to astronomy. But this is not enough to vindicate Kuhn's practical claim; the second and third questions must also be answered yes. But they cannot be.

There are many examples where competition between both metaphysical views and testable theories was fruitful. And the critical comparison of the different views led to progress. I need only mention two examples: The competition between the field and action at a distance views led to Faraday's and Maxwell's field theories. Helmholtz critically and undogmatically compared Maxwell's and Weber's theories of current. These are both testable, and they contradict each other. Helmholtz's critical comparison led to Hertz's demonstration of electromagnetic waves – a brilliant piece of applied science. Thus neither unity nor commitment are prerequisites to progress.

The final question, whether unity and uncritical belief are sometimes needed, is a tricky one. For it is not a purely historical question. It depends upon our ideas of what alternatives exist. Here we come to the heart of what I think is Kuhn's error. Kuhn thinks that to apply and develop a theory or metaphysics we have to *believe* in it *uncritically*. For instance, he says "... A commitment to the paradigm was needed simply to provide adequate motivation. Who would design and build elaborate special purpose apparatus, or who would spend months trying to solve a particular differential equation, without a quite firm guarantee that his effort, if successful, would yield the anticipated fruit?"[8] Kuhn's erroneous assumption that one must uncritically accept a theory or metaphysics as correct, to investigate it, was pointed out by Agassi in his review of Kuhn's book.[9] In fact, all he need do is accept the idea that working on the theory may help the advance of science. This is very different from believing uncritically in the theory. We may be highly critical of the theory, and believe that it is false, and yet think that it may lead to new discoveries, or that its development may help lead to a better theory. There are many examples of such investigations: I have already mentioned Helmholtz's development of Maxwell's theory. Another example is the reverse polarity theory of diamagnetism. Faraday invented it, and developed it, though he thought it false. He saw that it could explain many phenomena, and thought its refutation would lead to interesting results on field theory. Einstein helped create quantum theory, and applied it to 'puzzles' of radiation. Yet he always had serious reservations about the theory, and was the most outspoken critic of the theory.

The point is that commitment and suspension of criticism are not a prerequisite either of applied science, or of metaphysical research programs. Thus I conclude that Kuhn has no good argument for his practical advice of temporary suspension of criticism and search for alternatives.

But there are many arguments *against* such advice. I have already given the non-historical arguments of Popper. The historical examples abound. I mention only one: Maxwell, when he began his researchs on field theory was faced with the great success of the Newtonian theories of electricity: Wilhelm Weber had produced the first unified theory of static electricity, current attraction and repulsion, and electromagnetic induction. There were no experiments which were in open contradiction with Weber's theory, not were there any strong arguments on the basis of other testable theories. In such a case Kuhn would have to – for his advice ever to have force – say that Maxwell should not be critical, or create alternatives, but rather apply Weber's theory. Thus the acceptance of Kuhn's advice would have blocked the advent of a scientific revolution.

Thus my thesis on Kuhn's advice is this: It is never necessary as an aid to progress, and it will often be a block to progress. Of course I do not deny that a scientist might have uncritical faith in a theory, and that it might lead him to develop it successfully. What I deny is that the advice to believe in a theory uncritically is ever needed for progress. A related question is when one should work on a problem of application, rather than of basic theory. Here it seems to me that in principle we cannot give any general recommendation: We simply cannot predict, without knowledge beyond the existing, what avenues of research will be most fruitful.

I must admit that if a scientist shares the assumption of Kuhn that to work on a theory one must believe in it, then it might be fruitful advice to that scientist to believe in some theory. This would allow him to work at all. Yet I think that it would be far better to give the Popperian advice that uncritical belief in a theory is quite unnecessary. The belief in the necessity of suspension of criticism can cripple fruitful critical activity. It is especially harmful when coupled with the mistaken belief that refutation of a theory one has developed is a personal disgrace (which Kuhn seems also to hold). This mistaken idea has been beautifully criticized by Popper, who has shown its basis in the false theory that the truth is manifest.

Before commenting directly on Novakovic's paper, I would like to consider the question: If the practical consequences of Kuhn's theory are as rotten as I argue, why have many found his theory attractive, and why have especially Popperians felt that it is worthy of criticism? There are, I think, three main reasons: First is the widespread belief that belief in theories is essential. Second is the fact that applied science is today the most practiced sort, and no one before Kuhn has brought out so clearly that it is an integral and important part of science. Third is the fact that Kuhn's theory and his device have been the usually unconscious basis of our whole system of science education, and indeed education in general.

I have already elaborated the first point. The second point is one on which Popperians have not given Kuhn proper credit. Agassi has discussed the importance of competing metaphysical research programs in the growth of science. His account, as explained earlier, I think is superior to Kuhn's both in historical accuracy, and methodological progressiveness. But on applied research, Kuhn has been the first to bring out its importance so forcefully. Unfortunately he misconceived its nature in thinking it must be preceded by dogmatic commitment, and he also confused it with metaphysical research programs, which have most relevance to search for alternatives – that is, where 'revolutions' are made.

But I think Popper made a great mistake in reserving the name 'applied science' for 'uncritical science', in his recent comments on Kuhn's theory.[10] In fact applied science is an essential part of Popper's 'logic of research', a part which he did not emphasize: Applied science, as I am using the term, means work on the connection between theory and experiment, rather than directly on fundamental theory itself. In this sense, circuit theory is applied electromagnetic theory, solid state physics is applied quantum theory, etc. Such work, often the effort of a lifetime, is often essential for the testing of a theory, and for the application of the theory to explain the actual results of observation and experiment: These are both part of Popper's logic of research. The emphasis in such cases is on such problems as getting a good model of the experimental conditions, or solving the equations of the fundamental theory. Kuhn confuses the choice of a problem of application, with a suspension of criticism of the fundamental theory, or of search for alternatives. Outside the demands of time, both may be done, as I have mentioned in Einstein's case. And application may be for criticism or refutation of the theory rather than confirmation. But Kuhn's error does not make it less true that he has brought out an important side of scientific research, a part which is equally important for Popper, though for different reasons.

My final point on Kuhn is his accurate description of our science education, and its presumptions. This I think is a real service. But I, like Popper, regard his achievement here as the accurate identification of the worst features of our educational system, rather than the best. And it seems to me that here is where the greatest danger of Kuhn's recommendations lie. The physical sciences are looked to as an example of progressive methods. And it may well be true that in science classes a critical, imaginative spirit is more encouraged, or rather less discouraged, than elsewhere. Nevertheless there are many regressive tendencies. Our present system indoctrinates students in a textbook fabrication of a unified science. Conflicts and disagreements between theories, like between relativity and quantum theory, are neglected. The growth of science is presented as a series of formal modifications of equations, rather than as the battle of ideas that it is. The weak points of basic theories, like relativity and quantum theory, are glossed over. Application is constantly emphasized, while mention of criticisms and imaginative alternatives are rarely mentioned. The encouragement of imaginative search for alternatives and of sharp criticism of existing ideas is systematically sacrificed to training of competence in application of existing theory. And the application is restricted to relatively unproblematic cases.

Students are under great pressure to spend all their time becoming competent in application of theory, and are under similar pressure to do 'research' projects for their PhD's which are the most a foregone conclusion. Is it surprising that the 'products' of such a system almost invariably go into applied research? No, I think not; it is too much to expect large numbers of students to resist such pressure, especially when they are fooled into thinking that they are being trained for discovery. The best they can salvage is a critical and imaginative approach to problems of applied research. Of course much applied research is wonderful, but it is not the whole of science. And our present system of education does discourage the critical search for alternatives to the present basic theories. Naturally, I cannot say that this has prevented a revolution in physics. But I think I can say that it, like Kuhn's practical advice generally, blocks many possibly fruitful lines of research. We have an alternative to Kuhn's theory in Popper's, and I believe it most important that our educational system reform in the light of his suggestions, if progress in *all* the sciences is to be promoted.

Novakovic has given us some criticisms of Kuhn, Popper, Lakatos, and Feyerabend, and has offered some alternative suggestions.

My main point of agreement is with Novakovic's criticism of Kuhn. With Novakovic, I see the historical existence of more than one 'paradigm' as undermining some of the essential features of his system.[11] Let me just add one point to Novakovic's. One of the main problems Kuhn wanted to solve by his theory was to characterize the methods which separate physics, and for instance, psychology. If competition between theories has contributed to the growth of 'mature' physics, as Kuhn himself admitted in reply to Watkins, then the uncritical commitment to a single paradigm does not characterize physics as opposed to psychology. Thus a fundamental point of Kuhn's theory must be rejected. Kuhn himself has not acknowledged this consequence: it seems to me irresistible.

As to disagreements: I do not think the criticism of Popper is very telling: One can easily characterize scientific revolutions within his system: Revolutionary theories are those which involve a change in world picture, or metaphysics, as opposed to those in which laws within a given picture are changed. It is true, as Novakovic points out, that given this idea of revolution, we cannot say that science is a revolution in permanence; revolutions are not so frequent. Perhaps it would be better to characterize Popper's view as *attempted* revolution in permanence. Another minor disagreement is with Novakovic's claim that in social reality we have 'sudden and qualitative changes' as opposed to physical reality: Such changes also occur in physical reality: the weather. What is to be counted as qualitative and what as quantitative depends on our theories; common sense does not give any consistent distinction.

Now I come to my main disagreements: First, Novakovic seems to think that revolutionary periods in science are not favorable for the creation of new theories. This I think is historically false; the creation of quantum theory is one counterexample to it. Taken as a prescription it might hold up valuable work.

My second disagreement is with the treatment of Lakatos's and Feyerabend's ideas. Novakovic seems to accept their arguments that reason cannot be a strong guide to research. And he suggests, to prevent chaos in science, that we put some limitations on the proliferation of theories.

In the first place, the constraints of logic are stronger than Lakatos thinks. His claim that "No theory forbids some state of affairs specifiable in advance,"[12] is incorrect. Newton's law $F = ma$ forbids an acceleration of three units, given a unit force acting on a body of unit mass. In a case where we discover that the acceleration is actually 3, with a unit mass and force, then our alternatives are limited by logic to the following: (1) We have interpreted the initial conditions wrongly. (2) We have interpreted the experimental result wrongly. (3) The theory is false. (4) Our derivation of the result from the theory is invalid. These are the restrictions placed on us by logic, and they are, as Popper insists, a principle guide in research.

Feyerabend's phrases the 'principle of tenacity' and the 'principle of proliferation' are nice names for principles which are entirely consistent with Popper's theory, at least as revised by Agassi. Feyerabend's 'anything goes' claim does not follow from their adoption. As to logical incommensurability, I do not think this is a serious problem. Once we adopt the attitude that we want to make our theories as criticizable as possible, then we strive to bring theories into clear conflict. To make the conflict clear may sometimes be a difficult task, though it is usually a minor one compared with the task of bringing our theories in possible conflict with experiment. The real question is whether we want to make our theories criticizable. If we do not, then we can drop logic altogether. If we do wish to use reason, then 'incommensurability' is no insuperable difficulty.

Thus I think that Novakovic's restriction on proliferation is not needed and might be harmful. He says that theories have to "emerge out of" theoretical shortcomings, or conflicts between theory and observation. In a sense I agree that new ideas come out of particular problems; all research is problem-oriented. And all the research problems are somehow related to the overall aim of science, to explain the phenomena of nature. But beyond this very general aim, I think it unnecessary and probably harmful to place any restrictions on proliferation of theories. (It is not clear to me exactly what restrictions Novakovic wishes to place on proliferation.) I cannot see why theory proliferation is threatening: If the theories are not independently testable, they will not be interesting, unless they seem to provide an interesting research program; then pursuing it might lead to great discoveries. If the new theory is independently testable, all the better: We can discover where our error lies, and so advance our understanding of nature.

Massachusetts State College, Bridgewater

NOTES

[1] See T. S. Kuhn, *The Structure of Scientific Revolutions*, 1962, Preface; and K. R. Popper, *Conjectures and Refutations*, 1962, Chapter 1.
[2] K. R. Popper, *Logic of Scientific Discovery*, 1959, Chapter 2.
[3] Kuhn emphasized the anti-critical character of this theory in *Criticism and the Growth of Knowledge* (1970) (ed. by I. Lakatos and A. Musgrave): "It is precisely the abandonment of critical discourse that marks the transition to a science." (p. 6) "I hold that in the developed sciences occasions for criticism need not, and by practitioners ought not, deliberately be sought. When they are found a decent restraint is the appropriate first response." (p. 247) "... Confronted with the problem of theory-choice, the structure of my response runs roughly as follows: take a *group* of the ablest available people with the most appropriate motivation; train them in some science and in the specialities relevant to the choice at hand; imbue them with the value system, the ideology, current in their discipline (and to a great extent in other scientific fields as well); and finally, *let them make the choice*." (p. 237–8)
[4] *Op. cit.*, p. 246.
[5] *Op. cit.*, p. 262.
[6] J. Agassi, 'Sensationalism', *Mind* **75** (1966) 1–24; Kuhn, *op. cit.*, p. 14, also notes

that the corrigability of facts is a weak point in Popper's theory. Unfortunately, his discussion is vitiated by his assumption that Popper holds refutability to be a purely formal matter.

[7] J. Agassi, 'Scientific Problems and Their Roots in Metaphysics', in *The Critical Approach to Science and Philosophy* (1964) (ed. by M. Bunge).

[8] T. S. Kuhn, 'The Function of Dogma in Scientific Research', in *Scientific Change*, (1963) (ed. by A. C. Crombie), p. 362.

[9] J. Agassi, *Journal of the History of Philosophy* **4** (1966) 351.

[10] K. R. Popper, 'Normal Science and Its Dangers', in *Criticism and the Growth of Knowledge, op. cit.*, p. 51ff.

[11] Watkins and Popper also emphasize the importance of competing theories, *op. cit.*

[12] I. Lakatos, *op. cit.*, p. 130.

A. POLIKAROV

THE DIVERGENT-CONVERGENT METHOD –
A HEURISTIC APPROACH TO PROBLEM-SOLVING

> Though this be madness,
> yet there is method in't.
>
> Shakespeare, *Hamlet* (Act II, Sc. 2, 1.208)

I

How to approach the solution of an open scientific problem, i.e., one for which we do not know a determinate method or an algorithmic procedure for its solution. Which means exist for this purpose?

The most radical view on this topic denies the existence of scientific method at all. Thus, according to Bridgman, the scientific method "is nothing more than doing one's damnedest with one's mind, no holds barred."[1] In particular, creative thinking is grasped as something which excludes any limitations. In this sense one can believe that there is no other approach but the trial-error method, or that of testing hypotheses, and for the very formulation of hypotheses there are no rules.

Insofar as creative thinking is not separate from thinking in general, it is not the antithesis of the methodical approach, but it includes the latter at least as an element.

In the conception mentioned above the difficulties of problem-solving are absolutised and/or the requirements concerning similar methods are exaggerated. If we do not expect from the latter something like a magic wand to find the solution, but simply to contribute to problem-solving, then we can speak of such methods. They are known as heuristic methods.

A. *Heuristic Methods*

By heuristic methods we understand those complementary to the algorithmic methods of problem-solving; they presuppose a creative approach to the solution of open scientific problems. [As a matter of fact the very concept of 'scientific problem' may be defined by the impossibility of applying another approach.]

Boston Studies in the Philosophy of Science, XIV. All Rights Reserved.

Several heuristic methods are examined in the literature. In A. Osborn's opinion these are: adaptation, modification, substitution, addition, multiplication, subtraction, division, rearrangement, reversal and combination.[2] Dixon dwells on 'brainstorming', inversion, analogy, empathy, the study of new combinations.[3] According to Caude the fundamental heuristic methods are five, namely: 'brainstorming', matrix of discovery, teratological (by which the role of singular factors are consciously exaggerated), method of the advocate (when a given view is defended by all means), and of analogy.[4]

To these ones one should add the heuristic programming (artificial intellect) in cybernetics. By heuristic method, one understands here one which helps in discovering the solution of a problem by making plausible but fallible guesses as to what is the best thing to do next.[5]

It is wrong to think that, unlike algorithmic methods, the heuristic methods are always less perfect and reliable. As Minsky remarked, imperfect methods are not necessarily heuristic, nor *vice versa*.[6]

There are some attempts at classification of heuristic methods. Moles (1957) distinguishes three groups of heuristic methods: of application (use of a theoretical system), creative (of creation of a system), and *a priori* methods (intermediate between the first and the second groups). Properly heuristic are the methods of the second group. To this group belong the following six methods: of details, experimental disorder, matrix of discovery, recodification, representation and phenomenological reduction.[7]

In a later work Moles proposes a 'philosophical classification' of the methods of scientific creativity containing six groups. He demonstrates different possibilities of classification of scientific methods, namely in the following three categories: heuristic, artistic and general. Other criteria for classification, proposed by the working group for methodological studies in France (Moles, Caude *et al.*) are: purpose, similarities and ways used. The purposes may be heuristic, pedagogical, dialectical and pragmatic ones. Besides, one can establish groups of methods which are similar as well as ensembles of general (or super-) methods. According to the techniques applied, these may be classified in view of the relations between imagination and tradition, theory and experimentation, synthesis and analysis, abstract and concrete, certainty and risk.[8]

The known heuristic methods may be classified according to two pairs of opposite traits: spontaneous and systematic, on the one hand, and direct and indirect (not excluding transitional cases) on the other. The spontaneous methods, based mainly on nonrational (emotional) factors or on the deliberate use of these factors, are indirect in the sense that the solution sought can be obtained as a result of the discarding of irrelevant guesses. To the systematic direct methods (i.e. without choice) is to be related the matrix of discovery, which aims to establish all relevant possibilities. The systematic indirect methods include methods of detecting possible solutions, and methods of choice of relevant (optimal or true) ones. The teratological method and the method of constructing the set or field of possible solutions (fps) belong to the former. The method of reduction of the fps, as well as the heuristic programming refer to the second sub-group. Lastly, we can speak of combined methods. As a matter of fact, all indirect methods are combined with the procedure or method of reduction, i.e. of choice of solution. This is illustrated in Table I.

TABLE I

		spontaneous	systematic
direct			matrix of discovery
indirect	detecting of new possibilities	brainstorming synectics	teratology field of possible solutions
	choice of relevant possibilities		reduction heuristic programming

We should like to indicate that the confrontation between algorithmic and non-algorithmic (heuristic) means of problem-solving is conditional. The possibility to apply an existing theory or method to a given problem is not always clear and may in turn be related to supplementary problems. Moreover, there are cases when the same problems can be solved both algorithmically and heuristically.[9]

B. *Possible and Relevant Solutions*

Among the several investigations we shall examine concisely only a few, which have a closer bearing on the following exposition; about other works we shall speak only incidentally.

The matrix of discovery or the study of new combinations has as its remote predecessor Leibniz and even R. Lullus.[10] In the last century it was used successfully by such outstanding scholars as Faraday and Mendeleev.

This method – called *morphological* – was revived and further developed by Zwicky. Taking into account the nature (parameters) of the problem – alongside with the real or accomplished cases – we can outline all supplementary possibilities. In this way the method directs us to discover new objects, situations, relationships.[11] Zwicky has foreseen and established the existence of new astronomical objects – different kinds of super-novae (stars), compact galaxies, dispersed intergalactic matter, etc. The same method has found a successful use in a set of technical inventions. During the last few years, Zwicky and his collaborators have registered 16 patents. In technical problems, a similar method has been used in the works of J. Müller.[12]

In general, we can conceive a considerably greater set of possibilities, out of which the relevant ones have to be selected. Inasmuch as there exists no reliable direct way to the relevant possibilities, such an approach is justified and fruitful.

In fact, such a method is applicable in cases where we establish a set of possible solutions of a problem, among which – on the basis of certain criteria – we have to choose a definite solution. Such, for example, is the case in relativistic cosmology which allows (essentially) different cosmological models. In this connection the question of the investigation of various possibilities and of a choice among them acquires particular importance. The same refers to a set of astronomical and other problems, on which – for a long time – two or more competing hypotheses, views or trends coexist.

Historically, some of these conceptions can result from the change (improvement) of other ones, i.e. to be genetically connected. In a logical respect, however, they can be considered as variant conceptions (which tend to improve their accordance with experience and/or other requirements).

When the elimination of some of them and the acceptance of others cannot be accomplished on the basis of empirical data, general desiderata or methodological principles can be used for their comparison in order to establish their advantages. In connection with the question of the validity of the requirements for simplicity of the scientific conceptions, Bunge systematizes the desiderata of a scientific theory or the symptoms of truth in five groups, namely: syntactical, semantic, epistemological, methodological, and philosophical. The following 20 desiderata refer to them:

syntactical correctedness, systematicity or conceptual unity;

linguistic exactness, empirical interpretability, representativeness, semantical simplicity;

external consistency, explanatory power, predictive power, depth, extensibility, fertility, originality;

scrutability, refutability, confirmability, methodological simplicity;

level-parsimony, metascientific soundness, world-view compatibility.[13]

A properly chosen and completed subset of these requirements can be applied for the reduction of the fps (see below).

It is worth mentioning that a similar approach is used also in decision-making. Simon distinguishes the following three principal phases in this process:

(a) finding occasions for making a decision, or intelligence activity;

(b) finding possible courses of action, or design activity;

(c) choosing among courses of action, or choice activity.

The decisions are of two kinds: programmed and non-programmed, or with a 'good' or 'bad' structure respectively.[14]

Further, we shall characterize the so-called divergent-convergent method for solving a definite class of problems, which has been worked out and used in the examination of philosophical problems of modern physics and in other cases by the author and some of his collaborators in Bulgaria, in the German Democratic Republic and elsewhere.[15]

II

Essential elements of the method mentioned are subsumption of concrete problems under definite types, establishing the set of solutions of these types and reduction of this set on the basis of some methodological desiderata or principles.

A. *Types of Problems*

The existing attempts at typisation of problems can be related to the following three kinds of classification, namely:

(a) according to scientific branches (or disciplines);
(b) according to the (interdisciplinary) character of the problems; and
(c) mixed – according to branches and to character of the problems.

Bunge's attempt belongs to the latter category. He classifies the (non-elementary) problems as empirical, conceptual and strategic. The conceptual problems are problems of describing, arranging, elucidation, deducting, building and metalogical ones. To the strategy problems belong the methodological and valuational ones.[16] The enumerated metalogical and methodological problems are determined rather according to branches than to their interdisciplinary character as is the case with the other problems.

From the second kind is for instance the classification of Organ as well as those carried out in view of the solution of Zwicky, Ackoff, etc.

The problems in general (not only the scientific ones), according to Caude, can be distinguished according to two traits: essence (research and action problems) and degree of complexity (simple, complex and complicated). Some authors draw attention to the difference existing between problems to find and problems to prove.[17]

Organ concerns four types of problems: those of identification, of causation, of means, and of ends.[18] A difference between problems of explanation, description, proof, explication and definition is made by Parthey and Wächter.[19]

According to Zwicky, problems may be classified into three classes, namely: (1) for the solution of which we ought to substitute only a few known parameters; (2) when elements not yet known are required; and (3) problems of large numbers.[20] Ackoff distinguishes evaluative problems for selecting the 'best' solution and developmental problems involving the search for a better than any solution available at the time.[21]

The scientific problems may be characterized according to different traits (proper, relational, a.o.), among which the most appropriate seem to be the characteristics with regard to their solution.

Usually the division of scientific problems is made according to branches. We can speak about as many scientific problems as there are scientific disciplines (D), as for example mechanics, physics, astronomy, chemistry, biology, etc. This can be represented by the following scheme:

$$P_{11}, \quad P_{12}, \ldots \quad P_{1k}, \ldots \quad P_{1n} \qquad D_1$$
$$P_{21}, \quad P_{22}, \ldots \quad P_{2k}, \ldots \quad P_{2n} \qquad D_2$$

. . . .

$$P_{i1}, \quad P_{i2}, \ldots \quad P_{ik}, \ldots \quad P_{in} \qquad D_i$$

. . . .

$$(T_1) \quad (T_2) \qquad (T_k) \qquad (T_n),$$

whereby the first digit is related to the kind of problem, and the second to the species of each kind.

Sporadically the similarity between certain problems has been displayed, e.g. between P_{21} and P_{31} or P_{32} and P_{42}, i.e. in these cases they could be grouped in a new way, namely:

problems of the kind P_{k1} $(P_{21}, P_{31}, \ldots)$

problems of the kind P_{k2} $(P_{32}, P_{42}, \ldots)$ etc.

On this basis, in our opinion, a general classification could be carried out, which from the view point of the one existing represents a kind of cross-section of the sciences and a grouping of the problems of the same type $(T_1, T_2, \ldots)$ in definite general kinds (notwithstanding the branch to which they belong).

On the basis of the analysis of philosophical and scientific problems, the following preliminary (non exhaustive) classification of scientific problems can be outlined:

Problems of existence (detection and description);

Problems of essence (explanation) which can be related to the

structure or mechanism,

relation or connection (law),

causation or genesis,

specificity or classification,

reduction or deduction,

concept or definition (explication),

generality (validity) or interpretation;

Problems of construction, substantiation (proof), including

classification,

generality or validity,

uniqueness, necessary and sufficient conditions;

Problems of elimination of a contradiction (or paradox);

Problems of ascertaining criteria;
Problems of development; and
Problems of solubility (Entscheidbarkeit).

In this connection the question arises whether the set of problems according to branches can be transformed into a set according to types, and whether such a transformation has a real sense, i.e., can it facilitate the solution of problems, at least of some problems.

We can maintain that for some problems this transformation is possible, i.e., the problems in different branches can be arranged according to types. This leads to a decrease in the diversity of problems which already involves a simplification. At the same time this statement made in accordance with the general structure of solutions brings us nearer to the solutions sought. Herein we see the advantage of this approach.

B. *Structures of Solutions*

In the transition from the problem to its solution it is important to establish: (a) whether we are dealing with a problem which cannot be solved – in general or within a given scientific system; (b) whether the problem in question has a unique solution or, on the contrary, it admits of more than one solution. These two cases are characteristic for mathematical (and logical) problems: for some we possess a proof of their insolubility, and for others there is a theorem of existence of a unique solution.

At the other pole lie problems with – practically or *stricto senso* – an infinite number of solutions, for which heuristic methods of choice (Newell-Shaw-Simon, a.o.) have been devised.

The problems which do not belong to these two categories – we can admit a considerable network of scientific problems – are characterized by a comparatively small number of parameters, and these have a discrete and relatively small number of values.

In some cases we can establish the parameters and their possible values as well. In other cases, this is possible only by means of subsequent approximation, whereby the empirically shaped draft solutions are taken into consideration. In the first case we speak of problems of *regular* class and in the second of an *irregular* class. To the problems of a regular class belong the problems of the 'relationship' type and of the 'contradiction' type or paradox.

In general form the solution of scientific problems (including the ir-regular ones) can be presented in the following way:

(I) Stage of existence of a few initial design-solutions (hypotheses, inter-theories, conceptions), including those which are contradictory or mutually exclusive.

Here we can introduce the concept of *inverse* solutions, by which we mean the following: at a stage of scientific investigation a definite complex of phenomena is described or explained by a given theory. Thereafter, if one is confronted with a new phenomenon which cannot be encompassed in the theory, i.e. an exception, then two possibilities occur:

(a) Usually one tries to modify the theory in such a manner that the new phenomenon can be included, i.e., its anomalous status is eliminated.

(b) A more radical approach might be applied – and this is a special case worth noting – when the exceptions turn out to be the rule, and the old rule a special case. It is in this way that the revolutionary change in contemporary theoretical physics took place.

In this connection we can quote Mendeleev's interesting thought: "Be able to accept the opposite opinion. Therein consists genuine wisdom."

(II) Stage of development of the initial solutions involving their inter-action with new variant-solutions. The resulting increased set of design-solutions can be presented in the following way:

(1) Radical design-solutions which take the opposite poles in this set.

(2) Moderate design-solutions taking an intermediate stand (of course this classification can be carried out from different points of view).

(3) Combined solutions:

(3.1) without constraint: the radical-synthetic (or alternativistic) de-sign-solutions belongs here;

(3.2) with weak constraint of the diversity; i.e. by removing some variants;

(3.3) with a strong constraint of the diversity; this can lead to a unique combination.

It is essential to take into account the cases in which a problem is insoluble within the framework of a given set of design-solutions, i.e. to design supplementary possibilities of its solution both within a given conceptual system and in different (still unknown) conceptual systems.

In this respect it is important to distinguish the variations of certain conceptions or the different meanings of the concepts concerned. Then it

can be shown that alongside the known meanings and customary interpretations of these concepts, other concepts are possible. The latter for
one reason or another have not been taken into account so far but can
be reasonably maintained and might prove to be superior to other conceptions. For the regular problems, according to their parameters, the
f p s can be constructed. This construction for problems (of a given type)
presupposes supplementary (concrete) theoretical considerations.

The problem of relationship between two concepts has a 'standard'
structure of solutions. This kind of problems is a very general one. Inasmuch as general concepts and more exactly their denotations represent
actually a family of different concepts, the comparison between both
concepts is reduced to a comparison between two groups of concepts.
Hence their relationship involves a set of cases.

The number of possible relations is $N = \Pi A_i B_j$, where i and j take on
the values from 1 to m and n respectively. This formula can be easily
generalized for cases involving more than two concepts.

Figuratively speaking two general concepts do not intersect at one point
but involve a whole area and form something like a network. From this
we can deduce a methodological imperative: when you have to deal with
general concepts, you should expect different nuances to exist or be
possible. Try (if possible on the basis of theoretical considerations) to
distinguish them and, by taking all of them into account, make systematic
comparisons.

For illustration's sake let us take the actual dilemma: is the world space
finite or infinite. The concrete findings of the investigation of this problem
contribute to the clarification of the very concepts of finite and infinite
and of their relationship. First of all, it turns out that the concepts boundless and infinite do not always coincide and therefore, boundless but
finite spaces can exist (Riemann, Einstein). Furthermore, it comes out
that the concept of infinite is not as simple as it seems at first glance.
Rather it involves different concepts: practical infinity, cyclic infinity,
etc.[22] This implies that between finite and infinite a relation of exclusion
does not always exist; cases of compatibility are possible. Alongside the
well-known 'Aufhebung' of finite and infinite in cyclic infinity (Hegel), in
the relativistic cosmology – in the model of inhomogeneous Universe –
new combinations or unities of finite and infinite are quite possible. Among
the latter the inclusion of the infinite (relatively to one reference system)

space in the finite (relatively to another system) space is also possible. These possibilities were revealed by Schrödinger, Zelmanov, a.o.

In a special case we may have to do with a transition from one known solution to a second possible one (i.e. polar solution) or/and to a third (intermediate, combined or integral) or/and to more (radical or/and intermediate) solutions, etc. This is illustrated in the scheme of a problem with 12 possible solutions, of which three are known. Starting from them we have to reconstruct the whole set, i.e. to find out the 9 remaining ones.

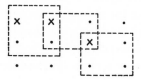

At the same time in this scheme three special cases of problems with 2×2 solutions are shown by dotted lines, among which one or two with a different position in the matrix are known.

In many cases the transition from some possible solutions to other ones is connected with the overcoming of the so-called 'psychological inertia' (Dixon).

Essential is the question of the structure of the fps, i.e. the existence of a relationship of subordination and of groups of solutions accordingly, which eventually may be presented in the form of a matrix and in the process of selection to be eliminated not singularly but as whole groups.

A number of requirements or methodological principles regarding inclusion in the fps are imposed on the design-solutions, namely: syntactical and semantic desiderata of consistency, completeness, solubility, etc. for the axioms (of the formal systems), scrutability of presumptions, verifiability, causality, restricted reducibility and metascientific soundness.

This first stage in the process of problem-solving may be called *divergent*. The next stage is concerned with the opposite process of *convergence* or reduction of the fps to a minimum of relevant (optimal, plausible) possibly one true solution, i.e. it is concerned with an option among the various design-solutions.

C. *Reduction*

The question of the truth of a theory, viz. of the preference of one theory to another is connected with open and disputable questions, which cannot be examined here. We shall merely note that recently Fine examined the different possible circumstances to carry over a term from one theory into another one. He states that under conditions which can be formulated in the new theory, the objects of the latter satisfying these conditions are suitable objects also for the old theory in various cases of a predicate term, an operation term and a term for a magnitude.[23]

In a particular case (for the physical theories) Strauss thinks that the comparison between two theories is physically meaningful if the cross section of the regions, where they are defined, is not empty.[24] Relevant for the comparison of two theories is the character of the pair of theories. Strauss suggests a classification of the boundary-relations between two theories according to (three) different points of view.

It deserves mentioning that already in the beginning of the nineteenth century Whewell put forward the conception that the core of the scientific discovery consists in a creative invention of hypotheses and the successful choice of the right ones. Accordingly, in his system refutation is a methodologically stronger procedure than confirmation.[25]

Nowadays this view has been elaborated by Popper, according to whom science progresses by forwarding and refuting hypotheses. The choice occurs in accordance with the characteristic of simplicity, and under simplicity he understands degree of falsifiability of the hypotheses. If the conjecture survives the tests, it is accepted provisionally, because its truth can never be known.[26]

Without doing a critical analysis of this conception[27] we can assert that a higher degree of falsifiability cannot be identified with a higher degree of substantiation. Besides, Popper's conception rests on an idealistic basis. In opposition to the latter, Nidditch rightly argues that scientific theories are essentially models of segments of reality.[28]

Within the framework of the inductive logic the questions of degree of confirmation (Carnap a.o.), degree of factual support (Kemeny-Oppenheim), degree of belief (Ramsey), etc. have been elaborated and applied to competing theories.[29]

The choice among competing conceptions is made on the basis of their comparative examination by dialectically combining the *exclusion* of non-plausible solutions and the *substantiation* of plausible ones. In connection with hypothetical reasoning Rescher speaks of a principle of rejection and retaining.[30]

This presupposes the introduction and elaboration of respective criteria and procedures for reduction of the fps. For this purpose, along with logical considerations we can use scientific data and methodological principles. To the principles of choice among the design-solutions belong: representativeness, simplicity, originality, explanatory and predictive power, depth, extensibility, external consistency and compatibility with the world-view. If we are concerned with opposite requirements the method of optimation can be applied.

The question about the validity of the methodological principles and the possibility of their hierarchical arrangement is relevant. At least at some stage particular principles may be considered as practically true. Some principles of symmetry have such a character. In the last resort their validity depends upon the extent of our knowledge, i.e. we cannot warrant against a reconsideration of the universal applicability of some of these principles.

It it important that notwithstanding revision of the general validity of some principles, there do exist a set of such principles. The question of the possibility to build up a consistent and relatively complete system of these principles remains open.

We can apply the procedure of systematic reduction beginning with a choice of the most plausible values for the basic parameter and continuing with a similar choice among the possible combinations with the next parameter, in a hierarchical arrangement of the parameters. This means to use a partially or completely ordered sequence for each parameter, or a lattice diagram.[31]

We shall illustrate this by an example related to the plausible conception of quasars and representing a sketch of a paper by M. Kalinkov and the author.[32] The following parameters: A–distance, B–source of energy, C–mass and D–kind of object play an essential role in the quasars hypotheses.

The parameter A may have the following values: A_1–extremely remote objects (cosmological hypothesis), A_2–remote (galactic hypothesis), A_3–near (local hypothesis) and event. A_4–another possibility (for instance some combination of the foregoing possibilities). The values of B in turn can be B_1–annihilation, B_2–gravitational contraction, B_3–nuclear energy, B_4–combined, and eventually B_5–an unknown source. For the mass we can accept at least three values, namely C_1–of the order of 10^2 solar masses ($M_\odot$), C_2–of the order of $10^7 \pm 10 M_\odot$, and C_3–of the order of $10^{11} \pm 10 M_\odot$, i.e. these can be D_1–star, D_2–superstar, D_3–galaxy and eventually D_4–some other object.

Accordingly the table of possibilities can be written down as follows:

$$
\begin{array}{lllll}
A_1 & A_2 & A_3 & A_4 & \\
B_1 & B_2 & B_3 & B_4 & B_5 \\
C_1 & C_2 & C_3 & & \\
D_1 & D_2 & D_3 & D_4 &
\end{array}
$$

Formally 240 combinations are possible. Let us note that this is the order of the already singularly forwarded hypotheses, which range generally speaking within the framework of these combinations.

If we accept (as plausible from a contemporary point of view) that we have to do with cosmological objects, i.e. A_1, the amount of possibilities is reduced to 60. Further, if we restrict ourselves with the known forms of energy, for B 4 possibilities (and altogether 48) are left. In the same time certain constraints on the next parameter are necessary. If the quasars exist more than 10^6 (until 10^9) years, then their minimal mass must be at least 10^3–$10^6 M_\odot$, i.e. we must prefer the possibilities C_2 and C_3, i.e. in this case we can have to do with superstars (D_2) or galaxies (D_3). The possibility B_3 is conceivable in combination with C_3–D_3, and C_2–D_2 in combination with B_1, B_2 or with the more general case B_4. Thus the amount of possibilities turns out to be 4, i.e. the former table is transformed into the following one: (see figure following page).

Here the two basic types of combinations are marked: B_3–C_3–D_3 and B_4–C_2–D_2. Under these assumptions, the last possibility could be regarded as plausible from a contemporary point of view. We shall repeat that this illustration – on account of an over-simplification – is providing only an idea for the application of the method.

In the process of reduction a preferable solution or (preferable) group of solutions might emerge. In view of the result, different situations are possible:

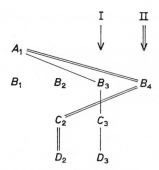

(1) The f p s includes a true solution and the remaining ones are false.

(2) All design-solutions are (partly) true, i.e. the combined (integral) solution, containing various solutions as special cases, eventually in some hierarchical relationship, is true.

These two extreme situations can be denoted as 'classical' and 'morphological' (according to Zwicky) respectively. Alongside those – as more general – intermediate (or 'mixed') cases may be presented when we have true, false and partially true design-solutions. This situation can be represented by the matrix of truth of conjunction in the polyvalent logic.[33]

In this case the true solution is obtained by eliminating the false solutions and elements and by integration within it (supposedly in a relationship of subordination) the true elements of the respective group of design-solutions. This can be achieved by subsequent approximations.

When a definite subclass of the f p s is false the reduction process is carried out by eliminating whole subclasses, without testing the individual design-solutions separately.

Among the various design-solutions debates are not only permitted but indispensable. They must be *sine ira et studio* (without rage and partiality) following the logic of the conceptions possibly in their optimal fashion. As a result of such debates and further development and improvement of the competing conceptions, one arrives at a plausible or optimal solution. Already Herodotus was aware that "if opposed opinions are not expressed, then there is nothing from which to choose the best."

From the point of view of the reduction of f p s, the solutions of the kind (3.2) and (3.3) (according to the above-adduced scheme) are usually more plausible. Herein are included the extreme cases and in this sense they – especially (3.2) – are concrete samples of solutions of a dialectical type.

Already in the transition from the first to the second stage (as mentioned above) some design-solutions were discarded. In comparison with (3.1), the solution (3.2) is a reduced one. The same can be said also about the solution (3.3) (compared with (3.2)), which in the limiting case is reduced to solution 1.

At the same time the possibility of improving rival conceptions makes the strife between them a continuous and difficult process. Generally speaking this process can go on with alternative success. These conceptions clash repeatedly and it is possible that the discarded conception may be revived again at a later stage or be included in a more general solution of the problem. Therein manifests itself the dialectical character of this process.

III

The validity of the proposed method can be outlined as follows: firstly for regular problems mainly, i.e. problems in which a definite connection exists between the type of problem and the type of solution; secondly, for problems which allow for more than one (but neither very many, nor in fixed number) possible solution; and thirdly, the choice among which is made mainly on the basis of methodological principles.

A. *Significance of the Method*

The possibilities of application and hence the significance of this method is related to certain factors which we shall try to enumerate:

(1) The method covers a wide class of scientific problems. Especially in our work, '*Relativity and Quanta*', we showed that seven of the philosophical problems of modern physics are reducible to two main types and their combination.

Besides, the procedure of problem-solving may also be applied in the process of decision-making.

(2) The dialectical unity of the processes of divergence and convergence is an illustration to the so-called 'inventor's paradox', according to which the more ambitious plan may have more chances of success. To put it otherwise, more questions may be easier to answer than just one

question; the more comprehensive theorem may be easier to prove, the more general problem may be easier to solve (Polya).

(3) The given statement of the question offers a model in the quest for a more-or-less systematic looking for new solutions.

The use of methods of this kind can assist in the transition from accidental discoveries to their designed realization.

In the same way a controlled change of f p s can result as a function of the increase in information.

(4) The choice of relevant solutions is the result of optimum argumentation and is subject to controlled correction on the basis of newly acquired knowledge. The search for optimal solutions proceeds according to the requirements or criteria of optimality. This permits to offer a (complete) set of solutions varying according to the different conditions.

(5) It is worthwhile noting that the method in question has not appeared as Pallas Athene from the head of Zeus, but that thereby the line of continuity of investigations can be pursued. The words of Goethe are applicable here, that all sensible has been already thought over and we must only try to think it over once more. The novelty is not in the general idea, but in the details, because "only in the details is there truth and originality" (Stendhal).

The method at issue has touching points with many contemporary branches and research trends such as psychology of productive thinking, logic of discovery, decision-making, interdisciplinary, methodological, historical, and other approaches.

This is one of the basic problems of the philosophy of scientific knowledge. According to such philosophers as Carnap, the main subject of the philosophy of science is the formal system. For other philosophers of science (Braithwaite, Hempel, Nagel) this is the problem of scientific explanation. Recently, the view is gaining ground that the question of problem-solving – which had remained in the shadow for a long time – is a central one for the philosophy of science. This view is supported by Hanson, Polanyi, etc.

As a matter of fact, this problematics has an important methodological significance. It has been elaborated by several scholars (mathematicians, physicists, biologists, psychologists, sociologists, historians of science) such as – besides those already mentioned – M. Aiserman, N. Amosov, S. Amarel, R. Banerji, L. Fogel, A. Ivakhnenko, B. Melzer, D. Michie,

N. Nilsson, A. Owens, J. Pitrat, E. Sandewall, O. Serebrjannikov, A. Ujomov, M. Walsh, etc.[34-43]

This bears witness to the fact that interest in the problems of scientific creativity is growing, in the field of several sciences, interdisciplinary studies and in the philosophy of science.

The heuristic approach like the one considered here corresponds to the dialectical requirement of thorough analysis of problems and their design-solutions. At the same time it is an antithesis of irrationalist and empiricist conceptions on these topics. This is related to the departure from the positivistic trend, in the spirit of which the process of acquiring new knowledge was neglected.

A similar approach is of a far-reaching importance and value for the applied sciences, such as technology and management.

B. *Some Applications*

The approach described here has a wide field of applications especially in the realms of technology, medicine, management, physical sciences and philosophy. Alongside with the already indicated applications of a similar approach, further applications of the latter in different fields and cases can be mentioned.

Let us recall that the well-known Soviet theoretical-physicist A. Z. Petrov obtained an invariant classification of *all possible* gravitational fields, based upon the algebraic analysis of the structure of the tensor of space-time curvature. This enables to outline the exact framework of the research of *every* problem of the general theory of relativity, at least by means of mathematical criteria.[34]

The possibility of alternative theories in all branches of physics is noteworthy. Earlier this possibility was thought of nearly as a curiosity. Now one begins to be aware of its fundamental significance. This can be seen for instance from the lectures of Feynman on physical laws and from his Nobel lecture.

"One of the amazing characteristics of nature", he writes, "is the variety of interpretational schemes which is possible."[35] As a matter of fact he bears in mind – and it should be said accordingly – the characteristics of the cognition of nature. In his Nobel lecture Feynman returns to these topics. Here he notes that different physical ideas can describe the same physical reality. According to him theories of the known which are described by different physical ideas may be equivalent in all their predictions. However, they are not identical when trying to move from that base into the unknown. Feynman thinks that a good theoretical physicist today might find it useful to have a wide range of physical viewpoints and mathematical expressions of the same theory available to him.[36]

Hoyle draws attention to the fact that the attempt to shorten the way by giving up the systematic investigation of the sequence of various theoretical alternatives and to choose intuitively the "most probable" one at each forking point, most frequently turns out to be wrong. For the non-investigated possibilities, taken together, contain the required answer with the greater probability (then the single investigated possibility) the more these subsequent possibilities are.[37]

A similar approach was used in the elaboration of a mathematical model of the Earth. Taking into account some parameters of our planet (mass, distribution of density, momentum of inertia, velocity of propagation of the seismic waves, etc.), the American geophysicist F. Press tested with the help of a computer five million possible models. Among them three turned out most adequate. The latter models introduce a correction in the Earth's radius, the density of the nucleus, etc.[38]

Lastly we shall note two articles concerning metatheoretical issues. For the construction of a physical theory V. Turchin suggests to go along the following lines:

(1) To determine a language for descriptions of the set of facts.

(2) To determine the set of admissible theories by introducing a universal language containing the languages of all admissible theories.

(3) To formalize the procedure by applying the theory to the facts and to introduce criteria for the 'quality' of the theory and hence

(4) To show how the best theory is to be found out.[39]

Another application concerns the possibility to construct a heuristics. For this purpose E. Alexandrov and A. Frolov compare in Table II the main mechanisms of the living organism with the sciences and scientific trends describing this system in different languages.

Thus according to these authors three dialectically related approaches toward heuristic programming are outlined, namely the *functional* approach, establishing a connection between the functions of the living organism and describing them in different languages; the *structural* approach, determining the transitions of the functional structures from a level of description to other ones (in the framework of the concrete functions of the living organism); and lastly the *applied* approach, warranting the solution of practical tasks on the basis of the singularly manifested laws.[40]

TABLE II

Main stages of elaboration of information processing in living organism	Main levels of descriptions			
	(1) physiological	(2) psychological	(3) Log. and math.	(4) technological
(1) perception	(1.1)	(1.2)	(1.3)	(1.4)
(2) thought	(2.1)	(2.2)	(2.3)	(2.4)
(3) behaviour	(3.1)	(3.2)	(3.3)	(3.4)

C. *Some Open Questions*

The method at issue is not universal, but – like nearly all scientific statements – it has a definite domain of validity, and this is very important to bear in mind. The very distinction of regular and irregular problems which can or cannot be formalized respectively, is an essential step forward in – and a prerequisite for – applying this approach. Nevertheless this approach can be successfully resorted to in several cases. It is another question whether it will be used spontaneously or deliberately, although the advantages of its deliberate use hardly need to be stressed.

We should remember that at every stage in the fps 'vacant spots' may be foreseen in a general fashion. Scientific progress leads to their being filled up and to the further modification of the fps, and thus in principle outstrips the (empirically) known design-solutions.

This general statement of the question opens up research possibilities; here, no doubt, it shows its advantage.

In connection with the procedure of reduction, the elaboration of the methodological principles for the choice of solutions as a coherent, simple and consistent system would be of great importance. It is essential to find out ways of preliminary reduction for the fps and in general for the choice of solutions without testing all design-solutions. To what extent is this feasible?

The suggested approach to problem-solving is based on a general experience with the solution of different scientific problems. Its gist consists in (1) typisation of problems, (2) divergent transition to the set or field of possible solutions, and (3) convergent transition or reduction of this set to the relevant (plausible, optimal) solutions. In this respect it is very important to pick out a group of problem-types with a definite number of solutions.

When there is a change in the available information about empirical facts or/and scientific methods, the set of possible solutions and its reduction have to be systematically reconsidered. The application of this approach reveals new possibilities over a wide range of problems in different branches especially in facing of complex and difficult problems, where in the course of solving them new possible solutions are revealed and accordingly, the choice of the relevant ones among them must be revaluated. In similar cases the advantages of the method described become obvious. Of course this, in turn, rises new problems. The point is to what extent they can be effectively solved.

Bulgarian Academy of Science, Institute of Scientific Information, Sofia

NOTES

[1] P. W. Bridgman, *Reflexions of a Physicist*, New York 1950, p. 370.
[2] A. F. Osborn, *Applied Imagination*, New York 1953, Chapters XXII–XXIV.
[3] J. Dixon, *Design Engineering*, New York 1966.
[4] E. Caude, *Comment étudier un problème*, Paris 1968, p. 91.
[5] Fr. M. Tonge, 'Summary of a Heuristic Line Balancing Procedure', in *Computers and Thought* (ed. by E. A. Feigenbaum and J. Feldman), New York 1963, p. 172.
[6] M. Minsky, 'Steps Toward Artificial Intelligence', *ibid.*, p. 408.
[7] A. Moles, *La création scientifique*, Genève 1957, Chapters IV–VI.
[8] R. Caude, A. Moles *et al.*, *Methodologie – vers une science de l'action*, Paris 1964, pp. 40, 45–46.
[9] H. Wang, 'Toward Mechanical Mathematics', *IBM Journ. of Research and Development* 4 (1960), No. 2.
[10] P. Rossi, *Clavis universalis*, Milano–Napoli 1960.
[11] F. Zwicky, *Entdecken, Erfinden, Forschen*, München–Zürich 1966.
[12] J. Müller, 'Operationen und Verfahren des problemlösenden Denkens in der konstruktiven technischen Entwicklungsarbeit', *Wiss. Z. Techn. Hochschule K. Marx-Stadt* 9 (1967), No. 5.

[13] M. Bunge, 'The Weight of Simplicity in the Construction and Assaying of Scientific Theories', in *Probability, Confirmation, and Simplicity* (ed. by M. H. Foster and M. L. Martin), New York 1966.

[14] H. A. Simon, *The Shape of Automation for Man and Management*, New York 1966, p. 54.

[15] A. Polikarov, *Relativity and Quanta*, Moscow 1966, Chapter VII (russ.).

[16] M. Bunge, *Scientific Research*, Vol. I, Berlin 1967, p. 185.

[17] G. Polya, *How to Solve It?* New York 1957, p. 154.

[18] T. W. Organ, *The Art of Critical Thinking*, Boston 1965, Chapter 2.

[19] H. Parthey and W. Wächter, in 'Problemstruktur und Problemverhalten in der wissenschaftlichen Forschung', *Rostocker philosophische Manuskripte*, Rostock 1966.

[20] F. Zwicky, l.c.

[21] R. L. Ackoff, *Scientific Method*, New York 1962, p. 24.

[22] G. I. Naan, 'The Problem of Infinity', *Voprosy Philosophii* (1965), No. 12 (russ.).

[23] A. Fine, 'Consistency, Derivability and Scientific Change', *J. Philos.* **64** (1967), 237–238.

[24] M. Strauss, 'Intertheoretische Relationen', *Deut. Z. f. Philos.* **17** (1969), 74.

[25] W. Whewell, *The Philosophy of the Inductive Sciences*, London 1847, quoted after P. Medawar, *The Art of the Soluble*, London 1967, p. 145.

[26] K. R. Popper, *Conjectures and Refutations*, London 1963.

[27] J. C. Harsanyi, 'Popper's Improbability Criterion for the Choice of Scientific Hypotheses', *Philosophy* **35** (1960), 332.

[28] P. H. Nidditch (ed.), *The Philosophy of Science*, Oxford 1968, p. 11.

[29] M. Foster and M. Martin, (See Note 13).

[30] N. Rescher, *Hypothetical Reasoning*, Amsterdam 1964, p. 21.

[31] A. Kaufmann, *The Science of Decision-Making*, New York 1968, pp. 61, 236.

[32] M. Kalinkov and A. Polikarov, 'On the Classification of Quasar's Hypotheses', *Isvestia sektiata Astronomia*, Vol. II (1967) (bulg.)

[33] A. Polikarov, 'On the Truth of the Philosophical Interpretation of Scientific Conceptions', *Comptes Rendus de l'Académie Bulgare des Sciences* **16** (1963), 569.

[34] *Machine Intelligence* (ed. by D. Michie *et al.*), Edinburgh 1967.

[35] *Artificial Intelligence*, Amsterdam, 1970.

[36] A. Newell and H. Simon, *Human Problem Solving*, Englewood Cliffs, N. J., 1972.

[37] N. Amosov, *Artificial Intellect*, Kiev 1969 (russ.).

[38] R. Banerji, *Theory of Problem Solving*, Amsterdam 1969.

[39] O. Serebrjannikov, *Heuristic Principles and Logical Calculi*, Moscow 1970 (russ.).

[40] A. Ujomov, *Analogy in the Praxis of Scientific Investigation*, Moscow 1970 (russ.).

[41] J. Slage, *Artificial Intelligence: the Heuristic Programing Approach*, New York 1971.

[42] N. Nilsson, *Problem Solving Methods in Artificial Intelligence*, New York 1971.

[43] A. Ivakhnenko, *Heuristic Self-Organizing Systems in Cybernetic Engineering*, Kiev 1971 (russ.)

[44] A. Z. Petrov, 'The Contemporary State of Development of Gravitation Theory', *Voprosy Filosofii* (1964) No. 11, 91 (russ.).

[45] R. Feynman, *The Character of Physical Law*, London 1965, p. 54.

[46] R. Feynman, 'The Development of the Space-Time View of Quantum Electrodynamics', *Physics Today* **19** (1966), 31.

[47] F. Hoyle, *Galaxies, Nuclei, and Quasars*, New York 1965, Chapter IV.

[48] F. Press, *Science News* **93** (1968) 551.

[49] V. F. Turchin, ' "Crazy" Theories and Metascience', *Voprosy Filosofii* (1968) No. 5 (russ.).
[50] E. A. Aleksandrov and A. S. Frolov, 'Heuristics is the Way to "Artificial Intellect" ', *Priroda* (1968) No. 11, 76 (russ.).

FRED SOMMERS

THE LOGICAL AND THE EXTRA-LOGICAL

> Thomas Hobbes, everywhere a profound
> examiner of principles, rightly stated that
> everything done by the mind is a *computa-
> tion*, by which is understood either the ad-
> dition of a sum or the subtraction of a
> difference. (*De Corpore* I.i.2). So just as
> there are two primary signs of algebra and
> analytics, $+$ and $-$, in the same way there
> are, as it were, two copulas 'is' and 'is not'.
>
> Leibniz

> If, as I hope, I can conceive all propositions
> as terms, and hypotheticals as categoricals,
> and if I can treat all propositions univers-
> ally, this promises a wonderful ease in my
> symbolism and analysis of concepts, and
> will be a discovery of the greatest impor-
> tance.
>
> Leibniz

1.   Medieval logicians called words like 'every', 'not', 'is' and 'some'
"syncategorematic." Today, such words – along with propositional constants such as 'and', 'if...then' and 'or' – are referred to as logical or
formative signs in contrast to extra-logical or descriptive signs.

The older label is essentially negative: the syncategorematic elements
of a categorical proposition are not its terms (predicates or names). A
negative characterization has the virtue of not pretending to define a concept governing the list of non-descriptive expressions. Contemporary
labels sound more positive but in fact there is no accepted principle for
the distinction between formatives and descriptive elements. What counts
as a logical sign remains arbitrary. Logicians will agree that 'is identical
with' ought to be included and that 'is faster then' ought to be excluded.
But they will hesitate over 'is greater than' or 'is between'.

2.   The question I wish to consider is: What is a formative element? To

Boston Studies in the Philosophy of Science, XIV. All Rights Reserved.

say that an expression is formative because it is not descriptive or not extra-logical is to appeal to the distinction that wants illumination. The appeal is not more illuminating than explaining the formative character of 'not' or 'some' by referring to its formative powers.

2.1.   Because formatives lack effective characterization, logicians usually give a list of them along with the stipulation that the list is exhaustive. This procedure works well enough. The number of formatives needed for adequate systems of logic is finite and small. Nevertheless this way of introducing the logical signs of a logically adequate language is defective in principle. One consequence is that the notion of logical necessity is made relative to initial and arbitrary stipulation. A related defect is discussed by Tarski who uses the formative-descriptive distinction to construct a definition of logical consequence:

> Underlying our whole construction is the division of terms of the language discussed into logical and extra-logical.... No objective grounds are known to me which permit us to draw a sharp boundary between these two groups of terms.... If for example we were to include among the extra-logical signs, the implication sign, then our definition of the concept of consequence would lead to results which obviously contradict ordinary usage.... It seems possible to include among logical terms some which are usually regarded by logicians as extra-logical without running into consequences which stand in sharp contrast to ordinary usage.
>
> LSM 419

Tarski suggests that the problem of characterizing the formatives may have no solution.[1] And, indeed, the attempt to find some recognizable feature possessed by all and only those signs we intuitively consider to be logical appears hopeless. What – apart from their "formative powers" – could signs as different as 'not' and 'every' have in common? There is no question that these signs are essentially logical; on the other hand, once we renounce this unexceptionable truism as an adequate basis for the distinction between formative and descriptive elements we intuit nothing that can be used for recognizing the formatives and distinguishing them from descriptive or extra-logical elements.

2.2.   The next sections will develop a simple and uniform account of the formatives that sharply distinguishes formative and descriptive elements. The discovery that the formatives can be defined by a common characteristic comes as something of a surprise. The account, which could not be anticipated, is the result of an investigation into the logical relations obtaining among the logical signs.

3.   Logicians properly distinguish two kinds of logical signs: those like 'and', 'if...then' and 'or' which function as propositional connectives belong in one group. We shall refer to these as propositional formatives. Formatives like 'is', 'some' and 'not' function inside the categorical proposition; they form a second group – the traditional syncategorematic elements.[2] We shall take up the syncategorematic formatives first beginning with some that clearly do have something in common.

3.1.   The words 'no', 'not', 'isn't' (and other forms of the negative copula), the prefix 'un-' (and other negative signs affixed to terms) are all negative elements. Of these negative formatives only 'isn't' has an explicit positive sign opposed to it. The formatives 'is' and 'isn't' are *opposed* elements. In the case of 'no', 'not', and 'un-' the opposed positive elements are tacit. The situation with negative and positive formatives is familiar in arithmetic. The numeral for a negative number has the form '$-x$' but the numeral for the corresponding positive is rarely given the form '$+x$'. For the opposition between positive and negative numbers we use the forms 'x' and '$-x$' thus marking the negative number by a sign and the positive one by the absence of a sign. And generally the opposition between positive and negative elements is marked by one sign; the lack of an explicit positive sign serves to identify the positive element. In language, for example, the word 'unwise' contains a negative formative analogous to the minus-sign in '-7'. The lack of an explicit positive prefix in 'wise' is a *tacit* positive element. But in arithmetic an explicit positive sign is available; we can, if we wish, write '$+7$' instead of '7'. In natural languages a positive sign corresponding to 'un-' is unavailable.

3.2.   At least some formatives then are *signs of opposition*. All negative formatives are explicit, most positive formatives are tacit. But it does not appear we can extend this characterization to all the formatives. For example, the pair 'some' and 'every' are in some sense contrasted but we do not think them opposed as positive and negative signs. The idea that all formatives consist of positive and negative signs seems even less

plausible when applied to the propositional constants. We do not think of
'if...then' or 'and' as opposed to some other logical element.

3.3.   These reservations seem to count very seriously against the suggestion that all formative signs are oppositional, analogous to plus and
minus signs in arithmetic. Nevertheless some formatives are quite clearly
oppositional and the suggestion is attractive enough to deserve further
consideration. We shall at first confine ourselves to propositions containing only formatives that are intuitively oppositional. We begin with
four propositions – equivalent to a traditional AEIO schedule. All formatives in the schedule are uncontroversially oppositional.

    I.A (1) No scientists were inattentive
    E   (2) No scientists were attentive
    I   (3) Scientists were attentive
    O   (4) Scientists were inattentive

(1) and (2) contradict (4) and (3) and the latter are weakly interpreted to
claim that scientists – some, anyway – were (in)attentive. In (3) the positive predicate 'were attentive' is affirmed and the claim is made that this
predicate is true of (some) scientists. In (2) the predicate is denied and
the claim is made that 'were attentive' is not true of (any) scientists. A
similar analysis holds for (4) and (1) where the negative predicate 'were
inattentive' is being affirmed (4) and denied (1).
    Two propositions equivalent to (1) and (2) are:

    (1.)   Not: Some scientists were inattentive
    (2.)   Not: Some scientists were attentive

I have avoided these formulations because they contain 'some', a formative which we do not intuitively think of as either positive or negative. In
contrast, all explicit formatives of Schedule I are positive or negative.
'No' and 'in-' are clearly negative. And 'were' is a positive copula. If we
represent positive formatives by a plus-sign and negative formatives by
a minus-sign our four propositions take the following forms:

    Ia. (1a) $-(S+(-A))$
        (2a) $-(S+A)$
        (3a) $S+A$
        (4a) $S+(-A)$

In (1a) and (2a) the scope of the initial minus-sign comprehends both the subject and the predicate. The minus-sign signifies that the predicate is denied of the subject. In (3a) and (4a) the positive and negative predicates are affirmed but there is no explicit sign of affirmation corresponding to the word 'no'. Also there is no positive prefix corresponding to the 'in'-of 'inattentive' or the 'non-' of 'non-scientists'. Using arithmetic notation we could make good these natural omissions by representing every tacit positive formative by a plus-sign. Schedule I would then look like this:

Ib. (1b) $-((+S)+(-A))$
(2b) $-((+S)+A))$
(3b) $+((+S)+(+A))$
(4b) $+((+S)+(-A))$

Ib is the more explicit schedule. But the propositions of Ia are direct transcriptions of the natural language originals economizing on positive signs in a natural and harmless way. Note also that Ia corresponds to arithmetical notational practice. We shall for convenience make use of abbreviated natural formulations like those of Ia. But we shall keep in mind that they are abbreviations of more explicitly formulated propositions like those of Ib.

3.4.   The formative 'not' is negative. It does not appear in I. However we notice that (1) is equivalent to

(1.)   Not: Some scientists were inattentive

The equivalence

No scientists were inattentive ≡ Not: Some scientists were inattentive could be represented as the equation
$$-(S+(-A)) = -(+S+(-A))$$

at the price of arbitrarily representing 'some' by a plus-sign. The assignment of 'plus' to 'some' seems harmless until we observe that 'some' and 'every' are now opposed. If 'some' is positive, 'every' must be treated as negative. Assigning 'minus' to 'every' may seem contrary to intuition. This isn't fatal. We can hardly hope for a uniform account of the formatives that is initially supported by intuition. Nevertheless the assigning of 'minus' to 'every' could only be justified if the negative quality of 'every' is a requirement of its logical relation to the other syncategorematic formatives. To test the negative assignment we consider a number of equivalences:

(a)   No S were non-A ≡ Every S was A
(b)   Not: Some S were A ≡ Every S was non-A
(c)   No S were A ≡ Every S was non-A
(d)   Every S was A ≡ Every non-A was non-S
(e)   Not: every S was A ≡ Some S were non-A

If we represent 'every' by *minus* and 'was' by *plus* we get

(a) $-(S+(-A)) = -S+A$
(b) $-(+S+A) = -S+(-A)$
(c) $-(S+A)   = -S+(-A)$
(d) $-S+A   = -(-A)+(-S)$
(e) $-(-S+A) = +S+(-A)$

All the equations representing the logically equivalent propositions are correct; the negative representation for 'every' thus checks out consistently. In addition the oppositional notation has provided us with a neat arithmetical method for testing the validity of immediate inferences from one categorical proposition to another.

Let us say that an immediate inference is "regular" if both its premise and conclusion are universal or if both are particular. If one is universal and the other is particular, we call the inference "irregular." Only regular inferences are valid and only those whose equations are correct. And a regular inference whose equation is true is valid. To test any immediate inference, 'P, hence C', we first see whether it is regular. If it isn't, the inference is invalid. If it is, we check whether '$P = C$' is a true equation. If it is, the inference is valid; if not, it is invalid.

Our investigation into the nature of formatives has come up with at least a partial answer. All the traditional syncategorematic signs are oppositional. Confining ourselves to the *explicit* non-propositional formatives we have found that the positive copula and the word 'some' are positive. All other explicit signs are negative.

If we make all implicit syncategorematic signs *explicit*, the general form of categorical proposition is:

$$\pm(\pm(\pm S)\pm(\pm P))$$

The left-most signs of opposition are signs of affirmation $(+)$ or denial $(-)$. The next opposition is that of 'some' $(+)$ and 'every' $(-)$. The signs prefixing S and P indicate whether these terms are positive or negative. The fourth opposition represents the positive and negative copula. With these assignments logical reckoning coincides with arithmetic reckoning: an inference is valid only if its equation is true.

3.5.   The view that the syncategorematic formatives are "oppositional" receives further confirmation from syllogistic logic.

A regular syllogism[3] is valid if and only if the *sum* of its premises is equal to the conclusion.

Thus let $P_1...P_n$ be the premises and C be the conclusion of a regular syllogism. Then '$P_1...P_n$, hence C' is valid if and only if '$P_1+...P_n=C$' is true.

An example from Lewis Carroll illustrates how we may use oppositional notation to calculate the conclusion from a set of premises.

(1)   Colored flowers are always scented
(2)   I dislike flowers not grown in the open air
(3)   No flowers grown in the open air are colorless

We reformulate these premises, transcribe them arithmetically, and add them up:

(1)	$-C+S$	Every colored flower is scented
(2)	$-(-G)-L$	Every flower not grown in the open air isn't liked by me
(3)	$-(G+(-C))$	No flowers grown are colorless
	$=-L+S$	i.e., Every flower liked by me is scented

The oppositional thesis is also confirmed by relational inference. Some examples are:

(i) Some laws fail-to-be-applicable to all men, hence some laws aren't applicable to some men.
$+L+(-A)-M=+L-(A+M)$
(ii) Every censor witholds some books from every minor. Every book is printed matter. Some minors are females
$\therefore$ Every censor witholds some printed matter from some females

We transcribe the premises and add them up to see whether the conclusion follows.

$$- C + W + B - M$$
$$- B + PM$$
$$+ M + F$$
$$= - C + W + PM + F \qquad \text{Q.E.D.}$$

4. We have argued that the syncategorematic formatives are oppositional signs. When these formatives are treated as positive or negative elements of categorical propositions the logical structure of such propositions is seen to be analogous to simple arithmetic expressions. An arithmetical notation for the oppositional signs is an effective logical instrument useful for testing inferences. As such it illuminates validity and logical truth. It also illuminates the nature of tautology and contradiction.

4.1. Let S be *scientists* and P be *Persians*. The valid inference

(1)   Some S are P hence some P are S corresponds to the tautology
(1T)   Every state of affairs in which some S are P is a state of affairs in which some P are S.

The phrase 'state of affairs in which' will henceforth be abbreviated by square brackets. (1T) may then be written

$$- [+ S + P] + [+ P + S]$$

(1T)   is very like the contradictory proposition:
(1C)   Some state of affairs in which some S is P is not a state of affairs in which some P is S.

This is represented as

$$+ [+ S + P] - [+ P + S]$$

Arithmetically (1T) and (1C) add up to zero. Logicians have remarked that tautologies and contradictions say nothing. In oppositional notation this is literally true.

To make clear the relation between a valid inference and the corresponding tautology and contradiction we generalize the above example. Let P be the premise or premises of a valid inference and let C be the conclusion. Then the true equation '$P = C$' can be read 'P, hence C'. Since '$P = C$' is true the equations

$$- [P] + [C] = 0$$
$$+ [P] - [C] = 0$$

will both be true. The left side of the first equation is a *universal* null proposition – a tautology. The left side of the second equation is a *particular* null proposition – a contradiction. Let us say that P and C are "homoscopic" when both are universal or both are particular. When one is universal and the other particular we call them "heteroscopic." For homoscopic P and C, a universal proposition, $\pm [P] \pm [C]$, that says nothing is a tautology. A particular proposition that says nothing is a contradiction.

5. We have yet to consider the propositional formatives. The oppositional interpretation which works so well for the syncategorematic formatives does not seem applicable to 'and', 'or', or 'if... then'. Propositions propositions appear to have a structure that is radically different from the structure of a categorical proposition. For example the conditional proposition:

(V)   If all humans are mortal, then some animals are vulnerable

has no subject and no predicate though its component propositions are of subject predicate form. The connective 'if... then' is not thought of as opposed. The whole proposition is neither universal nor particular, neither affirmative nor negative in 'quality'.

Most logicians before Frege were of the view that all propositions – including compound propositions – could be interpreted as categorical subject-predicate propositions. Modern logicians have rejected this traditional program. Moreover the tendency now is to reverse the traditional analysis: instead of analyzing compound functions as categorical propositions of subject-predicate form, the modern logician "decomposes" the subject-predicate relation of the categorical proposition into function and argument. For example, the function 'if A_x then B_x' is used as an element in decomposing the categorical form, 'Every A is B'.

Leaving historical considerations to the side, we may recognize that the classical or traditional doctrine is semantically simple and attractive. In-

terpreting all propositions as predications, it views them all as saying
something about something. For example, a conditional proposition, on
interpretation, is really indicative and this must become evident when we
analyze it as a subject-predicate categorical proposition. Oddly enough we
can find this sort of interpretation of the conditional in Frege. In his paper
'On Sense and Reference' Frege says the following of a statement of form
'if p then q':

A relation between the truth values of the conditional and dependent clauses has been
asserted, viz. such that the case does not occur in which the antecedent stands for the
True and the consequent for the False.

According to this, 'if p then q' *indicatively* asserts that no case of p standing
for the True is a case of q standing for the False.

We may also adopt an indicative interpretation of 'if p then q' without
committing ourselves to Frege's doctrine that every proposition stands
for the True or the False. Tarski, Austin and other philosophers favor a
different account of the relation of a proposition to its object: to every
proposition there corresponds a state of affairs. The proposition p corresponds to the $[p]$ – the state of affairs in which p. If we substitute '$[p]$'
for Frege's 'case of p standing for the True', we may regard 'if p then q'
as asserting that no $[p]$ fails to be a $[q]$ or, equivalently, as asserting that
every $[p]$ is a $[q]$.

A proposition about states of affairs will be called a state proposition.
We call the state proposition 'Every $[p]$ is $[q]$' a categorical transform of
'if p then q'. According to classical doctrine every proposition says something about something and compound propositions say something about
In particular, a proposition of form 'if p then q' says of every $[p]$ that it
is a $[q]$.

Every compound proposition has its categorical transform. The following table correlates conditionals, conjunctions, and disjunctions with their
respective 'state' categoricals.

if p then q	$-[p]+[q]$
p and q	$+[p]+[q]$
p or q = if not p then q	$-[-p]+[q]$

The truth function 'not' belongs among the syncategorematic expressions.
We have seen that these expressions present no difficulty for the thesis
that formatives are positive or negative signs. And now we see that the

oppositional thesis can be extended to the propositional formatives 'if...
then' and 'and'. The difference is that these binary formatives are represented by a *pair* of oppositional signs in a "state" proposition. Thus
'if... then' is the ordered sequence '[···]+[···]' and 'and' is '+[···]+[···]'.
The formative 'or' is defined by the equivalence of '*p* or *q*' to 'if not *p* then
q'. The further function 'not' has already been represented as a syncategorematic formative so that the definition of 'or' in terms of 'not' and
'if... then' introduces no new elements. Applying the table of transformation we can get a categorical transform of any truth function. For
example the fairly complicated formula

$$((pp \supset q) \lor s \cdot r) \supset \sim t$$

is arithmetically transcribed as the subject-predicate state proposition

$$- [- [- (- [p] + [q])] + [+ [ss] + [r]]] + [- t].$$

The systematic correspondence between compound forms and their categorical transforms is fundamental for logical syntax. The correspondence
shows how to represent the propositional formatives as ordered sequences
of signs wholly isomorphic to sequences of syncategorematic formatives
of subject-predicate propositions. This reveals that every molecular expression containing 'and', 'or', or 'if...then' has the *form* of a subject-
predicate proposition. Indeed, we could get along without propositional
formatives altogether if we were willing to reformulate all compound
propositions as "state" propositions. But even if we do not do this the
isomorphisms between compounds and categoricals remain. The following formulas exhibit these isomorphisms for compound terms as well as
for compound propositions. In each case the vernacular is followed by
its arithmetized transcription.

Molecular Form		Subject-predicate State Proposition
and	$\begin{cases} x \text{ is } A \text{ and } B \\ x + \langle + A + B \rangle \end{cases}$	Some [x is A] is [x is B] $+ [x + A] + [x + B]$
and	$\begin{cases} p \text{ and } q \\ + p + q \end{cases}$	Some [p] is a [q] $+ [p] + [q]$
or	$\begin{cases} x \text{ is } A \text{ or } B \\ x + \langle - (- A) - (- B) \rangle \end{cases}$	Every [x is not A] is not [x is not B] $- [x - A] - [x - B]$

or	$\begin{cases} p \text{ or } q \\ -(-p)-(-q) \end{cases}$	Every [not-p] is not [not-q] $-[-p]-[-q]$
if ... then	$\begin{cases} \text{if } p \text{ then } q \\ -p + q \end{cases}$	Every [p] is a [q] $-[p]+[q]$

5.1.   The difference between the syncategorematic formatives and the propositional formatives is now clear: A syncategorematic formative is a single sign of opposition, a propositional formative is an ordered sequence of oppositional/signs. This complexity of the propositional formatives accounts for the fact that we do not intuitively think of them as either positive or negative in the way we naturally do with most of the syncategorematic formative signs. The extension of the 'oppositional' thesis to propositional formatives accepts the traditional view of compound propositions as having the *form* of subject-predicate propositions. Leibniz realized that such a uniform analysis of compound propositions must simplify logical syntax.

In the next section we bring additional support to this traditional analysis of compound propositions by showing that it is logical efficient.

6.   We have seen that it is possible to express any function of two or more propositions as a function whose logical form is isomorphic to that of a categorical subject-predicate proposition. But a doctrine of logical form that does not facilitate logical reckoning cannot be taken seriously. For this reason our argument is not complete unless we can show how categorical forms of compound propositions can be used in inference. If the use of categorical form made reckoning difficult or impossible, the classical doctrine of logical form could hardly be acceptable. We showed earlier that the oppositional interpretation of the syncategorematic formatives is logically efficient. We must show this now for the propositional formatives.

6.1.   If p is a proposition the expression '[p]' is a term that applies to the state of affairs in which S. We refer to '[p]' as a state term. For convenience we can omit the square brackets and read 'p' as 'state of affairs in which p'. This notational license permits us to formulate state propositions with fewer brackets. Where no misunderstanding is possible we shall avail ourselves of this liberty, thus ' $-[p]+[q]$ ' will be written

'$-p+q$' with the understanding that 'p' and 'q' are state terms. Also, *anyone who reads '$-p+q$' as the compound proposition 'if p then q' will not go astray.*

If p is a state term then the state proposition '$-p+p$' is a *basic* tautology. Equally, if p is a proposition, the conditional proposition '$-p+p$' is a basic tautology. If BT is a basic tautology, then any proposition of form $-p+$BT is a *general* tautology. All tautologies are either basic or general. A basic tautology is a null proposition. A general tautology need not be null but its predicate term must be null. The following principles hold:

(1) Rule of Iteration   Standard Form
$$p = +p+p; p = --p--p \qquad p = p{\cdot}p = p \vee p$$

(2) Laws of Association
$$+p+[+q+r] = +[p+q]+r \qquad p{\cdot}(q{\cdot}r) = (p{\cdot}q){\cdot}r$$
$$--p----q--r = \qquad p \vee (q \vee r) =$$
$$=---p--q--r \qquad = (p \vee q) \vee r$$

(3) Laws of Commutation
$$+p+q = +q+p \qquad p{\cdot}q = q{\cdot}p$$
$$--p--q = --q--p \qquad p \vee q = q \vee p$$

(4) Laws of Distribution
$$+[--p--p]+[--q--r] = \quad (p \vee p)(q \vee r) =$$
$$=--[+p+q]--[+p+r] \quad = (p{\cdot}q) \vee (p{\cdot}r)$$
$$--[+p+p]--[+q=r] = \quad (p{\cdot}p) \vee (q{\cdot}r) =$$
$$= +[--p--q]+[--p--r] \quad = (p \vee q){\cdot}(p \vee r)$$

Because molecular expressions have the same logical forms as state categorical propositions, the aforementioned laws hold equally for state propositions, for compound propositions, and for compound terms. This may be seen from another standpoint. On looking over the traditional formatives we noted that 'is' is a positive sign while 'isn't', 'not', 'no' and 'un-' are negative. Beginning with this information we got an assigmnent for 'every' via the equivalence.

$$x \text{ is un-}y = \text{every } x \text{ is } y$$

The left side contains only formatives whose positive and negative quality is intuitively given. Since

$$- (x + (- y)) = - x + y$$

we discovered that 'every…is…' is represented by '$- \dots + \dots$' and, in particular, that 'every' is a negative sign.

The same method may be applied to get an oppositional representation for the propositional formatives. Looking over the propositional formatives we note that 'and' is often used as a synonym for 'plus' in the sense of addition. We therefore represent 'p and q' as '$p + q$'. We then take the equivalence not (p and not $-q$)=if p then q whose left side contains only formatives whose oppositional representation is intuitively given. Now we note that

$$- (p + (- q)) = - p + q$$

and so we discover that 'if…then…' is the ordered pair '$- \dots + \dots$'.[4]

The isomorphism of 'and' and 'some…is…' and of 'if…then' and 'every…is…' is reflected in modern predicate logic which uses the conjunctive form '$Ax \cdot Bx$' for "translating" 'Some A is B' and the conditional form '$Ax \rightarrow Bx$' for translating 'Every A is B.' More recently the isomorphism has figured in deep structure analysis which finds, for example, that the particular categorical sentence 'A man is wise' underlies the surface (conjunctive) expression 'a wise man' as it occurs in 'a wise man is honest.' The expression 'a wise man' is logically compound (wise and a man) and the sentence 'a man is wise' is logically particular. Transcribed in oppositional notation both expressions have the form '$+ x + y$'. By the same token when the logical structure is clearly revealed in a felicitous representation of the logical syntax of natural language the need for an analysis which treats one expression as deeper than the other is dubious.

With the exception of the notational rule (1), the four formulas preserve the analogy to arithmetic. Before proceeding to apply them we shall add to our notational license.

A function like 'p and (q and r)' is transcribed as '$+ p + [+ q + r]$'. This formulation still has the form of a categorical proposition. Given some license we can write this as '$+ p + q + r$' with the understanding that the terms of this formula are associated to form a two term (state) proposition. Strictly speaking the unbracketed formulation is ill-formed. So, too, speaking strictly, is the function 'p and q and r'. But this sort of abbreviation is common, useful, and – with proper care – harmless.

We may apply the above principles to any state or compound proposition in oder to determine whether it is tautological, contradictory, or neither. If a proposition is tautological it will either be found equivalent to a proposition of form $-S + S$ or to a proposition of form $-p + [-S + S]$. If it is contradictory it will be equivalent to the denial of a tantology. Thus if Q is tautological, $-Q$ will be contradictory. If neither is the case the proposition is contingent. Similarly the arithmetic notation can be used for testing the validity of propositional inference. For example, given the premises 'p' and 'if p then q' we may validly infer q. Using transforms in arithmetic notation we have:

$$p$$
$$-p + q$$
$$\therefore q$$

The corresponding tautology is '$p(p \supset q) \supset q$'.
We have:

$$- [+ p + [- p + q]] + q$$
$$= - - q - [+ p + [- p + q]] \text{ (Immediate Inference)}$$
$$= - - q - p - [- p + q] \text{ I.I., Assoc.}$$
$$= - [+ (- q) + p] - [- p + q] \text{ I.I.}$$
$$= - [+ (- q) + p] - [- (- q) - p] \text{ I.I.}$$
$$= - [+ (- q) + p] + [+ (- q) + p] \text{ B.T. a basic}$$
tautology.

A second example: Show that '$p \supset q \lor p$' is tautological. We have:

$$- p + [- - q - - p]$$
$$= - - (- p) - - [- - q - - p] \text{ I.I.}$$
$$= - - q - - p - - (- p) \text{ I.I., Assoc.}$$
$$= - (- q) + [- (- p) + (- p)] \text{ G.T.}$$

The last formulation is a general tautology.

These examples illustrate how to do propositional logic with oppositional formulas. The possibility of an 'oppositional calculus' vindicates our thesis that all the formatives – including proposition 'constants' – are analogous to plus and minus signs of arithmetic. The thesis applies directly to the syncategorematic formatives. Propositional formatives are represented by more than one syncategorematic sign in a proposition of categorical form. The thesis simplifies logical syntax: All propositions consist of descriptive signs and signs of opposition. In the concluding section we shall lightly touch on one or two topics affected by this account of logical form.

7. The distinction between logical and descriptive elements has never been seriously in question. But it is fair to say that logicians have not been clear about the nature of logical particles or "formatives". Moreover, it has been assumed that no common syntactical feature distinguishes the formatives from extra-logical elements. This negative assumption has appeared reasonable on considering such seemingly disparate elements as 'is', 'some', 'or', and 'no'. Nevertheless the idea that the logical elements are essentially disparate is formally unacceptable. Nor are we forced to accept it. A uniform account of logical signs was offered along with some indication of its logical power.

Any such account of the formatives bears on a variety of important matters in logical theory and the theory of syntax. We have discussed validity and tautology. The account given illuminates the notion of logical truth. But any useful interpretation of the formatives is also a theory of logical syntax.

In this connection it is worth saying a word about the recently much discussed idea of a universal grammar. There is a sense in which a universal grammar is nothing more and nothing less than the logical grammar inherent in the subject-predicate structure of all natural languages. In this admittedly narrow sense the recent claim of linguists on universal grammar must be rejected. For linguistics is or ought to be empirical science and its theories are or ought to be empirical theories. But the logical syntax of natural language belongs to no empirical science. Its universality is indeed innate in the way arithmetical or logical laws are innate and free of empirical confirmation.

That human language is logically structured is not an exciting new idea recently rediscovered by linguists. On this score, the current polemic against empiricists philosophies of language is surely misleading. With the possible exception of Mill, no empiricist denies that logical constraints are universal and even innate. Indeed the distinction between empirical and logical truth is an empiricist emphasis. Moreover, the idea that natural syntax is logically constrained is the central thesis of Aristotle's Organon. And the more specific suggestion that the logical syntax of natural languages is closely analogous to arithmetic goes back at least to Leibniz. In any case, untill modern linguists carefully disinguish between logical and empirical universality, the *linguistic* thesis on universal grammar, along with the innateness claim, must be accounted as too confused to be open to serious criticism.

Concluding Remark. Russell was of the opinion that logical discovery contributed to our understanding of the real structure often obscured by the apparent grammatical structure of sentences in natural languages. In his view, the syntax of logically naive grammarians required correction by an adequate theory of logical form. Russell saw logical syntax as a universal constraint on empirical grammar. He would probably have objected to assimilating the theory of logical form to any empirical science. The present paper is in essential agreement with the modest but importantly clear program implicit in Russell's view of the relation of logical syntax to empirical linguistics.

Brandeis University

NOTES

[1] Thus Tarski says:
Perhaps it will be possible to find important objective arguments which enable us to justify the traditional boundary between logical and extra-logical expressions. But I also consider it quite possible that investigation will bring no positive results in this direction so that we shall be compelled to regard such concepts as 'logical consequence' 'analytic statement' and 'tautology' as relative concepts which must on each occasion be related to a definite, although, in greater or less degree, arbitrary, division of terms into logical and extra-logical.

LSM 420

[2] The list of syncategorematic formatives consists of those elements that determine the logical form of a categorematical statement. The list consists of 'every', 'some', 'no', 'is', 'isn't' and 'un-' and other elements such as 'only', definable by these. (Only x is $y = $ No un-x is y.)

[3] I call a syllogism regular if it satisfies the rule of quantity. Thus a syllogism is regular if
   (a) its conclusion is universal and all premises are universal
   (b) its conclusion is particular and one (but no more than one) premise is particular.
Only regular syllogisms are valid.

[4] There is a direct argument for representing 'if ... then ...' by '$- \cdots + \cdots$'. the words 'and' and 'plus' are often synonyms. We therefore represent 'p and q' by '$p + q$' and 'not: p and not $-q$' by '$-[p + (-q)]$'. This latter form is logically equivalent to 'if p then q'. But it is arithmetically equal to '$-p + q$' which justifies our representing 'if ... then ...' by the ordered pair '$- \cdots + \cdots$'.

ERNEST SOSA

WHAT IS A LOGICAL CONSTANT?

The main question discussed by Sommers – what is a logical constant? – leads him to develop what he considers a simple and uniform account of the logical constants, one that sharply distinguishes logical and descriptive elements of sentences.

The first group of formatives, or logical constants, to be considered are the syncategorematic formatives, i.e., the logical constants of syllogistic logic. Consider the traditional AEIO schedule:

A	Every S is P
E	No S is P
I	Some S is P
O	Some S is not P.

We are offered the following as correlates of these:

A	$+(-(+S)+(+P))$
E	$-(+(+S)+(+P))$
I	$+(+(+S)+(+P))$
O	$+(+(+S)-(+P))$.

The left-most signs of opposition are signs of affirmation $(+)$ or denial $(-)$. The next opposition is that of 'some' $(+)$ and 'every' $(-)$. The signs prefixing S and P indicate whether these terms are positive or negative. The fourth opposition represents the positive and negative copula. With these assignments logical reasoning coincides with arithmetic reasoning: an inference is valid only if its equation is true.[1] (The equation of an inference has on one side the conclusion and on the other the sum of the premises.)

One might be tempted to directly convert the foregoing suggestion into a full-fledged arithmetical criterion of validity. But an obvious counterexample is provided by the fact that 'Every S is P' does not entail 'Some P is non-S' whereas the corresponding equation is indeed true:

$$+(-(+S)+(+P))=-S+P=+(+(+P)+(-S)).$$

Presumably in order to avoid such counterexamples, Sommers restricts the applicability of his principle to *regular* inferences, where an inference

254

(syllogistic or immediate) is regular if and only if (a) the conclusion and all premises are universal or (b) the conclusion and exactly one premise are particular. *A regular inference is valid if and only if its equation is true.*[2]

This account is then extended to cover propositional formatives by dint of further ingenious correlations and by use of such devices as the operation of bracketing a sentence (which produces a term that is true of states of affairs where the sentence holds true). Without suggesting that this extension is trouble-free, I will not try to follow Sommers into that question. What is already before us is debatable enough for the present occasion.

Consider the following passages from Tarski (quoted by Sommers):

> Underlying our whole construction is the division of terms of the language discussed into logical and extra-logical... No objective grounds are known to me which permit us to draw a sharp boundary between these two groups of terms... It seems possible to include among logical terms some which are usually regarded by logicians as extra-logical without running into consequences which stand in sharp contrast to ordinary usage...[3]
>
> Perhaps it will be possible to find important objective arguments which enable us to justify the traditional boundary between logical and extra-logical expressions. But I also consider it quite possible that investigation will bring no positive results in this direction...[4]

Sommers considers his investigation to have indeed brought positive results in this direction. Syncategorematic logical terms or formatives are oppositional, i.e., are the terms which can be correlated with + and − in the way indicated above so as to yield an arithmetical criterion of validity. (The account must be complicated to deal with the propositional formatives, which are correlated not with single oppositional signs, + and −, but with sequences of oppositional signs.)

The main thesis before us appears therefore to be this: All and only formatives can be correlated with + and − in the way shown by Sommers so as to yield arithmetical criteria of validity.[5]

There are questions of detail that one might raise about Sommers' criteria of validity. For instance, they appear incapable of accounting for the validity of the inference from (a) Somebody likes everybody ($+B+L-B$) to (b) Somebody likes somebody ($+B+L+B$).[6] In addition, it is hard to see how they can account for cases of inconvertible immediate implication, such as that of (b) by (a). For arithmetic equality *is* of course convertible: if the premise is equal to the conclusion, then the conclusion is equal to the premise.

Let us put aside such questions of detail, however, in order to concentrate on a more basic question that will come into prominence gradually as follows.

Let us begin with some definitions: A *temperature proposition either* (a) predicates 'is hot' or 'is cold 'with respect to some singular term, *or* (b) is the negation of such a predication, or the negation of such a negation, etc. A *temperature inference* contains a premise and a conclusion, each of which must be a temperature proposition. A temperature inference is *regular* iff (a) both premise and conclusion are positive (i.e., each has an even number of negation signs) or (b) the conclusion is negative (i.e., has an odd number of negations signs). Let us correlate 'It is not the case that' with ' − ', and let us correlate the predicate 'is hot' with ' + ' and the predicate 'is cold 'with ' − '. The *arithmetical correlate* of a regular temperature inference is an equation obtained by (a) replacing the elements of the inference that have arithmetical correlates with their correlates, and (b) letting the resulting correlate of the premise stand on one side of the equality sign and the resulting correlate of the conclusion stand on the other side of the equality sign. Finally, we can accept the following arithmetical criterion of validity for regular temperature inferences: *A regular temperature inference is valid if and only if its arithmetical correlate is true.*

Thus consider the following inference:

The motor is hot
_____
Therefore, the motor is not cold.

Its validity is shown by the truth of the following equation:

$$+ \text{(the motor)} = - (- \text{(the motor)}).$$

According to Sommers' account, therefore, 'is cold' and 'is hot' will count as logical constants.

It may be objected that the discussion of simple temperature inferences is implausibly *ad hoc*. The applicability of the arithmetical criterion may be thought to be implausibly restricted to certain special types of inference involving the terms 'hot' and 'cold'. However, I can detect no difference in kind here between our regular temperature inferences and Sommers' regular immediate inferences. If either arithmetical criterion is implausibly *ad hoc*, then surely they both are.

It may be objected that the criterion for temperature inferences fails to cover valid inferences covered by Sommers' criterion. But the converse is equally true. Moreover, the criterion for temperature inferences could easily be extended to cover weight (heavy – light), height (tall – short), color (black – white), and any other of the many types of polar opposition.

Concerning Sommers' arithmetical criteria of validity, I have argued that they do not work since (a) there are valid inferences that the criteria would pronounce invalid, and since (b) the criteria seem unable to account for any valid but inconvertible immediate inferences. Even if such arithmetical criteria did work, however, we still could not use them to obtain the account of the logical constants proposed by Sommers, or so I have argued. My main reason is that it is possible to correlate the polar opposites 'hot' and 'cold' with '$+$' and '$-$', and to define a notion of regular temperature inference for which one can then formulate an arithmetical criterion of validity quite parallel to Sommers' arithmetical criterion for his regular syllogistic inferences. Therefore, if the syllogistic arithmetical criterion shows the oppositional words 'all' and 'some' to be logical constants, then equally the temperature arithmetic criterion shows the oppositional words 'hot' and 'cold' to be logical constants.[7]

Brown University and
The University of Michigan

NOTES

[1] Fred Sommers, 'The Logical and the Extra-Logical', this volume, p. 241.
[2] *Ibid.*, p. 240.
[3] Alfred Tarski, *Logic, Semantics, Meta-Mathematics*, Oxford University Press, London, 1956, pp. 418–419.
[4] *Ibid.*, p. 420.
[5] Thus Sommers tells us that he has provided a "...uniform account of logical signs" as oppositional. Only by advancing the biconditional ("All *and only*..."), moreover, does he answer Tarski. There are many properties that formatives have in common (e.g., they are all symbols). What Tarski wants is an interesting property or combination of properties which is not only necessary for being a logical constant (formative), but also sufficient.
[6] It is clear that Sommers wishes his criteria to cover relational inference. An example he provides in favor of his thesis is the following: Some laws fail-to-be-applicable to all men, hence some laws aren't applicable to some men. ($+L+(-A)-M=+L-(A+M)$).
[7] Thanks to Jaegwon Kim and Stephen Leeds for helpful comments.

GOTTLOB FREGE

ON THE LAW OF INERTIA*

*Translation and notes by Howard Jackson and Edwin Levy** [1]*

It will undoubtedly seem strange to many that a law as long regarded as unquestionable as that of inertia should again receive exhaustive examination and that a new conception for it be sought. 'In the absence of outside forces, a body at rest remains at rest and a body in motion retains its velocity'. This has been established in innumerable cases; and what is meant by 'a body is in motion' or 'is at rest' appears to be so clear that nothing remains to be explained. The purpose of the worthwhile work mentioned below [i.e. Lange's book] is to upset this false security and inspire further speculation. It is well-known, and the author [Lange] brings it out in detail, that the philosophers of antiquity found it difficult to answer the question whether a given body move or not. I am thinking of the ship anchored in the current; and of the man moving toward the rear of a ship under sail, whose position with respect to objects on the shore does not change. In such cases our question may easily receive different answers, according to the weight one places on this or that relative position; a general criterion is wanting. All of these disputes would of course be quite simply settles if one recognized the incompleteness of an expression like 'a moves' and in its place wrote 'a moves with respect to b'. The sentences, 'a moves with respect to b' and 'a does not move with respect to c' need not contradict one another. Indeed the physicists will admit that the motion of a body is never absolutely ascertainable, but rather is only so in relation to another. One thereby acknowledges the defectiveness of the above statement of the law of inertia; for the reference therein is to absolute motion and rest. And the bad part of this is that this defect cannot be corrected by including in the law a reference to another body; for which [reference body] should we choose? A given body would appear to be at rest, or to move in rectilinear or curved paths, uniformly or nonuniformly, depending on the choice of the reference body. The nature of the law of inertia prohibits reference to a particular body, since none exists which deserves this distinction; but at the same time absolute motion remains unrecognizable. This is the difficulty. How

Boston Studies in the Philosophy of Science, XIV. All Rights Reserved.

is it then that it is so little heeded by the physicists? The incomplete expression, '*a* moves' is so convenient and sanctified by ordinary usage that it is used all too often even in physics. The theoretical impropriety of this expression is all the more readily forgotten as it easily helps to overcome many a difficulty. Whenever one cannot answer a question, one can at least allow it to disappear behind the cloud of imprecise speech – which in the present case is especially agreeable. For if one considered the matter quite openly, the entire foundations of physics would seem to totter. For this reason, one has unconsciously avoided throughout using the complete expression '*a* moves with respect to *b*'. Further, the law of inertia has become such unassailable common knowledge that we do not readily notice when we tacitly presuppose it in order to prove it. In this way we easily make use of laws of motion and of expressions like 'mass' and 'force' although the law of inertia is the foundation of all laws of motion and lends meaning to these expressions. How is it then that physics makes such sure progress despite this deficiency? Indeed, astronomy does introduce a coördinate system that suffices for practical purposes. When we express the law of inertia in terms of motion with respect to this system we find that all consequences are sufficiently in accord with experience. Theoretically, however, we gain nothing by this; for no one doubts that the fixed stars, which we need to set up our coördinate system, are only apparently at rest with respect to one another, and that this appearance is the consequence of our imprecise observations. Added to this is the fact that reference to specific bodies is contrary to the concept of a natural law, which requires generality. [2] On the other hand, no one would want to doubt that the completeness with which our coördinate system satisfies the requirements of scientific explanation indicates a regularity without which that satisfaction would be unexplainable.

In general one still holds to the Newtonian position, in which motion is relative to absolute space and absolute time, however little Newton's theological considerations are acceptable to modern tastes and although a region of absolute space is not recognizable, so that it is impossible to specify the velocity of a body with respect to absolute space and time. The difficulty is not avoided by eschewing the expressions 'absolute space' and 'absolute time' in favor of true and apparent motion. The author [Lange] asks, "With what justification does Newton assert that the paths of unacted upon bodies describe straight lines relative to absolute space,

whose parts, as he [Newton] himself admits, cannot be observed?... Newton could not refute us in the least if we advanced the contrary thesis that
the paths of unacted upon particles are spirals." "How does Newton know
that the oscillation of the perfect pendulum is isochronic with respect to
absolute time?" Newton's justification of the law of inertia is obviously
circular, something still frequently encountered today. The circularity is:
to specify true motion and to distinguish it from apparent motion, one
uses laws of motion which depend on the law of inertia; and only then,
on the basis of this true motion, can one establish the law of inertia.
Lange does not even regard Newton's absolute space and time as necessary evils. He considers them superfluous products of an *esprit méta-
physique*. In this instance he [Lange] appears to me to have overshot the
mark. This is due to his considering the hypotheses individually, while they
only have meaning taken as a whole. When we consider the hypothesis
of an absolute space by itself, we obviously have something that goes
beyond any experience; motion relative to absolute space is not recognizable as such, and for this reason no laws of motion relative to absolute
space are derivable from experience. The matter is different if we combine
the hypothesis of absolute space and time and the law of inertia into a
single hypothesis. In this way [the hypothesis of an] absolute space is
connected with observable phenomena. Under this hypothesis propositions such as Newton's concerning absolute motion are possible and can
be confronted with experience. Newton himself did not clearly recognize
this fact. The deficiency of his presentation may be explained by his reluc-
tance to put forth an hypothesis. He wished to derive each step directly
from experience or from earlier (to him) self-evident truths, and in this
way was led to consider individually principles which could only be confronted with experience as a single totality. To me, the difference between
Newton's theory and the author's [Lange's] is not as great as Lange takes
it to be. Though indeed I do not fail to recognize that Lange's work
represents some progress in this matter. [3]

Lange relates motion to 'inertial systems'. [4] He considers three freely
moving point-masses not on a straight line which simultaneously depart
from a spatial point. He calls a coördinate system an 'inertial system' if
the paths of these particles relative to that coördinate system are straight
lines. A coördinate system of this sort can always be set up; of course its
state at each instant must be determined by reference to the three particles.

That in such a system the paths of these particles are straight lines is not a truth of experience, but follows from the definition of inertial system. However, that a fourth freely moving particle likewise moves in a straight line in that inertial system is not a consequence of the definition, but if it does so nevertheless, and every other freely moving particle does so, then that is a natural law. [5] For the time component of the law Lange introduces – following the procedure of C. Neumann – an 'inertial time-scale', that is, a method of time measurement according to which a freely moving particle proceeds uniformly in the inertial system. [6] That every other unacted upon particle also proceeds uniformly is also not a consequence of the definition, but is a natural law.

What is gained by this? The author [Lange] regards his reference system as an ideal one while Newton's absolute space is transcendentally real. This seems to follow from Newton's words. But, were Newton's absolute space really transcendental, it could not have served in explaining nature, as it has indeed done for a long time. Space is also connected to experience by the law of inertia; but indeed in an unobvious manner. It is no small accomplishment on the part of the author [Lange] to have clearly formulated this connection. Newton's assumption of a single absolute space contains more than is necessary for the explanation of natural phenomena. Of the infinitely many possible inertial systems moving uniformly with respect to each other without rotation, none is in any way distinguished so that one might consider it rather than another to be at rest in absolute space. Newton can therefore not distinguish rest from uniform motion with respect to absolute space, because there is no ground for this distinction given in experience. This designation of a single inertial system, worthless for explanation and going beyond experience, was happily avoided by Lange, who was right to this extent in finding fault with Newton's transcendentalism. [7]

By no means do I consider the matter closed. One could similarly reproach the author as he did Newton. The question whether or not a particle is 'unacted upon' (freely moving) goes beyond experience just as much as the question whether or not it is at absolute rest. With Newton the question was: How do we distinguish real from apparent motion? Here the question is: How do we distinguish influenced motion from that of a freely moving particle? In order to answer the question in Newton's case we required knowledge of an absolute space, which we do not have;

here we require knowledge of an inertial system which we equally do not have. For in order to know whether a given coordinate system is an inertial system, we must already have answered our question. In the same way with Newton, we could not know a given coördinate system were absolute without first knowing whether the origin of the coördinate system is at rest in absolute space. This fault is in both cases the result of considering the hypotheses individually. As hypotheses, the basic laws of dynamics can only be compared collectively with experience and thereby verified. Thus my venerated teacher K. Snell formulated the law of inertia in somewhat the following manner: 'A particle has an acceleration only as a consequence of its interaction with other particles'. What is to be understood by 'interaction' can only be more precisely determined by means of the other dynamical principles. In this way the law of inertia immediately becomes linked with these other principles. [8]

Still, Lange's thesis requires supplementation from yet another point of view. [9] It may at first appear strange, but a little reflection will confirm that we have no means of observing whether and how much the length of an object changes through time. With every such judgment we make we always assume the invariance of a scale. What we observe is therefore not the change in length itself, but only the change in relation to another length.[1] If all lengths were reduced simultaneously by one half, we would have no means at all to notice it. [10] For the angles of vision in which objects appeared to us would remain the same, as would the parallax in relation to our eyes and all such relations to the lengths of our body parts, since our own body would participate in the shrinking. One might say that the receptivity of the eyes would have to change; but one cannot say anything about this, for by doing so something dynamical would be introduced, namely the elastic forces in the ether. We must assume here the position that we do not yet know anything about force. What force is can only be defined subsequently. Here we ask: what is observable, without the introduction of an hypothesis concerning the movement of matter? What is purely empirical? And we cannot deny that we are without means of observing the constancy of the distance between two bodies, just as we have no means of recognizing a point in space after the passage of time, and just as we have no means of deciding whether there is an inertial system with respect to which a particle is at rest. Agreed, we have no means of addressing these problems without hypotheses. I do not intend to say that there is no difference between the uniform and accelerated motions of a particle; or between the constancy and change in the distance between two particles. But those distinctions can only be recognized after we have acknowledged a set of hypotheses. Just as the assumption that the earth is at rest forces itself on us because of the overwhelming preponderance of relatively motionless phenomena the earth presents; and, on a higher level, as the fixed stars are taken to be motionless because they appear to be motionless with respect to each other; we can therefore hardly avoid calling a length unaltered if it belongs to the overwhelming totality of lengths which do not appear to have changed with respect to each other. In all these cases, the observation of an extended relative permanence leads us to the assumption of an absolute permanence, although from a purely geometric point of view a non-relative account of rest makes just as little sense as a non-relative account of rigidity. In a passage quoted by Lange[2] Leibniz says: "For this reason motion is by its nature relative. But this holds only in a rigorous mathematical sense. On the other hand we assign motions to bodies in conformity with the appropriate hypotheses; in this way appearances are explained in the easiest way and there is no difference between a true hypothesis and one that fits." Lange here rightly objects to the expression 'hypothesis' and proposes 'convention'. One could also say 'definition'. Conventions are actually neither true nor false but rather useful or pointless. One will always advocate that manner of speaking in which the laws of nature are most simply stated. It is the same here; one cannot assert that a length remains the same without first saying how one is going to compare two intervals that are not introduced simultaneously; and only after this has been determined will one be able to say whether or not a given interval has changed. But not all such possible stipulations are conducive to short statements of the laws of nature. Therefore, if one wishes to introduce a coördinate system to describe motion, one must specify the unit of length for each instant. [11]

    As a further illustration, let me add the following consideration: imagine a rectangular coördinate system and a standard of space and time measurement such that three freely moving [12] particles move uniformly along three straight lines passing through the origin of the coördinate system, and further that they move as if they had all been at this origin at the same time. The instant determined by this coincidence we will call the zero point of the time scale. Then their coördinates are proportional to the time and the triangles formed at any given time by the particles remain similar. [13] We now relate everything to a new coördinate system, whose only difference from the first is that its unit of length is proportional to the time as measured by the unit of length in the first system. [14] Let the units of length in the two systems agree at time $t=1$. Let us now call a coördinate in the old system x and the corresponding one in the new system ξ, so

$$\xi = \frac{x}{t}$$

at time t. With reference to our new coördinate system our three points are at rest. The old coördinates of any point moving uniformly in a straight line with respect to that old system are linear functions of time, thus of the form:

$$x = a + b \cdot t$$

From this we derive the coördinates of the new system:

$$\xi = a \cdot \frac{1}{t} + b;$$

that is, in the new system the spatial coördinates are linear functions of $1/t$. In the new system the [a fourth] particle [which is moving uniformly with respect to the old system and does not pass through the zero point] would not move uniformly according to the time system employed thus far. We can however introduce a new time system with respect to which the particle does move uniformly. We need only let,

$$1/t = \tau$$

where now τ is the number which in the new system denotes the same instant as t did in the old. [15] Thus the particles which have uniform rectilinear motion with respect to the old coördinate system as measured by the old time scale will again have uniform rectilinear motion with respect to the new coördinate system, as measured by the new time scale. Both systems are inertial systems according to the definition: for we assumed that the original three particles considered were moving freely, without any external influence; and for both it holds that a fourth freely moving particle also describes uniform rectilinear motion. Of course these paths [i.e. our descriptions of them] are in general different and the previous

time origin has been moved out to infinity, and conversely, the former
infinitely remote time point has been shifted to the time point $\tau = 0$ of the
new system. Thus, what was before, in a matter of speaking, envisaged
as an ideal goal that could never be fully realized is now actually reached,
but as soon abandoned. We now have the equation:

$$\xi = x \cdot \tau$$

from which it follows that ξ can go to infinity only when x or τ does so.
The latter would agree with what could happen in our familiar system of
space-time measurement, namely that in the course of time a body fades
farther and farther away beyond every boundary. But that a particle
within a finite time should disappear into infinity and immediately thereafter reappear in the finite strikes us as absurd; and *prima facie* it might
seem possible in the new space-time scale for ξ to go to infinity, even for
finite τ, if x were to go to infinity. Suppose that our old system of measurement were the familiar one; then x could go to infinity only for infinite t and according to our knowledge of nature we may indeed assume
that x can only go to the same degree of infinity as t or remain finite.
Then x/t or ξ would remain finite even for infinite x, and it would hold
also on our new scale that a particle cannot disappear into infinity in a
finite time but only in an infinite time.

The acceleration of a particle with respect to the new scale has to go
to 0 at the same time as it does with respect to the old scale; for we have
seen that an unaccelerated, that is, a uniform motion in the one system is
also a uniform motion in the other. It is therefore no surprise that the
calculation yields the following relation between the two accelerations:

$$\frac{d^2\xi}{d\tau^2} = t^3 \frac{d^2x}{dt^2} \quad \text{or} \quad \frac{d^2x}{dt^2} = \tau^3 \frac{d^2\xi}{d\tau^2}.$$

From this it follows that at a given instant, all accelerations in the new
system have the same relationship to each other as the corresponding ones
in the old system have to each other. Therefore, all the [general, dynamic]
laws [of motion] relating accelerations within a system at a given time
hold equally well in the new system as in the old. As examples of such
laws I mention the proposition concerning the parallelogram of accelerations; the proposition that the accelerations of two mutually interacting
particles are directed along the line joining them and directed in opposite

directions and is always constant for those two particles; and let me further mention, the proposition that the acceleration ratio of particles B and C is a simple function of the acceleration ratios of particles A and B and A and C. All this remains unchanged if all simultaneously occurring acceleration components are increased or decreased in the same proportion; also from these laws one cannot discern which system of measuring space and time we should choose. It is otherwise when we consider the dependence of the acceleration on the distance between the interacting particles. In Newtonian mechanics the acceleration of a particle whose acceleration is caused by [gravitational] interaction with another particle is inversely proportional to the square of the distance of separation:

$$p = \frac{a}{r^2}$$

where p is the acceleration, r the distance between the two particles, and a is a time-independent constant, this is all with respect to the familiar coördinate system. In the new system [in contrast], let π correspond to p and ρ to r. Then we have $p = \tau^3 \pi$, $r = \rho/\tau$, and we obtain:

$$\pi = \frac{a}{\tau \cdot \rho^2}.$$

Here a direct dependence of acceleration on time would enter, or the constant a would be replaced by the time-dependent variable a/τ. Our familiar space-time system is thereby distinguished in this way: with respect to it Newton's law [of gravitation] is time-independent. This we demand of a natural law. When identical circumstances arise, we expect the same consequences to occur irrespective of when they happen. If other consequences occur, we infer that we failed to notice all of the relevant circumstances; we do not put the blame on the time difference as such. [16]

Therefore it seems to me that for now the only thing to be said is: it is possible to set up a coördinate system and a method of measuring length and time such that the movement of particles in the universe in relation to that system is such that the acceleration of a given particle obeys the parallelogram law according to which accelerations can be decomposed into components, each of which corresponds to an interaction with another material article, whereby the above-mentioned laws [of interaction] hold, and whereby the laws according to which the magnitude of the acceleration for the single interaction is determined, contain neither the time nor the position of the material particles with respect to the coördinate system. Thus, one can change the origin of the time scale, and replace the [original] coördinate system with a closely related congruent one, without the analytic expressions for these laws being altered, except by replacing the old symbols by corresponding new ones.

That there exist infinitely many such coördinate systems which move without rotation or alteration of distance-scale uniformly with respect to one another is a mathematical truth.

It seems to me that this has the effect of threatening to make the question of the reality of motion degenerate into a verbal dispute. The only question can be whether the distinction between accelerated and non-accelerated motion, or as Lange puts it, between inertial rotation and inertial rest (p. 56) and the distinctions between various sorts of acceleration are real. The following decision may be the most relevant: in the same sense in which one calls the invariability of a length (e.g. a usual meter stick at a fixed temperature) real, the distinctions among the various sorts of motion are also real. In both cases arbitrary stipulations are involved, which are however so bound up with the lawlike aspect of nature that they are distinguished from all other logically and mathematically possible stipulations. If one wishes to express the intimate connection our familiar system has with the lawlike aspect of phenomena by calling it 'real', one must do this in both cases. But perhaps the word 'objective' would be more apt. [17]

I should make a few remarks here concerning the expressions 'concept' and 'conception'. The first, it seems to me, belongs to logic, for logic after all has the oldest claim to it, and needs such an expression to enunciate its laws. What is required of concepts for this purpose are sharp boundaries but on no account consistency. Whatever fails to draw such clear distinctions cannot be regarded as a concept of logic, any more than a point can be accepted in geometry unless it is dimensionless; for otherwise it would be impossible to set up axioms for geometry. In every science the point of view for the development of the technical language must be authoritative in order that the regularities be expressible in the most simple and exact manner. From this point of view I must deplore the current use of the word 'concept' which is incompatible with its logical sense. There is no development, no history of a logical concept, at least in the way one cares to talk about it. I cannot agree with the author that there is a great need to be able to talk about the history of a concept; and I find that there is ample ground to avoid such usage. If one said instead, 'the history of the attempt to comprehend a concept' or 'the history of the grasping of a concept', that would appear to me much more appropriate; for the concept is something objective which we do not imagine, and which does not impress itself on us; rather it is something we try to grasp, and hope in the end to have grasped correctly, if we have not mistakenly looked for something where there was nothing to find. 'The number three falls under the concept prime number' is an objective truth. When I say it I do not mean, I have within myself a conception which I name 'three' and a conception which I name 'prime number', such that these conceptions stand in a particular relation to each other; and it is yet to be determined whether other people have similar conceptions which stand in the same relation to each other. I cannot know whether the conception I call 'prime number' can alter itself so that it no longer stands in the same relation to the other conception [i.e. of the number three]; this can only be determined by experience.

If one were to talk in this manner, one would completely miss the real meaning of the sentence; similarly if one said instead, 'I imagine for myself these concepts', rather than 'I find these conceptions within myself'; one would in this way still be talking in terms of a private mental process. On the other hand we want to assert something with our sentence, which is valid independent of our waking or sleeping, our living or being dead, which has always held and will hold regardless of whether or not there are or will be creatures who recognize this truth.

Lange thinks 'that a concept which is still developing is by its very nature not free of internal contradictions; if it were, a motive for further development, would be lacking'. This seems to me a totally false representation, which in particular fails to cope with the above case. A contradiction in a concept is certainly no motive for its further development. The concept of being reflexively non-identical contains a contradiction and yet remains what it is and always was, and no one thinks of its further development. There is good reason to acknowledge this concept in logic, for makes the sharpest possible distinction, and is useful for defining the integer 0, as I have shown in my book on the *Foundations of Arithmetic*. And similarly in our case there are no contradictions in the concept of motion which demand to be further developed. In any case the contradictions have arisen not because we tried to treat something as a concept which was not such in the logical sense, but because the clear-cut boundary is absent. One looks for a boundary and notices because of the contradictions that arise, that the assumed boundary is still vaguely defined or is not even the one we were looking for. In this way contradictions have always pushed the inquiring mind further; but not contradictions in the concepts, for these always carry sharp distinctions with them. One knows, indeed that nothing falls under a self-contradictory concept, as any doubt whether or not a given object falls under it is impossible as soon as the contradiction is recognized. [18] What drives us on is the observation of vague boundaries. Thus in our case all efforts have been directed at finding a sharp boundary. One can now say that these efforts failed because a boundary was sought in the wrong place. We have therefore found a new boundary, not between motion and rest but between inertial rest and non-inertial rest; and it is to the author's merit to have first clearly seen this boundary. With this he can console himself for having failed to 'more quickly reach the end of the development of the concept of motion'.

I cannot agree with the current use of the word 'conception' any more than with that of the word 'concept'. Just as the latter belongs to logic, the former belongs to psychology. In this way one remains not only as close as possible to ordinary linguistic intuitions but also to the tradition in psychology. We say, 'I imagine something' and mean thereby an internal mental process and we understand by 'conception' an internal picture. Therefore one should never use the word 'conception' in physics, mathematics, or logic; or at least regard it as useless. Physics, for example, is concerned with bodies and, like all other sciences, with concepts, but never with conceptions, they should be left to psychology. In science one should never talk about conceptions without at least specifying the point of view of the person conceiving, or at least indicating whose conception it is.

For the conception of one person is not the conception of another any more than one person's nose is another's even if they were congruent. The use of the word 'conception' without an indicated frame of reference is just as scientifically inappropriate as that of the word 'motion' without a frame of reference.

Lange says for example: "We overlook the fact that in our judgment of 'real' motion the conception of the earth is involved", and in another place: "the conception of the configuration of a point system is the conception of the location of all its points." Can the author conceive of the locations of the aggregate of atoms in a piece of paper? But the system does not depend on that; if in fact the configuration of a system of points were the total of the locations of its points, that would suffice. Why should we be concerned with the conceptions that anyone might form of that? When are people going to stop mixing up the psychological and the logical, conceptions and concepts? Of course the author could cite common usage against me and show me hundreds of books and papers to that effect. Unfortunately he could do this, and therefore my remarks are not directed particularly at him and his book. I base the demand that the necessity of a distinction be acknowledged, and that separate expressions be allotted to logic and psychology, on the nature of the case and out of a scientific necessity, and I appeal to common usage only when there is a question concerning a choice of expressions. But common usage can never justify obscuring existing distinctions.

All the same I would not like to leave this stimulating book by bringing digressions into prominence, but rather by agreeing with the criticism in Lange's first appendix of the so-called 'absolute translation of the sun' and agreeing with the sentence, 'the elementary concepts are not given at the beginning of scientific experience'. Or as I would prefer to say, they must be discovered by logical analysis. Similarly, the elements were not known at the beginning of chemistry, but rather were discovered at an already highly developed state of that science. The logical and technical starting point is not the psychological and historical origin.

NOTES

* With reference to Ludwig Lange, *Die geschichtliche Entwicklung des Bewegungsbegriffs und ihr voraussichtliches Endergebnis*, W. Engelmann, Leipzig, 1886.
** University of British Columbia. This translation first appeared in *Studies in History and Philosophy of Science* 2 (1971), 195–212; we are grateful to the editors and the publisher, Macmillan Journals Ltd., for permission to print it here. (Editor's note.)
1 It is therefore completely false that a fixed scale belongs to the foundations of geometry. Whether a length changes through time is completely inconsequential to geometry, for from the point of view of pure geometry this question does not even make sense. The comparison of lengths which are no given simultaneously does not belong to the domain of geometry. This discipline does not deal with time and consequently not with that constancy that can only be spoken of with respect to the passage of time. That

belongs to physics. Should the phosphorous content of the brain and the temperature of the sun also belong to the foundations of geometry?

² *Leibnizens mathem. Schriften* (ed. by Gerhardt), Vol. VI, p. 507.

TRANSLATORS' NOTES

We wish to thank S. Straker, P. Belluce and M. Jackson for their critical reading of the translation and notes.

[1] The following is a translation with notes of Gottlob Frege's 'Über das Trägheitsgesetz', which originally appeared in the *Zeitschrift für Philosophie und philosophische Kritik* **98** (1891) 145–161. This is Frege's single published writing in non-deductive science, and it illuminates his genius from a slightly different view. However, the main reason for publishing an annotated translation of the paper is that it deserves a place, we feel, in any history of physical theory leading up to Einstein's epochal 1905 paper; possibly because Frege has not been thought of in this connection it has, so far as we know, gone unnoticed by historians of relativity.

As for these notes, they are merely an attempt to elucidate Frege's philosophical position. We have not attempted the extensive historical analysis required to put Frege's contribution in proper perspective; we hope that this will be done.

Reference to our notes in the text is indicated by square brackets; the same device is also used to mark additions to the text where we felt such ammendment clarifying while not changing the intended sense. In several places we have broken lengthy paragraphs into shorter ones; we felt it unnecessary to indicate such changes.

[2] This is the first of several criteria relating to general laws, conventions and assumptions, For the sake of convenience, all these Fregean characteristics are listed below:

(a) A general law does not contain a reference to specific entities.

(b) A natural law is synthetic (p. 260) see note [5].

(c) General laws are time-independent. (p. 265)

(d) One will always advocate those conventions which result in the laws of nature being most simply stated. (p. 262)

(e) In those cases involving arbitrary stipulations or conventions, some of these conventions – e.g. our familiar ones – "...are however so bound up with the lawlike aspect of nature that they are distinguished from all other logically and mathematically possible stipulations. If one wishes to express the intimate connection our familiar system has with the lawlike aspect of phenomena by calling it 'real', one...[can do so]. But perhaps the word 'objective' would be more apt." (p. 266)

[3] In this paragraph Frege delineates three views of absolute space, time and motion:

(1) The Newtonian position. Actually this view includes contributions from Newton, Lange and Frege. Although it would be interesting to disentangle these elements, we shall not attempt to do so. According to the Newtonian approach, absolute motion cannot be recognized by empirical means. Yet, due to theological and transcendental considerations, absolute space and absolute time are admissible into the conceptual framework of mechanics.

(2) Lange's position as interpreted by Frege: absolute motion is a superfluous notion. Lange offers (see text below) an empirical method which, allegedly, will enable us to identify a class of inertial reference systems. But the method renounces

as empirically impossible, all attempts to divide this class into frames which are moving absolutely and those which are absolutely at rest. In later portions of the article Frege criticizes Lange's method for distinguishing inertial reference frames (see [5] and [8] and relevant text). However Frege claims that Lange's approach is an improvement on (1): "I do not fail to recognize that Lange's work represents some progress in this matter."

(3) Frege's own view, which we shall call 'collective verification':

(a) Absolute space, and thus absolute motion, are not, by themselves, ascertainable by empirical means:

> When we consider the hypothesis of absolute space by itself, we obviously have something that goes beyond any experience; motion relative to absolute space is not recognizable as such, and for this reason no laws of motion relative to absolute space are derivable from experience.

(b) However, a collective hypothesis can be formed by combining the law of inertia with hypotheses of absolute space, time and motion. This collective hypothesis has empirical content, i.e. it can be tested by experimental means.

(c) The observational character of the collective confers empirical content upon the totality, and perhaps the components. "Under this [collective] hypothesis propositions such as Newton's concerning absolute motion are possible and can be confronted with experience".

(d) Newton did not recognize this 'collective' method for admitting statements into the body of empirical science because Newton "...wished to derive each step directly from experiment or from earlier (to him) self-evident truths, and in this way was led to consider individually principles which could only be confronted with experience as a single totality".

(e)   Although there is a distinction between the Newtonian position (1) and Lange's (2). Frege believes that this difference is not so great as that between the Newtonian view (1) and the Fregean claims contained in (3a) through (3d). (See note [8].

[4] In what follows, Frege presents Lange's method of ascertaining and of employing inertial reference frames. Later, beginning on p. 260, Frege criticizes Lange's analysis. In discussions of relativity theory today, Hans Reichenbach's method for ascertaining inertial systems is the one generally adopted (see, for example, articles by W. Salmon and by A. Grünbaum in *Philosophy of Science* 36 (March, 1969)). Reichenbach's method is sketched in §27 of his *Philosophy of Space and Time* (translated by Marie Reichenbach and published by Dover in 1957); this presentation relies heavily on more detailed discussions in his *Axiomatization of the Theory of Relativity* (translated from the 1924 volume by Marie Reichenbach and published by the University of California Press, 1969.)

[5] We are given three 'freely moving' point masses P_1, P_2 and P_3 which simultaneously depart from some point 0 and which describe non-collinear paths $\bar{P}_1$, $\bar{P}_2$ and $\bar{P}_3$ respectively. According to Frege, Lange defines a class of 'inertial reference systems', IRS, such that with respect to any one of these the paths $\bar{P}_1$, $\bar{P}_2$ and $\bar{P}_3$ are straight lines. In other words any coördinate system is inertial if in that system the paths described by the 'freely moving' particles P_1, P_2 and P_3 are straight lines. Further, it is claimed that it will always be mathematically possible to find some inertial frame for any three (non-collinear) freely moving particles. Having found some inertial frame R, a fourth freely moving particle, P_4, is introduced. Whether

or not P_4 moves in a straight line with respect to R is a matter for experiment to decide. If P_4 and all other freely moving particles move in straight lines with respect to the inertial system R, then Lange will have found inductive support for the following *synthetic* natural law: All freely moving particles move in straight lines relative to inertial systems.

There are at least two major difficulties involved in this approach. Since these two issues form the backbone of Frege's paper, we will discuss them in some detail.

(A)   The notion of 'freely moving' or 'unacted upon' particles has to be de-fined independently of and prior to that of 'inertial system'. No such definition is offered.

On p. 260 (see [8]) Frege points out that "The question whether or not a particle is 'unacted upon' ['freely moving'] goes beyond experience just as much as the question whether or not it is at absolute rest." Frege's remedy for this difficulty is what we have called 'collective verification' in note [3]. According to Frege's scheme, we form an hypothesis which designates some particles as freely moving. This hypothesis, taken individually, is not empirically verifiable. "As hypotheses, the basic laws of dynamics can only be compared collectively with experience and thereby verified" (p. 261)

(B)   The kind of reference frame and metric which qualify for membership in IRS will be extraordinarily varied.

Some of the effects of this variety can be seen by considering an example.

Case 1

Imagine a 3-dimensional reference frame V whose origin is fixed at 0 and whose axes are $\bar{P}_1, \bar{P}_2$, and $\bar{P}_3$ as depicted in Figure 1. (NB: According to the text, P_1, P_2 and P_3, should not lie on the same straight line. We believe it necessary to im-pose a stronger condition: P_1, P_2 and P_3 are not coplanar). If we assume that differential forces are negligible (an assumption we shall make throughout these notes) and if we assume the standard conventions of congruence (e.g. a rigid rod remains congruent to itself under transport), then suppose $\bar{P}_1, \bar{P}_2$ and $\bar{P}_3$ are geo-desics in V. Then V is a member of IRS as defined by Lange-Frege (and the geom-

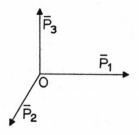

Fig. 1.

etry of V is Euclidean). Under these circumstances, there are clearly other reference systems which are also members of IRS and whose geometry is Euclidean but which employ nonstandard metrics. For example Frege introduces (see note [14] and relevant text), an inertial coördinate system, suppose we call it V', in which the distance metric is inversely proportional to time. The purpose of that exercise is to show that there are, in Frege's view, unacceptable consequences to admitting V' into IRS; viz., with respect to V' general laws become time dependent. That is, Frege argues that IRS as defined by Lange is much too wide; he then offers some conditions which would exclude some types of reference systems. (See notes [16] and [17] and relevant portions of the text.)

In the preceding discussion of Case 1, we have alluded to some conditions which Frege explicitly adds to Lange's scheme in order to narrow the class of inertial systems. However, it may be the case that Frege, as well as Lange, has made other *implicit* assumptions which tend to narrow IRS in a different way. The issue is this: does Frege make implicit assumptions about the behavior of 'freely moving' particles? Consideration of this matter is facilitated by examining a case in which, in contrast to Case 1, the particles move arbitrarily.

Case 2

Imagine a 3-dimensional reference frame S whose origin is fixed at 0 and whose axes are $\bar{P}_1$, $\bar{P}_2$ and $\bar{P}_3$ as schematized in Figure 2. (NB: There should be a condition limiting the number of times a particle can cross its own and other paths).

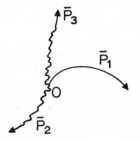

Fig. 2.

Now the nature of physical space may be such that if we employ standard conventions of congruence, then $\bar{P}_1$, $\bar{P}_2$ and $\bar{P}_3$ turn out to be geodesics in S. But in this case the geometry of S will, in general, be non-Euclidean. On the other hand, $\bar{P}_1$, $\bar{P}_2$ and $\bar{P}_3$ may be geodesics in S only if we adopt a radically non-standard definition of congruence. In this case also, the geometry of S will, in general, turn out to be non-Euclidean. Now given S and some congruence convention such that $\bar{P}_1$, $\bar{P}_2$ and $\bar{P}_3$ are geodesics in S, is there a three-dimensional reference system S' such that $\bar{P}_1$, P_2 and $\bar{P}_3$ are geodesics with respect to S' and the geometry of S' is Euclidean? It seems to us that in general it will not be possible to find such an S'. Thus if freely moving particles are defined independently of inertial systems and behave in a manner similar to that depicted in fig. 2, it will always be possible to construct an inertial system in the sense of Lange-Frege, but such a system may require a radically non-standard metric and its geometry may well be non-Euclidean. Frege says that "... a coördinate system of this [inertial] sort can always be set up". This suggests a high degree of generality, i.e. his statement implies that no matter how three, non-coplanar, freely moving particles behave, an inertial system can be found. But as we saw in Case 2, if we are free to ascribe any sort of behavior to the particles, then we can conjure up some paths such that there is no member of IRS whose geometry is Euclidean. Since Frege does not mention non-Euclidean geometries, we believe that he implicitly assumes that the 'freely moving' restriction circumscribes conditions so as to eliminate situations like Case 2. This implicit assumption does not obviate Frege's analysis since he could explicitly incorporate such a condition into his hypotheses about 'freely moving' particles, i. e. he could hypothesize that three non-coplanar particles which simultaneously depart from a point are freely moving only if there is some reference frame W with respect to which the paths of the particles are straight lines and the geometry of W is Euclidean. In Lange's approach W would be inertial (and the schema circular), but since Frege introduces additional conditions for membership into IRS, W would not necessarily be inertial in his approach (and circularity would be avoided). Of course this hypothesis must be 'collectively verified'.

[6] Having obtained an inertial reference system for $\bar{P}_1$, $\bar{P}_2$ and $\bar{P}_3$, the Langean scheme now attempts to introduce a time-scale T' – i.e., an 'inertial time-scale' – with respect to which the particles' speed is uniform. Here again we are faced with difficulties similar to those discussed at the end of the preceding note. For if the speed of the particles were entirely arbitrary with respect to an inertial system, it is not clear that a time-scale can be found such that all freely moving particles move uniformly with respect to it. Again, the 'freely moving' restriction must circumscribe conditions in special ways.

[7] In this paragraph Frege makes two main points: (i) Newtonian absolute space is not transcendental because a transcendental (absolute) space "... could not have served in explaining nature, as it has indeed done for a long time." This claim seems to be in accord with Frege's method of collective verification (see note [3]) which was designed to admit absolute space into empirical science: by employing collective verification' "... absolute space is connected with observable phenomena."

(ii) Lange is applauded for having shown that a preferred inertial system – and thus absolute motion – is "... worthless for explanation and [goes] ... beyond experience".

It thus appears that Frege is claiming, on one hand, that Newtonian absolute space does have empirical content and is useful for explanation and, on the other hand, Lange is correct in saying that Newtonian absolute space goes beyond experience and is useless for explanation.

This apparent inconsistency is removed (or ameleriorated) in the next paragraph of Frege's text and note [8].

[8] Frege believes that both Newton's approach and Lange's contain hypotheses which, when considered individually, are not amenable to experimental test. In the Newtonian case, there are, for example, the hypotheses which deal with absolute space or absolute motion; similarly, in Lange's analysis, there are hypotheses about 'freely-moving particles'. Nevertheless Frege believes that in both cases these hypotheses might be admissible into a scientific system if the total scheme is empirically verifiable – i.e., if collective verification is operative.

This then is the 'resolution' of the apparent inconsistency discussed in note [7]:

Frege regards Lange's analysis as an improvement on the Newtonian one *even though both require collective verification*; the superiority of Lange's analysis resides in the preferability of those Langean hypotheses whose empirical content is due solely to collective verification. Frege believes that Lange has 'made progress', because while the hypothesis of freely moving mass points is not empirically testable by itself, it is better than the similarly non-empirical Newtonian assumptions about absolute space. The Newtonian postulates are less desirable in Frege's eyes, because one gains nothing by singling out some members of IRS as being absolutely at rest; in contrast, Lange's hypothesis about freely moving particles is a starting point for constructing a system of mechanics.

It should now be clear why Frege regards his view as more divergent from Newton's than Lange's is. See note [3]. After all, Lange rejects the Newtonian approach because it allegedly contains non-empirical elements. Frege points out that Lange's thesis has components similar to the ones Lange found distasteful in Newton. But Frege does not reject theories simply because they have constituents which, considered individually, are non-empirical.

So much for Frege's positive views of Lange. Frege also has two main criticisms of Lange. These correspond to the two difficulties discussed in note [5]; the first is contained in the present paragraph of the text and the second begins in the following one.

(A) Lange's method of identifying inertial systems involves a prior identification of 'freely moving' particles. (B) Lange failed to consider reference systems which involve non-standard space and time metrics. By considering such non-standard systems, Frege shows that the class of reference frames which qualify as inertial under Lange's conditions is much wider than Lange assumed.

[9] This further supplementation is difficulty (B) in note [8].

[10] For recent treatments of this suggestion, see for example G. Schlesinger, 'It is False that Overnight Everything Has Doubled in Size', and A. Grünbaum, 'Is a Universal Nocturnal Expansion Falsifiable or Physically Vacuous?', both of which appear in *Philosophical Studies*, **15** (October 1964).

[11] It is not clear to us what distinctions Frege intends to draw with the terms 'hypothesis' (*Hypothese*), 'convention' (*Konvention*) and 'definition' (*Definition*). On one hand he applauds Lange for rejecting Leibniz's use of 'hypothesis' in favor of 'convention'. On the other hand, Frege himself uses 'hypothesis' in places where it would appear to be open to the same criticism.

[12] Because of 'collective verification', Frege believes that his use of freely moving here is unproblematical. That is, Frege does not consider himself vulnerable to the same charge as he leveled against Lange on p. 260.

[13] Clearly, Frege is also assuming that the particles are moving with constant velocity with respect to the coördinate system. Is this assumption made because freely moving particles move with uniform velocity (with respect to the indicated coördinate system) by definition?

[14] See note [5].

[15] Operationally there appears to be an asymmetry between t and τ. We have an operational definition of t, since $t = 0$ when the particles were at the origin and the scale is specified by the distance from the origin. In contrast, it appears that the only way to ascertain τ is first to obtain t. However, even if there were a problem in this case, Frege's general argument would be threatened only if it could be shown that all non-standard (time) metrics exhibit this operational asymmetry.

276   GOTTLOB FREGE

[16] Having introduced a coördinate system and time scale, V' (see note [5], Case 1), using non-standard space and time metrics, Frege has shown that with respect to it Newton's law of gravitation is time-dependent. Since Frege requires a general law to be time-independent, this result has given him grounds for excluding the non-standard system V' from the calss of inertial systems.

[17] The interpretation of this paragraph and the two preceding ones is facilitated by reviewing Frege's second objection to Lange's analysis – this objection appears as (B) in notes [5] and [8]. Frege contends that the class of reference systems which qualify as inertial under Lange's scheme is far wider than Lange imagines. As an illustration, Frege examines the non-standard system V' and concludes (see note [16]) that such systems can be rejected on the grounds that they result in time-dependent general laws. Now the point of the three paragraphs in question is to show that even *after* eliminating systems such as V', the class of inertial systems is still very wide. So wide in fact that it might include systems with respect to which some motion which we normally regard as accelerated turns out to be non-accelerated.

The question now arises, how do we whittle down this class even further? Frege answers that the distinctions between accelerated and non-accelerated motion are as real as the invariability of length. That being so, we can reject any reference system, V'', which blurs the distinction between accelerated and non-accelererated motion just as we could eliminate coordinate systems like V' which issued in time-dependent laws. In both cases the rejected reference systems qualify as inertial in Lange's scheme.

We must emphasize the nature of Frege's reasons for excluding from the class of inertial systems both (i) reference systems, such as V', which employ non-standard metrics and which result in time-dependent general laws and (ii) coördinate systems such as V'' which blur the distinction between the two kinds of motion. In both cases the grounds are extra-theoretical, though Frege appears to vacillate between these two related reasons: (a) such systems violate our notion of 'reality'; (b) such systems lead to unacceptable formulations of general laws. It seems to us that Frege's reasons are philosophically important because, among other things, they locate the issues in the proper perspective. That is, Frege clearly realizes that a debate about non-standard metrics, for example, must take place at the metatheoretical level. If, for example, one could produce non-standard metrics which did *not* lead to time-dependent general laws, then the acceptability of these metrics will hinge on their squaring with extra-theoretical motions of reality. And even if our notion of reality is violated by such a non-standard metric, there may well be other metatheoretical notions – e.g. mathematical or physical economy – which would counsel for their acceptance. (Of course, acceptance would entail a redefinition of reality.) It was on this extra-theoretical plane that Einstein conducted many of his arguments about non-standard metrics. We do not suggest that Frege anticipated Einstein's arguments, but we do believe Frege would have been ready to listen.

[18] Another way of getting at Frege's point in these paragraphs is by considering concept-*words*. For Frege a proper name is inappropriate for science if it is without denotation; the same applies to concept-names. However, a concept-name with an empty extension is not denotationless. Rather, it falls to denote if it is not possible to say of a given object that it does or does not satisfy the concept-name. When this is the case Frege speaks of 'imprecise boundaries'.

YEHUDA ELKANA

SCIENTIFIC AND METAPHYSICAL PROBLEMS:
EULER AND KANT

1. INTRODUCTION

Historiographical preoccupation is typical for fast developing fields, and the history of science is developing and changing rapidly. Fifteen years ago, when the professionalization of the history of science started, the then Young Turks were fiercely critical of the nineteenth-century positivistic influence and extolled the serene beauty of pure history of ideas which does not view the growth of knowledge as an item-by-item accumulation of positive contributions. In 1963 Agassi could still write that historians of science "paint people as well as ideas black or white",[1] and that their "criterion for whiteness is the up-to-date science textbook".[2] This criticism does not apply any more. By the early 1960's history of ideas in broad cultural setting, relying on metaphysical conceptions and 'Weltanschauung' as influencing factors, had won the day, and Alexander Koyré's work became the hall-mark of perfection. A few years ago, under the impression of rich and numerous historical studies, and with the weakening of the anti-Marxist sentiment among the intellectuals, the conclusion that among the decisive factors on the development of knowledge one had to consider socio-political and economic ones, became inescapable. The historiographical issue now became 'internal' as against 'external' history.[3] Under the banner of 'external' history the social history of science became dominant. What this meant was revising Marxist studies[4] written in the 1930's and writing institutional history of science, i.e. historical studies on the emergence, growth and structure of academies, universities and learned societies.[5] Generally the presupposition behind these works is that institutions have influence on the way scientists act and thus have an indirect impact on the scientific product. Some philosophers of science, like I. Lakatos[6] accept happily the internal- external dichotomy because it seems to them that it saves the applicability of logic as the unique tool for appraisal of scientific theories. At the same time the Young Turks of nowadays presuppose that there is a direct social influence on knowledge in general and on scientific knowledge in particular. These are groping towards a 'Mannheim Revisited' (or perhaps Scheler) yet to be written.[7]

What is common to all these historiographies is that they all need and rely on a demarcation criterion between science and metaphysics. The problem of demarcation was formulated by Popper[8] who calls it Kant's Problem, and put great emphasis on it. For the logical positivists of the Vienna circle, as for the nineteenth-century historians of science, science was the body of empirically testable statements; all statements which were not so, or could not be reduced to atomic descriptive statements were metaphysical i.e. for them pure gibberish. For Popper metaphysics was not meaningless nonsense, but it had to be sharply demarcated from empirically refutable statements which constituted the body of science. The great historians of scientific ideas described scientific and metaphysical ideas in one lump, and as they generally did not articulate their historical methodology, the result often looks confused.[9] The typical social historians of science account for metaphysical ideas as part of a 'Zeitgeist', i.e. a social phenomenon, and as such these can influence the institutionalization of a given brand of science or the choice of some problems as having priority – for them metaphysics plays no direct role in the birth of ideas.[10] Lakatos, in his recent publications[11] advanced a methodology of Scientific Research Programmes according to which the growth of knowledge is due to a continuous critical dialogue between competing Scientific Research Programmes. At the hard core[12] of such a Research Programme, Lakatos puts scientific metaphysics, i.e. those untestable hypotheses which deal with the structure of the physical world and which direct scientists in their research. For him metaphysical hypotheses are pure ideas and he does not inquire into their genesis. As a result of this inclusion his 'internal' history deals with scientific thoughts whether testable or not, and these thoughts are not influenced by anything but by other scientific thoughts. All other influences, according to Lakatos are 'external' influences not really relevant for a rational reconstruction of the growth of knowledge.

2. PROBLEM AND THESES

The central problem of historians and philosophers of science in the 1970's is: 'how does knowledge grow', i.e. what causes change in the con-

tents of knowledge, and what is that part which 'grows' i.e. serves as
nucleus of accumulation and continuity. My approach to this problem is
based on the following theses:

(a)   All analysis of changes that cause the growth of knowledge
        has to rely on three different kinds of interacting factors:
        (i) problems in the body of knowledge which emerge from
        the scientific ideas themselves and which point to possible
        directions of change. Such problems can both originate in
        hypotheses, as well as create testable as well as untestable
        hypotheses.
        (ii) the image of science in a given place or community at
        a given time: whatever people in general and scientists in par-
        ticular *think* of science, its role, its ethos etc. This influences
        heavily the choice of problems from among the enormous
        range of open problems as provided by the body of knowledge
        itself; i.e. by factor (i). The image of science shapes the formu-
        lation of selected problems, determines what is called the
        'frontiers of science' and determines the *reasons* for scientists'
        promotions. It is sometimes so influential, that it not only
        emphasizes some of the open problems but even totally ob-
        scures others. It decides what is legitimate science and what is
        pseudo-science, it shapes the demarcation criteria between
        science and metaphysics.
        (iii) Social and political factors which interfere directly with
        the lives of the scientist and the scientific institutions and thus
        influence the development of science through the scientists.

These three kinds of factors interact. Social, political factors, in addition
to influencing directly the lives of the scientists, influence also the image
of science, i.e. what people *think* of science. All scientific metaphysics is
heavily influenced both by development in science and by cultural and
social environment which it is impossible to disentangle.

On the other hand new scientific ideas, insights and products certainly
influence both the image of science and the socio-political developments.
These interactions are so interpenetrating that one can easily draw the
irrational conclusion that no analysis of factors can be undertaken and
rather look for hidden personal motives and look for the one, under-

lying, unifying principle which makes humanity tick. This conclusion is so strange to all rational scholars that they prefer to rush into absolute dichotomies like science vs. metaphysics, or internal vs. external history, and write 'rational reconstruction' which they admit has very little to do with actual history.

My justification for the above analysis into three kinds of interacting factors is that it seems to me much more satisfactory than the internal vs. external historical explanation and that it obviates the necessity of a demarcation criterion between science and metaphysics; it helps to realize a reconstruction of the changes in the past which is both rational and historical.

How to support this thesis? I do not know of any other way but that of the historical inductivist: to pile up case-histories and to allow the burden of evidence to carry its own weight. I myself tried three such case-histories.[13]

The bulk of this paper has to serve as a fourth case history with the following historical thesis.

(b)   Euler was a Cartesian in his metaphysics, a Newtonian in his methodology; his image of science was heavily influenced by Leibniz and by the tone of the Enlightenment as it was expressed by the Berlin and St. Petersburg Academies, and he was a major influence on Immanuel Kant.

Leonhard Euler is known as a great mathematical physicist and mathematical astronomer, but he is very little known as a physicist; his theory of matter is rarely mentioned either for its intrinsic interest or because of his influence on other natural philosophers in the late eighteenth and even in the early nineteenth century. The reason for this neglect is in the historian's attempt to adopt a demarcation between distinct and indistinct ideas, or between science and metaphysics. Because of the prevalent, Victorian image of science, so far historians write only history of 'distinct' ideas, and only of science, and not of metaphysics. A clear demarcation between distinct and indistinct ideas has been discussed by most philosophers of science, and the demarcation between science and metaphysics (even the Popperian demarcation, which is more tolerant of metaphysics) or the Lakatosian approach (which includes metaphysics in the hard-core of a Research Programme) has only resulted in a new dichotomy: rational reconstruction vs. historical narrative.[14]

3. EULER'S THEORY OF MATTER

I.   A substantial part of Euler's enormous volume of published works is 'metaphysical' in character, and specially deals with his theory of matter. When I say 'metaphysical' I refer to the usual view according to which scientific metaphysics consists of empirically untestable views about the structure of the world. The important influence of such views is not on the kind of answers that science gives, but on the choice of problems that science chooses to deal with. Moreover, it has to be remembered that what is considered as science and what is considered metaphysics is time-dependent; when dealing with any historical question it is the contemporary criterion of what is metaphysics that is usually applied. The Leibniz-Clarke debate was considered a scientific debate by its participants, and as such should be considered as relevant to any discussion of the Newtonian-Leibnizian critical dialogue. Newton's views on the Athanasian creed, or Leibniz's views on the unification of the Churches were not considered by them as directly relevant to their science, while any social historian today takes them seriously into account. That which in the past was not considered directly relevant to science can be, and indeed should be, investigated by us as to a possible indirect influence on science; however that which in the period being studied was considered relevant to their science must be in the domain of straightforward history of science.

Here the crucial question is, how does the historian of science know *a priori* what was for Newton, Leibniz, Euler or anybody else directly relevant to science? And the answer is that *a priori* he does not. For every single philosopher-scientist, for every period and place, he has to think anew. In other words, this approach is historically motivated philosophical a priorism combined with methodological inductivism, and it is just the kind of attitude which characterized historians like Emile Meyerson.[15]

II.   An example: In order to better explain the kind of historical analysis I have in mind, I should like to bring one example which will also explain why Euler's theory of matter had been neglected by historians. The case is Leibniz's concept of the monad.

There is no serious historical treatment of Leibniz's monad as a scientific concept, attempting to understand it as a generalization of his concept of force; yet Leibniz made it very clear that this generalization was the connection between the two concepts.

To deal with the monadology in terms of the history of metaphysical (not scientific metaphysics is meant) systems will simply not do. Even if Leibniz had not pointed to the connection between force and the monad (which he did)[16] we still have to presuppose that there is coherence among the various aspects of man's thought. This is not a coherence of a strict logical[17] kind – it may be that the choice of the term 'coherence' is unfortunate. It is rather like the coherence between different works of art by the same artist.[18]

As a good counter-case to the negligence of the Leibnizian coherence one could mention the accumulated result of fifty years of Newtonian scholarship. Newton's metaphysics and its influence on science was investigated in the thirties.[19] Recently, examinations of Newton's metaphysics and Newton's theology appeared; even a psychoanalysis of Newton has been undertaken.[20] Newton's various activities as a mathematical-physicist, as a historian and as a theologian, are compared; the influence of his dialogue with the Cambridge Platonists on his physical concepts was revealed. Koyré and I. Bernard Cohen showed Descartes' influence on Newton.[21] From the work of Guerlac, McGuire, Rattansi, Heimann and others, the 'other Newton' emerged. For us the important point is that when reading all this diverse critical material, written from so many points of view and each going through the *whole* of the Newtonian corpus, a *coherent* body of ideas is clearly conceived.

If we wish to understand the Leibnizian concept of force, we should look into all the works which Leibniz considered relevant to this concept. There are very few of his works which he did not consider relevant to his physics. One could ask whether this has not been done, and admirably so, by men like Russell, Cassirer or Schmalenbach?[22] Certainly these erudite scholars had studied very carefully all the relevant works by Leibniz; yet it seems to me that, from the historian's point of view, they missed the interesting connections between the different works of Leibniz. In their books a logically impeccable system was set as an ideal. In all of these studies of Leibniz, whatever was not consistent with the picture was ruthlessly expurgated. Cassirer considered Leibniz an early Kantian, Schmalenbach considered him a historian-politician, while, as is well known, Russell looked at all Leibnizian ideas in relation to his logic and mathematics. It has to be admitted that as a return for the price he paid, Russell at least was eminently successful. Nevertheless it seems to me that Meyerson was right when he claimed that relinquishing the rigour of logic in seeking understanding in natural philosophy is a smaller price than saving rigour while eliminating most of the philosophy of nature.

The search for a unitary principle goes through the work of Leibniz like a 'Leitmotiv'. In his early dynamical works [23] he spoke of such a fundamental principle as conservation of force. What is this force to which the Leibnizian universe can be reduced? It is nothing like the Newtonian force-vector. It is certainly not energy as some modern commentators would like to translate it. [24] What Leibniz actually tells us himself is that force has *an effect mv^2*, or *mv*, or the height reached by a body thrown upwards, as the case may be, while in later works he says that it is 'a metaphysical entity', 'the essence of matter 'or 'the main attribute of a monad'.

The monad serves as a final generalization of his concept of force, now uniting in it not only all the physical effects of this fundamental entity which is conserved in nature but also the physical and the spiritual: mind and matter. To us, scientifically-minded, logic-trained moderns all this sounds very confused; but the importance of vague concepts like Leibniz's force of monad for creating science is enormous. [25] Most scientific concepts at early stages of their evolution as well as theories in which they occur defy any attempt to decide whether they are distinct or indistinct, whether they belong to science or to metaphysics. It seems to me that historical research on the evolution, importance and influence of such concepts and theories can be usefully done only if we follow them in every work of the same author; or again if we choose a limited period during which we follow these developments in the work of all important authors.

III.   Concepts like the monad, discussed above, could be called for the sake of brevity, *half-distinct concepts*, and it is my suggestion that viewing such concepts in a special category can become a useful philosophical tool in the hands of historians of science.

At the centre of our discussion of Euler's theory of matter, the half-distinct concept of impenetrability will be found. Euler was trying to construct a physics which would suit his scientific metaphysics. He took for granted an attitude towards science and towards metaphysics which was new and which later cost Kant a lifetime of hard labour to put on a strong foundation. Euler discovered a discrepancy between his metaphysics, the central pillar of which was the tenet that there are no forces acting-at-a-distance, and the usual interpretation of the successful Newtonian physics; this constituted a serious problem for him. In trying to overcome the difficulty he reached back to the Cartesian and Leibnizian heritage which he transmitted to the coming generations. From Descartes and Leibniz he took the rational rejection of action-at-a-distance and from Leibniz the methodology of harmonizing and synthetizing. Though he failed in his theory of the ether and in his theory of gravitation, both resulting from his attempted synthesis, he was so eminently successful in his mechanics of solids and mechanics of fluid bodies that he became a central figure in the development of that kind of science which flourished in Germany in the nineteenth centruy and harked back to Eulerian and Leibnizian syntheses of conservation ideas and unitary principle. He was also an ardent supporter of the *wave* theory, or rather of the anti-corpuscularean theory of light. Moreover, in the development of that attitude to science which is best described as 'essentialistic',[26] Euler serves as an important link.

The concept of impenetrability occurs from Descartes onward almost in every treatise on the theory of matter – be it corpuscular or dynamic in approach, or a combination of both. 'Corpuscular' theories treat mass and motion as fundamental entities; the 'dynamicist' theories use force and motion, while there is a third approach which uses at once the concepts of force and mass.

This division is advocated by Meyerson. He calls Descartes, Boyle, D'Alembert, Lagrange, Laplace, Kirchhoff, Hertz and others corpusculareans; Leibniz, Boscovich and Kant are the arch-dynamicists; the third category is represented by those whom Meyerson simply calls the 'physicists', such as Newton, Maxwell or Helmholtz, who "preferred less absolute solutions, departing still farther from rigorous logic, but offering more of a hold on our imagination."[27] This approach is appealing; yet it seems to me that this classification leads into difficulties. It is appealing because Meyerson treats together whatever we nowadays call physics and metaphysics; it was following Meyerson that I formulated above my second qualification to the thesis that 'metaphysics influences physics'. But difficulties occur, because Meyerson concentrates not on the problem-situations of his subjects but rather on their answers. I must digress at this point.

IV.   It is a widely accepted tradition to call Descartes, Leibniz, Boscovich and Kant philosphers and not physicists, but what shall D'Alembert, Lagrange and Laplace be called? Was Hertz less of a physicist than Maxwell? These are seemingly only labelling difficulties but there is more to it. Meyerson's corpuscularean theory (if seen against a dynamicist theory) views force as a derivative quality; it is not inherent in matter, it is a result of collisions or of pressure which is a kind of impact [28] between particles. In Meyerson's words:

Action by impact constitutes the esential element... for every corpuscular theory.[29]

Contrarily, according to Meyerson, it is in the character of every dynamicist theory to assume action-at-a-distance, This leaves us with the following questions open: how is it then that Leibniz, who was the father of dynamicism, was so violently opposed to action-at-a-distance? How is it that Euler, who was a Leibnizian, at least in his opposition to action-at-a-distance,[30] did not give to force the same central and primary importance as to the (quality of) impenetrability?[31] How is it that Kant, at one stage of his life accepted Newtonian gravitation, and at a later stage almost rejected it, and yet always rejected the corpuscularean view?[32] These questions suggest that a different classification of philosopher-scientists is necessary.

If successful, such a classification would show that there is a scientific development on the Continent, parallel to the Newtonian, rooted in a metaphysics which is different from the Newtonian, and equally responsible for the birth of twentieth-century science, not less than the Newtonian.[33] This would also answer the historical questions: 'Why is Euler considered a Newtonian?'; 'why were Euler's philosophical works neglected?'; and, assuming as I do, that Euler had considerable conceptual influence on Kant, 'why has Euler's influence on Kant been so little emphasized'?[34]

In distinguishing between the various kinds of philosopher-scientists, it may be a better strategy to look at the questions they asked, and then to classify according to that, and not according to the answers they gave.

Those whom the tradition labelled physicists are also Meyerson's 'physicists' – his third category. Newton, Maxwell and Helmholtz were not system-builders; they were not, at least not explicitly, asking questions of the kind 'what is the very essence of matter?' (Helmholtz was different in his epistemological papers). Newton's problems, at least in the *Principia*, and Maxwell's and Helmholtz's, were the problems of the mathematical-physicist: they took some concepts for granted, and asked for the relations between them.[35] While Leibniz was waging his lifelong campaign to prove Descartes wrong by elevating the concept of force to central importance, Newton simply presupposed force, and worked with it. The physicists are those who replaced the 'absolute solutions' by more imaginative answers. As a result they are not committed to a rigid system. Newton did not claim that all forces were of the same kind; he could accommodate all kinds of forces in his physics. Contact forces explained some phenomena, while action-at-a-distance forces explained others. It seems to me that the extensive Newton literature has shown that Newton's denial of action-at-a-distance came after he had actually used it in his work. But I admit that this is open to debate.

V. Descartes, Leibniz, Boscovich, Kant, as well as Euler and many others on the other hand, began asking 'what is the essence of matter?'. Some of them were corpusculareans, others were dynamicists, but all were system-builders, concept-creators. These are the men in whose work half-distinct concepts play a major role. I must repeat their answers to the question 'what is the essence of matter?' and juxtapose them to the answer given by Euler. Descartes' choice was that extension was the fundamental property. His answer was soon discarded. Leibniz went through three stages: at first he thought that motion was the fundamental property of bodies; this was when he still acknowledged the jurisdiction alone in regard to material things. Later he became the dynamicist as which he is known. As a last stage he developed his theory of the monad. His monad, the way I read it, is a direct generalization of his concept of force; i.e. all that Leibnizian force did in his system, the monad does too, but it does more: the Leibnizian force was the essence of matter, and all phenomena of the physical world were related to it. Now the monad does all that, but in addition unites the material with the spiritual.

Euler and Boscovich[36] were followers of Leibniz. They too asked first of all what was the essential property of matter, and both considered the Leibnizian 'impenetrability' and 'force' as the essential properties of matter. But their solutions were complementary.

Euler found that the fundamental quality was impenetrability. It was fundamental insofar as Euler thought he could 'derive' – in those days 'derive' had a much less mathematical meaning than nowadays – all other properties and fundamental laws from it. Force was for Euler only an effect of impenetrability.

For Boscovich the essence of matter was force, and impenetrability was its effect. As a result of these different metaphysics, for Euler all action was by impact; that is, only contact-forces existed, while for Boscovich, all action was due to forces acting-at-a-distance. Yet he was not a proper corpuscularean for his system dispensed even with the concept of mass as a fundamental property.[37]

Kant, as we shall see, was influenced by Euler and opposed the corpuscularean view. As mentioned above, he first accepted action-at-a-distance, while in his old age he became even more Eulerian and he rejected action-at-a-distance. It was only Newton, and even more his followers in England and later in France, who, not having asked 'what is the essence of matter?' could say that some forces were contact forces while others acted-at-a-distance.

In view of all this, I shall now summarize the Eulerian theory of matter. It is a coherent system with the concept of impenetrability at its foundation.[38]

What is the essence of matter? Is it extension? Extension certainly is an essential property of matter, but not the most fundamental one. All bodies are extended, but not only bodies. Space is extended too. Let me interject here, that for Euler there are three different kinds of entities: bodies, spirits and space and time. (As for Kant after him.) The fundamental property for which Euler is looking must be a property of all bodies and only of bodies.

Is mobility an essential property of bodies? The answer is no: motion does not change anything in a body. A body at rest is not different from a body in motion. The property of mobility only serves to distinguish between bodies and space, for mobility is not a property of space. Yet it is not only bodies which are moving; an image in the mirror exhibits also the properties of mobility. One wonders whether Euler would categorize an image in the mirror as a spirit.

We come now to inertia, and with it to the first serious breach with Newtonian tradition. In the conceptual framework of Euler's days, the first problem on which he had to take a stand was: 'is inertia an innate property?' or in Newtonian terms: 'is inertia an innate force?'. Euler was imbued with the Leibnizian spirit and terminology: qualities were either passive or active. Inertia is the very essence of passivity. Force on the other hand is the very essence of activity. Thus Euler insisted that inertia was not a force.

Euler, who was very much language-conscious, took great trouble to eliminate any connection with Newtonian 'force of inertia' and he proposed a change in terminology. He suggested that inertia should rather be called 'Standhaftigkeit' – perseverance.[39]

Let me draw here a brief comparison with Newton. There is a widespread discussion in the historical literature whether Newton himself considered inertia as a force; the fact that he used the expression *vis insita* cannot be considered as conclusive evidence either for or against this view.[40]

In view of this, the real issue should be formulated in the following way: 'while it is doubtful whether Newton wanted to consider inertia a force like all other forces, he considered it an active resistance to change. He calls it a 'power' or sometimes an 'efficacy' – both these terms point to an active property. Euler, on the other hand, considered inertia not a force but a totally passive quality, even if Euler expressed himself that inertia is 'resistance to change'.

This difference between Newton and Euler is borne out in a very significant difference in their treatment of the law of inertia. As is well-known, Newton formulates his first law thus:

Every body continues in its state of rest, or of uniform motion in a right line, unless it is compelled to change that state by forces impressed upon it.[41]

That is, the two states, of rest and of uniform motion, are given a unified treatment. This is a law of conservation of state. External forces are necessary to change this state of the body, whether the state be a state of rest or of uniform motion, because inertia acts like a force to resist change.

Yet Newton did not draw a further conclusion from his own view of inertia. In spite of the fact that for him inertia was an active force, Newton did not consider it commensurable with external forces – it was different in kind. Euler on the other hand made for the first time a pure dimensional analysis in which it turned out that all forces can be added or substracted from each other. When Kant wrote his first work on *The Living Forces...* he had reviewed the entire literature on the subject (except the one correct solution, namely that of d'Alembert). Kant accepted the Eulerian view of fundamental conceptual equality of all kinds of forces but at first tried to keep the Newtonian view that force of inertia was an active force. Later, and especially in his opus posthumum, he too switched to the Eulerian view that impenetrability was the primary active force and *vis inertiae* was only a derivation of it.

In Euler's *Mechanica* – his first important work in Mechanics, and the one that established Euler's fame as a Newtonian – we find a surprisingly different treatment. Here two separate laws of inertia are introduced.[42] The one deals with bodies at rest, the other with bodies moving uniformly. Why is there such a difference in treatment? The answer is that both Newton and Euler accepted an absolute space. It is only with respect to absolute space that there is a difference between rest and uniform motion. Yet their treatments are different, and this creates a problem. One solution to the problem is offered by those Newtonian scholars, who in Rouse-Ball's wake, treat the Newtonian absolute space, unlike Newton, as a mathematical device and not as a physical reality. Here there is no problem, this would be a case of rational reconstruction which is historically untenable. As to Euler, there is little doubt that he attributes to absolute space physical reality. In Euler's words: "One would then be obliged to admit, as had been in reference to space, that time is something, which exists outside of our mind, or that time is something real, as properly as space."[43]

Yet it seems to me that there is an additional, though not independent reason for treating rest and uniform motion separately. Even if, with respect to absolute space, the two states are distinct, in Newton's treatment they are united by an active resistance to change of motion. By active, I mean that the body does something instead of something merely happening to it.[44] This falls away in Euler's treatment. Inertia is not a force, it is not an active resistance to change. It is a passive quality of all matter. In fact, the law of inertia is explained by Euler with reference to the principle of sufficient reason which underlines the passive nature of this property:

If a body is in a state of rest, and there is no external cause present which would act on it, it is inconceivable how it could be brought into motion. If it began to move, this would have to be in some direction; but there is no reason why it should move more in this or in the other direction. In the absence of such a sufficient reason, we can conclude that a body at rest will remain so forever as long as no external causes make it move.[45]

Is this kind of inertia an essential quality? It is shared by all bodies, and probably, though Euler does not say it, it is a property only of bodies. Yet it is not the essential property for it is derivable from another property.

Moreover, one cannot derive the second and third laws of motion from the concept of inertia alone, while these can be derived, according to Euler, from the concept of impenetrability.[46]

Impenetrability is the fundamental property of matter, it is the essence of matter. Euler calls it a metaphysical concept. It is not measurable, it has no degrees: there is no partial impenetrability. It is a property of all bodies and only of bodies. It is connected with important metaphysical principles like the conservation of state or the principle of least action. Conservation of state is the law of inertia, which is derivable from impenetrability. That nature chooses its phenomena to take place according to the principle of least action is a corollary of the property of impenetrability.

I shall give here one of these Eulerian derivations to illustrate what Euler means by 'deriving' one concept from another: for example Euler derives the concepts of extension, mobility, and inertia (or rather 'Standhaftigkeit') from that of impenetrability. The argument is as follows:

Without extension, impenetrability is meaningless, for a thing which has no extension cannot occupy a place. Then the question whether another thing can occupy the same place at the same time is nonsensical. Thus an impenetrable object is necessarily extended. An object which occupies a place can change its place at least potentially – and we have to ascribe mobility to this object; but it occupied place as we have seen, due to its impenetrability, and thus impenetrability implies mobility. But the 'Standhaftigkeit' – the Eulerian 'inertia' – is inseparably connected with mobility:

if an object is mobile, then also 'Standhaftigkeit' has to be attributed to it, for otherwise any change would occur without a sufficient reason.[47]

And thus,

In impenetrability all other properties like extension, mobility and 'Standhaftigkeit', are included.[48]

It is on similar lines that an attempt is made to 'derive' the third and second laws of motion, in this order, from the concept of impenetrability.

What about forces? All contact forces are effects of impenetrability. When two bodies collide, or when one body exerts pressure on another, the forces of opposition are just such as to prevent penetration, never more and never less. Can universal gravitation be explained by this fundamental property? Yes, but not on Newtonian lines. It is rather on Cartesian lines that Euler attempts to do it. For in Euler's system, as is well-known, there are no action-at-a-distance forces. All action is by contact – either between two particles of matter, or between the particles of matter and of the all-pervading ether. 'All-pervading' means all space where there is no matter.

What is called universal attraction is the result of the pushes and pulls of the contact forces between particles of matter and the particles of the subtile, tenuous ether. In other words, these contact forces are again the effects of impenetrability. Thus gravitation is not an essential property of matter.

The mechanism of Euler's theory of gravitation where the effect is transmitted by the ether is rather sophisticated; it seems to me that it is more ingenious, though not more successful, than Descartes' and Huygens' mechanism.

Descartes and Huygens operated with pushes between particles. Euler, on the other hand, operates with the new concept of pressure – a concept which, according to Speiser and Truesdell,[49] was introduced by him. A push is unidirectional, while pressure acts on a body from all sides and perpendicularly to the surface at every point. In other words, Euler applied his successful theory of the mechanics of fluid bodies to the ether.

The theory of contact forces unites all physical pehnomena. All the forces which act between bodies, whether the phenomena be mechanical, electrical, magnetic, or of light, are contact forces. And thus all phenomena can be deduced from the concept of impenetrability.

I conclude thus that there are no other forces in the world but those which originate in the impenetrability of the bodies.[50]

4. EULER'S INFLUENCE ON KANT

Euler's essentialism and theory of matter influenced Kant and Helmholtz
in Germany, and Thomas Young and through him Faraday in England.
Euler proved that important and successful work can be done when motivated by his metaphysics, and this became one of the important sources
for the continuum theory of matter which developed in the late eighteenth
and nineteenth century, very often through the mediation of Kant. It
founded the scalar formulation of mechanics, the mathematical study of
fluid mechanics; it motivated Helmholtz in developing the theory of energy, and gave a scientific support to the revival of the various vortex
theories in the mid-nineteenth centruy. Finally it was to Euler's metaphysics that the anti-atomistic tradition in the late nineteenth century
referred.

Euler's fame rests mainly on his mathematics and on his books on
Mechanic and on the *Theory of Motion of Rigid Bodies.* It is on the ground
of these treatises that he was labelled a Newtonian. If, however, these are
read in conjuction with his important philosophical papers, even these two
works will turn out to be much less Newtonian than generally considered.

Where does Kant come into this picture?[51] Why is it that Kant's name
so seldom occurs together with that of Euler? Is it at all possible to say
something about Kant which has not yet been said? Let me begin the
outline of an answer to these questions, not forgetting that I am dealing
here with half-distinct concepts in science and metaphysics, and that my
historical sub-thesis is that Kant's metaphysical or scientific concepts
were deeply influenced by Euler's metaphysics. This thesis is not new. It
was adumbrated by several late nineteenth century historians. I do not
pretend to know why it never gained momentum, but one answer lies in
our deep-seated prejudices, that scientists are not metaphysicians. Euler's
philosophical works have been totally ignored by the historians of science
and of philosophy of the nineteenth and twentieth centuries. In all probability even if Euler had not been neglected, his influence on Kant would
have been. The kind of influence that many historians love to cultivate
– and I am drawing a caricature on purpose – is something of the following: 'The greatest scientist of all times, Newton, exercised the deepest
influence on the greatest philosopher of all times, Kant'. It is hard to

warm up to the suggestion that an admittedly 'bad' philosopher like Euler, could have exercised an influence on the great Kant.

But, it seems to me that there is no direct and simple correlation between the field of influence of a philosopher-scientist on his contemporaries, and the field in which we assign to him greatness in our time. In Euler's case, this is true perhaps, more true than elsewhere. While his highly technical mathematical, astronomical and physical papers were written for the few great mathematicians of his age, his philosophical *Memoires* read at the Berlin Academy in the 1740's were widely read and discussed in all intellectual circles. For us he is mainly a mathematician, astronomer, or mechanician, – but that was yet another reason for not connecting Kant with Euler for so long.

Kant mentions Euler very often and in highly praising words, but it so happens that in the all-important first *Critique* Euler is not mentioned at all. The *Critique* so overshadowed all other works of Kant, that even the *Metaphysical Foundations* was mostly ignored, at least until the Kant-renaissance which began in the 1870's. Kant's last work, the posthumously published *On the Transition from the Metaphysical Foundations of Natural Science to Physics*,[52] which is the least Newtonian and most Eulerian of all his works, has been ignored as being a work of senility.

Yet it seems to me that Kant was directly influenced by Euler's concepts of space, time, impenetrability – in short by Euler's theory of matter. These concepts, half-distinct as they were in Euler's system, became the building blocks of that physics, on which Kant, by a further step of abstraction, constructed his metaphysics.[53] By its very formulation, and choice of words, Kant's metaphysics, as is well-known, became the chief support of the hostile attitude towards all metaphysics which characterized philosophy of science until the days of Einstein. This attitude generated a kind of history of science, which, with the possible exception of Cassirer, important as he was, looked for the scientific roots of Kant only in positive distinct science; it is small wonder that they thought they discovered this in Newton's work. The renaissance which rescued Newton's philosophy from the strait-jacket of positivism has not yet reached Euler and Kant. Kant's half-distinct ideas were made distinct by hindsight, after having been amputated from the body of a coherent system on which they had grown.

As I wish to concentrate here on Euler's influence on Kant's theory of

matter, I mention only briefly that Kant's theory of absolute space and
his theory of time are based on Euler's. The way I read Kant it seems to
me that he was impressed by Euler's complete separation between the
state of rest and the state of uniform motion. It was this separation which
made his absolute space more absolute than Newton's; his further abstraction of it, by elevating it into the world of the physically unattainable
ideas, did not make it less real than Euler's absolute space from which it
started. The influence can be detected in Kant's theory of physical time,
which is simply 'a measure of the duration of things'. The Kantian conception of place is again based on Euler. The idea here is that the concept
of place is not formed except by removing the body which occupied it.
It was on this that Kant based his well-known and erroneous theory of the
difference between a body and its mirror image.

From an early age Kant accepted the Eulerian notion that the answer
to the all-important question 'What is the essence of matter?' lies in the
concept of impenetrability.

But there are several things in it (in the universal science of nature) which are not quite
pure and independent of empirical sources, such as the concept of motion, that of impenetrability (upon which the empirical concept of matter rests...).[54]

But before that: mobility too, is a property of all bodies: Material substance is whatever is mobile in space. Quantity of matter is the mass, and
the measure of mass is the quantity of mobile substance. This is an attempt to identify mass with the concept of matter as mobile substance.
What was said above about extension, inertia and impenetrability in
Euler's system could be almost *verbatim* repeated for Kant's theory of
matter. The major difference between them is in the relative importance
of impenetrability and force. While for Euler impenetrability was the
cause of force, and this had been formulated unambiguously, for Kant it
very often was the force which was the cause of impenetrability. This was
the attitude which Meyerson called Kant's dynamism; it originated with
Leibniz and was shared to different degrees by Euler and Kant. It relied
on an anti-corpuscularean view in both of them.

As to the nature of these all-important forces, Kant went through a
whole cycle of opinions. While a young man, he hesitated between the
Newtonian and the Leibnizian conceptions of forces. Later, he tried to
unite the two by giving a Leibnizian, metaphysical importance to the concept of force, but as its physical nature he chose the Newtonian action-

at-a-distance. This is the well-known Newtonian element in the *Meta-physical Foundations* which led into internal difficulties mainly because of Kant's opposition to ultimate atomic units of matter. Here, in the Meta-physical Foundations, which could be called the second stage in Kant's theory of matter, he accepted the Newtonian gravitational force as action-at-a-distance. But even here Kant did not reject completely the possibility of contact forces, and when he talks about these, not surprisingly it is in connection with impenetrability:

Contact in the physical sense is the immediate action and reaction of impenetrability.[55]

Later, Kant tended to reject the Newtonian type of gravitation and adopt-ed the Eulerian theory of contact forces. For both Euler and Kant, whatever the nature of the forces that are active in the world, it is im-penetrability which, expressed through forces, makes a body capable of filling space. The empirical concept of matter, for both of them, rests on the concept of impenetrability.

The list of influences could be continued: Kant's theory of the ether, Kant's concept of infinity, Kant's *Monadologia Physica* are all Eulerian in spirit, might have influenced and probably were influenced by Euler. Even Kant's discovery of the retardation of the rotation of the earth caused by tidal friction came as an answer to the Berlin Academy Competition of 1754, which was probably formulated by Euler. There is even evi-dence,[56] though not conclusive, that Kant knew of Euler's theory of the gradual shortening of the course of the year. The importance of this is that Kant, the abstract metaphysician, worked all his life on an attempted unified system between physics and metaphysics. His first work dealt with the problem of living forces, his last work with the transition from meta-physics to physics. As the physical basis of his system he chose a half-distinct scientific concept extracted from Euler's metaphysics – the con-cept of impenetrability.

5. EIGHTEENTH CENTURY IMAGE OF SCIENCE

We set out to review one chapter in the history of scientific metaphysics from the point of view which combines (a) the history of (disembodied) scientific ideas: how Euler and his contemporaries dealt with the theory of matter and its open questions as it was left open by Newton and

Leibniz; (b) the influence of the image of science in the eighteenth century
or the kind of knowledge which is aspired; (c) the influence of institutional and other socio-political external factors which influence science.
(a) was covered in some detail. (c) has not been done satisfactorily in
detail by anybody. There will be many questions raised: what was
the impact of the fact that the Berlin Academy consisted mainly of
French, Swiss, Russians and other non-Germans? How did Frederic the
Great's German nationalism (created by him, its charismatic leader)
harmonize with the cosmopolitan 'Enlightenment' attitude to science and
culture in general? Why did Euler write his *Anleitung zur Naturlehre* in
German (instead of the usual French)[57] and why was it never published?
What was the institutional difference between the Berlin and St. Petersburg Academies and how did this influence Euler's work who was in
contact with both? Why was Germany the centre of growth of the strong
synthesizing and reconciliatory tradition best represented by Leibniz,
Euler and Kant? How did Kant deal with his own German nationalism,
pro-French revolution attitude and the changing Prussian-Russian occu-
pancy of his beloved Königsberg? Finally, how was the social and political
Rousseau-influenced philosophy of Kant (constructed on his scientific
metaphysics) influenced by the absolute autocratic rule of Frederic the
Great's successor? This I leave for the present.

However I shall attempt here at least to adumbrate an answer to (b).

The conclusion drawn from Newton's unprecedented success was that
certainty of non-revealed knowledge is possible. This was the starting
point of that quest for certainty which exemplifies our science, philosophy,
literature, and politics ever since the seventeenth century and which since
Freud has been described as an innate (i.e., non-cognitive and non-en-
vironment-influenced) need shared by all human beings (if not by all live
creatures). Due to the Baconian influence this certainty centered around
the concept of 'fact' in England; in France (due to Descartes!?) it looked
to mathematics to provide that certainty; Germany, since Leibniz, sought
certainty in consensus: such metaphysical synthesis which would harmonize between the differing schools and would discover underlying unitary principles (conservation laws in nature) which apply to everything.
This became the basic task of the scientist-philosopher and no problem
looked worthwhile to be worked upon if it was not directly contributory
to the broad metaphysical view. This image of science and this view of

the role of the scientist kept alive the Leibnizian tradition through Euler, Kant, Hegel, Fichte, Helmholtz, Fechner, Freud, Marx, Einstein and Bohr. This tradition has been engaged ever since in a critical dialogue with the competing Newtonian and Cartesian traditions.[58]

In the Introduction to Euler's *De Curvis Elasticitis, Additamentum* I to his *Methodus Inveniendi Lineas Curvas Maximi Minimive Proprietate Gaudentes* (this is the work where Euler solves the isoperimetric problem, and invents the calculus of variations which became so famous during the fight between Jacob and Johann Bernoulli), Lausanne and Geneva, 1774, we find a passage which is most typical of his approach to physical problems and casts also some light on his image of science and methodology:

Wherefore there is absolutely no doubt that every effect in the universe can be explained as satisfactorily from final causes as it can by the aid of the method of maxima and minima from the effective causes themselves. Now there exists on every hand such notable instances of this fact, this, in order to prove its truth, we have no need at all of a number of examples; nay rather one's task should be this namely, in any field of Natural Science whatsoever to study that quantity which takes on a maximum or a minimum value, an occupation that seems to belong to philosophy rather than to mathematics. Since, therefore, two methods of studying effects in Nature lie open to us, one by means of effective causes, which is commonly called the direct method, the other by means of final causes, the mathematician uses each with equal success. Of course, when the effective causes are too obscure, but the final causes are more readily ascertained, the problem is commonly solved by the indirect method; on the contrary, however, the direct method is employed whenever it is possible to determine the effect from the effective causes. But one ought to make a special effort to see that both ways of approach to the solution of the problem be laid open; for thus not only is one solution greatly strengthened by the other, but, more than that, from the agreement between the two solutions we secure the very highest satisfaction. Thus the curvature of a rope or of a chain in suspension has been discovered by both methods; first, *a priori*, from the attractions of gravity; and second, by the method of maxima and minima, since it was recognized that a rope of that kind ought to assume a curvature whose center of gravity was at the lowest point.[59]

Thus Euler's methodological demand is for theoretical proliferation; moreover, his image of science is such that whatever we do in our mathematics in a way of revealing actual reality in Nature and by necessity whether we proceed by the direct method which is the method of minima and maxima from efficient causes or by the indirect method of proceeding from final causes, i.e., from *a priori* principles, we must reach the same conclusions. This mutual reinforcement is the real satisfaction for the natural philosopher. The sharp distinction between final, *a priori* causes, and effective causes is Leibnizian rather than Newtonian and in this Kant turns out to be a few decades later Euler's most faithful disciple.

These views on the role of science and metaphysics combined with the strongly Puritan view of German Pietism which stressed the habit of self-renouncing labour, of singleness of purpose. This was a *sine qua non* for developing scientific metaphysics on unshakeable foundations on which a whole system of science, ethics, social theory and an anthropology had to be built.

In addition, Euler, and later Kant, developed a theory of growth of knowledge which rejects the view of knowledge-by-accumulation and comes much nearer to the view that knowledge grows by a continuous critical dialogue between competing metaphysics. The connecting nineteenth-century link in this tradition was William Whewell. Its twentieth-century main representatives are Popper, Agassi, Lakatos and their followers.

Euler's realistic attitude and his opposition to the growth-by-accumulation theory of the growth of scientific knowledge is beautifully illustrated by his systematic, detailed description of trials and experiments that failed, and allows us to gain a true insight into his method of discovery. In all the literature on Euler, the only source that I know that draws attention to this at all, without however drawing any conclusions as to Euler's theory of the growth of knowledge, is an essay by Fuchter in 1948.[60] All this in complete harmony with the above-mentioned theoretical pluralism.

The Eulerian-Kantian view of science and the resulting scientific-metaphysics determined for the next generations a choice of problems dealing with general principle in Nature. Their work complemented the Newtonian achievements and the French mathematization of mechanics: the principle of conservation of energy, the continuum approach which gave birth to the field concept, the new, specifically biological approach, both in its vitalist formulation and the 'hard-approach' cell-theory-reductionism grew on Kantian soil.

Finally, it is this same attitude of the Leibniz-Euler-Kant tradition which is at the core of twentieth century philosophy of science (from whatever school we happen to be) as expressed by Meyerson:

Science only accomplished finite progress, and all the mechanical hypotheses which it forms being contradictory in themselves – that is, absurd at bottom – it always remains separated by an infinite distance from the logical conception toward which it seems to tend.

For doing, understanding and evaluating science a demarcation between science and metaphysics is misleading and superfluous.

Department of History and Philosophy of Science
The Hebrew University, Jerusalem

NOTES

[1] J. Agassi, *Towards an Historiography of Science*, Mouton & Co., 's-Gravenhage, 1963, Beiheft 2, *History and Theory*, p. 3.

[2] *Ibid.*

[3] As typical of this stage one could point to Thomas S. Kuhn's article, 'The History of Science' in *International Encyclopaedia of the Social Sciences*, MacMillan 1968.

[4] In 1931 a conference was held in London where some of the greatest Russian theoreticians of the philosophy of science participated: Bukharin, Joffe, Rubinstein, Colman, Hessen, Vavilov, etc. The proceedings were published under the title *Science at the Crossroads*, Kniga, England, 1931. Other important writers in this vein were Lancelot Hogben, J. C. Crowther, Bernal with his monumental *Social History of Science*, S. Lilley's papers and especially the volume *Essays in the Social History of Science* (ed. by Lilley), which appeared in *Centaurus* 3 (1953) No. 1, 2. It was recently reprinted with an introduction by P. G. Wershey and a foreword by J. Needham, Frank Cass 1971.

[5] For example see J. Ben-David's recent book *The Role of the Scientist in Society*, Englewood Cliffs, N.Y., 1971; Roger Hahn *The Anatomy of a Scientific Institution: The Paris Academy of Sciences 1666–1803*, Calif. Univ. Press. 1971, and also numerous articles in *Minerva*, a journal dedicated to this topic, edited by Professor Edward Shils.

[6] I. Lakatos, 'Falsification and the Methodology of Scientific Research Programmes', in *Criticism and the Growth of Knowledge* (ed. by I. Lakatos and A. Musgrave), Cambridge U.P., 1970. Also, I. Lakatos, 'History and Its Rational Reconstructions', in *Boston Studies in the Philosophy of Science*, Vol. VIII, pp. 91–136, D. Reidel Publ. Co., Dordrecht-Holland, 1971.

[7] For this see E. Mendelsohn and A. Thackray (eds.). *Science and Values*, Humanities Press, in print, and J. E. Curtis and J. W. Petras (eds.), *The Sociology of Knowledge*, N.Y. 1969.

[8] K. R. Popper, *The Logic of Scientific Discovery*, Hutchinson of London, 1959, p. 34.

[9] Lakatos calls even the great A. Koyré confused for that reason: "For example, for non-inductivist historians Newton's *Hypotheses non fingo* represents a major problem. Duhem, who unlike most historians did not overindulge in Newton-worship, dismissed Newton's inductivist methodology as logical nonsense; but Koyré, whose many strong points did not include logic, devoted long chapters to the 'hidden depth' of Newton's muddle." In 'History and its Rational Reconstruction', *op cit.*

[10] S. Lilley, *op. cit.*

[11] *Op. cit.*

[12] The hard core of a Research Programme according to Lakatos is: "All scientific

research programmes may be characterized by their 'hard core'. The negative heuristic of the programmes forbids us to direct the *modus tollens* at this 'hard core'. Instead, we must use our ingenuity to articulate or even invent 'auxiliary hypotheses', which form a protective belt around this core, and we must redirect the *modus tollens* to *these*". Lakatos (1970), p. 133.

[13] The first case deals with the emergence of the concept of energy and the principle of its conservation in the work of Helmholtz. I attempted there to bring together the problems that occurred in the body of physical and biological science in the 1840's, the 1840's image of science and the institutional and cultural environment in which Helmholtz lived. This paper 'Helmholtz's Kraft: an Illustration of Concepts in Flux' appeared in *Historical Studies in the Physical Sciences*, Vol. 2 (ed. by McCormmach), University of Pennsylvania Press, 1971, pp. 263–298, and forms the main theme of my forthcoming book; *The Discovery of the Conservation of Energy*, Hutchinson's Educational, in press.

The second study is of the discovery of conservation of energy in Germany by Mayer, Helmholtz and others and the discovery of the motion theory of heat in England by Joule, Thomson and others. Two discoveries that at first, because of differences in scientific problems, in the image of science, and in the socio-cultural setting, looked unconnected until in the 1860's they were conflated into one major discovery. 'The Conservation of Energy: a Case of Simultaneous Discovery?' in *Arch. Inter. d'Histoire des Sciences*, No. 90 (1971), pp. 31–60.

The third case history is 'Boltzmann's Scientific Research Programme and the Alternative to It', to be published in *Chapters in the History of the Interaction between Science and Philosophy* (Proceedings of the International Symposium held at the Van Leer Jerusalem Foundation, January 1971), ed. by Y. Elkana, Humanities Press, in press.

[14] Such distinguished modern works as Dijksterhuis', *The Mechanization of the World Picture* or Dugas' *History of Mechanics* (it is purely incidental that both deal with mechanics) treat only such concepts which can be translated into the terms of a modern science textbook. For example, Dijksterhuis does not consider the 'conatus' of Hobbes nor Leibniz's general concept of force which was influenced by it. The same is true of Dugas who does not mention Hobbes at all; nor does he relate at all to Euler's 'impenetrability'. Both books ignore the concept of the monad. But these writers would still be considered old-school inductivists, and it is not my purpose to offer another sweeping attack on the inductivist-positivist historians of science since relatively few extreme inductivist-positivist histories of science have been written lately. Professionalization has freed the history of science from the positivistic strait-jacket. History of science is not being written any more by distinguished professors emeriti of science, and most of us historians have heeded the message of the new tradition inaugurated by writers like Hannequin, Metzger, Burtt, Pagel and Koyré. The common assumption in their work is, that there is a far-reaching influence of scientific metaphysics on science; it is also widely accepted that every scientist has a metaphysics. This has resulted in a welcome change. They realize that scientific concepts and theories are often transferred from the indistinct world of metaphysics to the distinct world of science. But it is our modern image of science and metaphysics which underlies our determination what is science and what is metaphysics. Thus it is still true that an idea is considered scientific if it has a genetic relation with a modern scientific idea. It is assumed that what is science and what is metaphysics is now clear, and that modern scientific ideas are distinct.

To prove them wrong, to show that the neglected theories or concepts do belong to the domain of the historian of science, it would be necessary to show that what they thought was metaphysics is actually science; i.e. a demarcation between science and metaphysics seems to be necessary. But this is paradoxical because it is precisely these 'modern' historians who claim to know where the demarcation is, while I happily admit total ignorance of such a demarcation.

15 Meyerson laid down the methodological foundation for a new kind of philosophy of science and indeed of a kind of history of science which, to the best of my knowledge, has not yet been written in the form of professional history of science. Meyerson himself considered historical research not an end but a tool for understanding the nature and growth of knowledge. Such an understanding can be achieved according to him by treating every concept used by a philosopher-scientist in the context of all the ideas which he thought relevant to it. I am naturally referring to his famous *Identity and Reality*, Dover. N.Y., 1962 and also to works like Meyerson's three volume, *Du Cheminement de La Pensée*, Paris 1931, and also his *De L'Explication dans les Sciences*, Paris 1927.

16 Leibniz himself says in his letters to De Volder: (G. W. Leibniz, *Philosophical Papers and Letters*, ed. by Leroy E. Loemaker, D. Reidel Publ. Co., Dordrecht-Holland, 1969) "...you are already tacitly assuming what matter would be except through monads, since it would always be an aggregate or rather, the result of a plurality of phenomena, until we arrived at these simple beings" (p. 534), and "...It is essential to substance that its present state involves its future states and vice versa. And there is nowhere else that force is to be found or a basis for the transition to new perceptions...It is also obvious that in actual bodies there is only a discrete quantity, that is a multitude of monads or of simple substances, though in any sensible aggregate or our corresponding to phenomena, this may be greater than any given number" (p. 539). In his 'Reply to the Thoughts on the System of Pre-Established Harmony' contained in the second edition of Mr. Bayle's *Critical Dictionary, Article Rorarius'*, 1702, Leibniz repeats: "And I do in fact regard souls, or rather monads as the atoms of substance since there are no material atoms in nature according to my view and the smallest particle of matter still has parts. Since an atom such as Epicurus imagined has a moving force which gives it a certain direction, it will carry out this direction without hindrance and uniformly if it encounters no other atom", *op. cit.*, p. 579. Finally in *The Monadology* (1714) he says "It follows from what I have said that the natural changes in monads come from an *internal principle*, since an external cause could not influence (influer dans) their interior", *op. cit.* pp. 643–644. Loemaker himself (*op. cit.*) also draws attention to the close relationship between monad and force: "All monads are temporal series of active force and passive content, representative of the universe and striving toward the purposes defined in the individualized law from which they proceed", (Introduction, *op. cit.*, p. 45) and in his annotation to 'matter' he says "...secondary matter, as matter is phenomenal, and only the aggregated monads whose passive force is expressed as resistance or inertia are substantial" (*op. cit.*, p. 508, note 12).

17 Anthropologists wrote extensively on coherence which is not of a logical kind. Lévy-Bruhl in his *Les Carnets de L. Lévy-Bruhl*, Paris 1949, writes that mystical thought is 'organized into a coherent system with a logic of its own' (p. 61). According to Evans-Pritchard in a review essay on Lévy-Bruhl (E. E. Evans-Pritchard, 'Lévy-Bruhl's Theory of Primitive Mentality', *Bull. of the Faculty of Arts* II (1934), 1–36) mystical thought is "intellectually consistent even if it is not logically consistent and primitive thought is eminently coherent, perhaps over-coherent. ... Beliefs are coordinated with other beliefs and behaviour into an organized system". Both quotations are brought

302   YEHUDA ELKANA

by Steven Lucas in his paper 'Some Problems about Rationality' which appeared in
Rationality (ed. by Bryan P. Wilson), Basil Blackwell, Oxford, 1970, p. 202. See also
G. Holton, 'Science and New Styles of Thought', *The Graduate Journal* 7 (1967),
399–422. The motto to this article is "not in logic alone."

[18] There is a sense in which the man who composed the *Razumowsky* quartets must
be the same man who wrote also *Fidelio*. The question is how to translate such a vague
insight into practical terms – into fruitful historical research. What a commitment to
such a belief in coherence implies is that the works of the philosopher-scientists must
be treated as a whole.

[19] E. A. Burtt, *The Metaphysical Foundation of Modern Physical Science*, N.Y. 1927.
J. M. Keynes, 'Newton the Man' (1836) in *Essays in Biography*, Horton & Co., N.Y.,
1963. E.W. Strong, 'Newton's Mathematical Way' in *Roots of Scientific Thought* (ed. by
P.P. Wiener and A. Noland), Basic Books, N.Y., 1957.

[20] Frank E. Manuel, *A Portrait of Isaac Newton*, Harvard U.P., Cambridge, 1968.

[21] Even in these works it is presupposed that, when investigating the influence of
metaphysics on physics, it is always known which is which, the emphasis being on
'always'. For example, Koyré in his *Newtonian Studies* (Harvard U.P., 1965) examines
in detail the influence of Newton's metaphysical idea of space on his physical idea of
inertia. He also adds that Leibniz's different metaphysics of space prevented him from
ever really understanding inertia.

[22] Herman Schmalenbach, *Leibniz*, München, Drei-Masken, Verlag, 1921.
   Ernst Cassirer, *Leibniz's System in seinen wissenschaftlichen Grundlagen*, Marburg
1902.
   Bertrand Russell, *A Critical Exposition of the Philosophy of Leibniz*, London 1900.

[23] The *New Physical Hypothesis* (1971) which appeared in two parts: (1) The Theory
of Abstract Motion and (2) The Theory of Concrete Motion.

[24] On the problem of translation of terms into modern terminology see my 'Helmholtz'
Kraft', *op. cit.*

[25] There is a famous quotation by H. A. Kramers which I had already used as a motto
to my paper on Helmholtz (*op. cit.*) but it is so poignant that it is worth repeating:
In the world of human thought generally and in physical science particularly, the most
fruitful concepts are those to which it is impossible to attach a well-defined meaning.

[26] K. R. Popper, 'The Aim of Science', *Ratio* 1 (1957), 24–35.

[27] Emile Meyerson, *Identity and Reality*, Dover Publications, 1962, p. 77.

[28] "We know that pressure is here represented as resulting from the impact of manifold
particles against one another." Meyerson, *op. cit.*, p. 68.

[29] *Ibid.*

[30] Euler in his *Theoria Motus*, Rostock 1765, p. 51, stated that the assumption of
action-at-a-distance contradicts the principle of inertia, according to which bodies
modify their motion only by impact.

[31] "This property of all bodies known by the term impenetrability, is then not only of
the last importance relatively to every branch of human knowledge, but we may consider it the master-spring which nature sets a-going to produce all her wonders."
D. Brewster (ed.), *Letters of Euler on Different Subjects in Natural Philosophy addressed
to a German Princess*, Letter No. LXX, N.Y. 1833, p. 236.

[32] Kant in his *Metaphysische Anfangsgründe der Naturwissenschaft* dealt a death-blow
to all corpuscularean theories but still accepted the Newtonian view of gravitation.
In his opus posthumum he followed Euler in most details of his matter-theory.

[33] See my review essay, 'Newtonianism and anti-Newtonianism in the 18th Century',

B. J. Phil. of Science 22 (1971), 297–306.
34 A rare exception to this is H. E. Timerding's 'Kant und Euler', Kantstudien 23 (1918/19), 18–64.
35 See my 'Helmholtz's Kraft', op. cit., p. 50.
36 As to Boscovich and Kant see: M. Oster, R. J. Boscovich als Naturphilosoph, Cöln 1909 and, my op. cit., p. 267 on Kant.
37 Boyle, who introduced the term 'corpusclearean', characterized all hypotheses which employed mass and motion as 'corpuscular'. Dynamic is applied to hypotheses which rely on force and motion only. Boyle, Works 3, V, London 1772.
38 Coherent is again meant in a non-logical sense as explained above.
39 In his Recherches sur les origines des forces..., Euler explains that the conservation of state is a fundamental property of bodies. The state can be changed only by an external force, and it should never be looked for in the body itself. To prove this point, Euler does now a 'Gedankenexperiment', saying that we should imagine the whole world annihilated and only one single body remaining extant in infinite and empty space. Then, he says, the principle of conservation of state will spring to the eye. Interestingly, Euler thinks that his conclusion will remain valid, even if somebody opposes his idea of empty space, because what he says, in his opinion, is only a result of abstracting all strange and external causes and annihilating them. This abstraction would be the same even if we imagined that the other bodies stayed around, provided they would be deprived of all force, by which they can act on other bodies. This is in itself a very strange idea. Then if all these other bodies have been annihilated, there would not be the slightest reason why the remaining body should change its motion. Thus, the law of sufficient reason is applied here, with the greatest simplicity. It will be interesting to see how one would apply this principle to a world with circular inertia, or how one would apply it to the general theory of relativity. In any case, this property of conserving its state is called inertia, and it is an essential attribute of all bodies. By 'essential' Euler means that it would be impossible to imagine a body deprived of this property. In other words, if we think of a body, we think of something which has the property of inertia. Euler's conception of inertia is the old idea of laziness, and he emphasizes his point, saying: "For we are accustomed to consider the continuation of motion as an action, and the word inertia in its natural significance, marks a clear opposition to all action." To avoid this old Aristotelean conception, he, very much aware of linguistic significance, suggested that a different name should be given to it.
40 "The vis insita or innate force of matter, is a power of resisting...", I. Newton, Principia, Vol. I, translated by Motte-Cajori Univ. of California Press, 1962. Def. III, p. 2.
41 Op. cit., Law I, p. 13.
42 In the German translation of Euler's Mechanica by J. Th. Wolfers, Greifswald 1846, we find:
Satz 7 §56: "Ein absolut ruhender Körper wird beständig in Ruhe verharren, wenn er nicht durch eine äussere Ursache zur Bewegung angetrieben wird."
Satz 8 §63: "Ein Körper welcher eine absolute Bewegung hat, wird sich stets gleichförmig bewegen und mit derselben Geschwindigkeit in jeder frühern Zeit bewegt haben, wenn nicht eine äussere Ursache auf ihn wirkt oder gewirkt hat."
43 Leonard Euler, Reflexions sur l'Espace et Temps, parts of it had been translated into English and occur in Arnold Koslow (ed.), The Changeless Order: the Physics of Space, Time and Motion, N.Y. 1967, pp. 115–125.
44 Here we see the perseverance of Aristotelian arguments.

[45] L. Euler, *Anleitung zur Naturlehre worin die Gründe zur Erklärung aller in der Natur sich ereignenden Begebenheiten und Veränderungen festgesetzt werden, Op. post.*, 1862, Chap. 4, §26, pp. 449–560.

[46] For details to this derivation see again the *Anleitungen*, Ch. 4, *op. cit.*, p. 45.

[47] *Op. cit.*, §38.

[48] *Op. cit.*

[49] The editors of Euler's collected works.

[50] *Op. cit.*, §49.

[51] Euler's influence on Kant was mainly through the Berlin Academy. To this I shall come later. The Eulerian metaphysical works, the impact of which can be easily seen in the work of Kant are:

(1) 'Anleitung...', *op. cit.*, p. 45. There is little doubt that such an all-embracing philosophy of Nature was widely discussed and debated like all other works of Euler.

(2) 'Recherches physiques sur la nature des moindres parties de la matière', *Mem. de l'académie des sciences de Berlin* I (1745, 1746), 28–32. The paper was read at the academy already on 18 June 1744.

(3) 'Reflexions sur l'espace et du temps', *Mem. de l'académie des sciences de Berlin*, 1748.

(4) 'Recherches sur les origines des forces', read before the Academy in 1750.

(5) The famous *Lettres à une Princesse d'Allemagne sur divers sujets de physique et de philosophie* (referred to in note 31), which was published in three volumes by the St. Petersburg Academy, the first two volumes in 1768 and the third in 1772. They were soon translated into many European languages and this is the work most quoted, also by Kant. (Into English 9 times, into German 6 times, Dutch twice, Swedish twice, into Danish, Italian and Spanish.) That Kant was acquainted with Euler's work has been definitely shown by Alois Riehl in the first volume of his *Der Philosophische Kritizismus*, Leipzig 1876.

[52] J. Kout, 'Vom Übergange von den Metaphysischen Anfangsgründen der Naturwissenschaft zur Physik', in A. Krause: *Das nachgelassene Werk J. Kout's*, Frankfurt 1888.

[53] In Methodology there is positive progress from Euler to Kant. Euler was one of the first scientist-philosophers who became very uneasy about the Leibnizian mathematical ideal. His certainty that metaphysics is harmonious with mathematically demonstrated truths originating in *a priori* principles was shaken. Kant, his true disciple in this too, gave up the Leibnizian ideal totally and he was after the genuine scientific metaphysics which links our transcendental notions with the experimental data in a unique way. That methodologically Euler influenced Kant so much is easily seen when we deal with Kant's attitude towards the Berlin Academy of Sciences of which Euler was the unofficial but most important spokesman. Moreover, as we shall see below, their image of science and their view on the role of the scientist were very similar.

[54] I. Kant, *Metaphysical Foundations of Natural Science* (transl. by E. B. Bax), London 1883, Explanation IV, p. 174.

[55] I. Kant, *op. cit.*

[56] See Willy Ley's revised edition of Hastie's translation of Kant's *Cosmogony*, New York 1968.

[57] The directives of Frederic the Great were that (a) all publications of the Academy should appear in French and (b) parallel to the French translation the work could be printed in any other language as desired by the author.

[58] See again, *op. cit.*, note 33.

59 The translation of the above passage is taken from the annotated translation by W. A. Oldfather, C. A. Ellis, and D.M. Brown, *Isis* **20** (1933), 72–160.

It is worthwhile to correct a curious mistake in the translation. On p. 77 we find "...every effect in the universe can be explained as satisfactorily from final causes, by the aid of the method of maxima and minima, as it can from the effective causes themselves."

This should read:

"...every effect in the universe can be explained as satisfactorily from final causes, as it can be by the aid of the method of maxima and minima from the effective causes themselves."

60 Dr. R. Fuchter, *Leonard Euler*, Basel 1948, pp. 1–24. On p. 14 here it is noticed that Euler's style of topics starts with a thorough analysis of the problem at hand. This characterizes his mathematical, physical and epistemological-philosophical work.

W. W. BARTLEY, III

THEORY OF LANGUAGE AND PHILOSOPHY OF SCIENCE AS INSTRUMENTS OF EDUCATIONAL REFORM: WITTGENSTEIN AND POPPER AS AUSTRIAN SCHOOLTEACHERS

> For the most part an experiment about es-
> sentials will not occur to anybody unless a
> good problem leads to it. And a problem
> arises in a theoretical context. Moreover,
> just why should we have merely facts, not
> theories and explanations?
>
> Wolfgang Köhler[1]

I

When we begin to study the immediate historical background of contemporary philosophy we encounter a curious fact: one of the most important parts of this background is precisely the *disappearance* of this
background from our field of vision.

II

No one needs to be reminded that several English and American writers
– Bertrand Russell above all – influenced the philosophical and scientific
movements of Vienna, Prague and Berlin between the wars. Yet many
ideas and developments closely associated with these movements – for
example, Wittgenstein's early and, even more so, his later thought, and
Popper's theories of induction and demarcation – were developed in soil
foreign to English and American philosophy, and under the influence of
German and Austrian thinkers whose names and views are unfamiliar
to the majority of English-speaking philosophers.

That this background should have disappeared from our historical perspective is itself a most intriguing fact, one that cries for explanation.
Such an explanation cannot be given in the present paper, which is intended only to suggest the extent of our ignorance about these matters,
and to indicate some of the corners and bypaths of 20th century philosophy into which we should have to look if we were to begin properly to
understand ourselves. For it is doubtful that the thought of, say, Witt-

genstein, Popper, or the members of the Vienna Circle can be properly understood or appreciated without better knowledge of the social and intellectual milieu in which it was developed. Moreover, this hidden background is well worth studying for its own sake, since it is an important part of one of the most fertile and exciting periods which any culture has enjoyed.[2] If this essay will, then, do no more than probe a few small parts of a rich terrain, I nonetheless hope that my remarks will stimulate others to begin to view the ideas of our contemporary philosophical heroes in a somewhat broader context than is usual.

The topics which I have chosen to discuss are: (1) the once famous but now virtually forgotten school reform movement which was developed in Austria by Otto Glöckel immediately following the collapse of the dual monarchy and which managed to survive until the Dollfuss dictatorship of 1934; (2) the psychological school whose ideas undergirded this school reform: namely, Bühlerian child psychology, a critical version of Gestalt psychology, difficult to classify precisely, but perhaps closer to the thought of Piaget than to that of Wertheimer, Koffka, Köhler, or Kurt Lewin; (3) the personal participation in this movement by Ludwig Wittgenstein and Karl Popper; (4) the development of Wittgenstein's thought construed as that of an amateur child psychologist turning – partly as the result of his experience in schoolteaching at this particular time – from an essentially associationist psychology to a configurationism or contextualism close to that of the Gestaltists; (5) the thought of Popper viewed as that of one chiefly a schoolteacher and neo-Kantian Gestalt psychologist, a man far removed from the essential ideas of logical positivism, who virtually stumbled into his relationship with the Vienna Circle and the consequent development of his hobby – namely, the philosophy of science – on which his reputation came to rest but which cannot properly be understood without some knowledge of his earlier research interests and permanent anti-positivist outlook; (6) the thought of the later Wittgenstein and the early Popper viewed as far more closely linked in spirit one to the other than to that of the Viennese positivists whom they influenced.

III

We should begin with an account of Austrian school reform, for the role it played is crucial. The character of the debate over the Austrian school system, and some of its implications for both philosophy and psychology, are reflected in the title of a pamplhet which Otto Glöckel published in 1928: *Drillschule, Lernschule, Arbeitsschule* (or Drill School, Learning School, Working School). The year 1928 was a rather late date in the debate about these various kinds of schools; in fact, the debate began long before Glöckel's own birthdate in 1874. For many decades, even under the Habsburgs, the Austrians had enjoyed one of the most progressive school systems in Europe. Yet, however well the pre-war Austrian system may have compared with that of other European countries, it was hardly a paradigm of progressive thinking: instruction, largely in the hands of the Roman Catholic Church, was mechanical and as uniform as was practical. As Count Rottenhan, royal advisor, defined its aims, the purpose of the lower schools was "to make thoroughly pious, good, *tractable*, and industrious men of the laboring classes of the people." The constitution of the common schools issued by the emperor in 1805 was unequivocal: "The method of instruction" it decreed, "must endeavor *first and foremost to train the memory*; then, however, according to the pressure of the circumstances, the intellect and the heart. *The trivial schools will strictly refrain from any explanations other than those exactly prescribed* in the 'school and method book'..."[3] If any educational psychology lay behind this approach, it was a version of associationism like that propounded by Johann Friedrich Herbart (1776–1841). After the revolution of 1848 most chairs of philosophy in Austria were filled by followers of Herbart,[4] who viewed the human mind as neutral and passive, lacking innate faculties for producing ideas. The theory of the human mind, as presented by Herbart, rather resembles what Popper in his early article, 'Die Gedächtnispflege unter dem Gesichtspunkt der Selbsttätigkeit',[5] was to describe as 'the bucket or tub theory of the mind', an expression that Popper was often to repeat in his later work. According to such theories, ideas themselves might be active; but they lead their lives in passive storehouse minds. To a Herbartian, whose aim above all is *moral* education, teaching consists in feeding students those ideas which it has been decided should dominate their lives. At no time, according to Herbart, should a teacher debate with his students on any matter. As he explained in his *Outlines of Educational Doctrine*: "Cases may arise when the impetuosity of the pupil challenges the teacher to a kind of combat. Rather than accept such a challenge, he will usually find it sufficient

310 W. W. BARTLEY, III

at first to reprove calmly, to look on quietly, to wait until fatigue sets in."[6]

It was such a doctrine, such schools – Drillschule and Lernschule – and such a school system, that Otto Glöckel (8 February 1874–23 July 1935), not to mention his perhaps somewhat better-known subordinates, Ludwig Wittgenstein and Karl Popper, was to combat. Before turning to Wittgenstein and Popper, a word need be said of Glöckel. Himself the son of a public schoolteacher, Glöckel in effect made his entry into the school reform movement by being dismissed from his teaching post due to his political activities. This act, the work of Karl Lueger, a famous and controversial Christian Socialist mayor of Vienna, occurred in early September 1897; and it was not until nearly twenty years later, in 1916, that Glöckel reappeared in a position of educational significance. At that time Glöckel put together a programme of educational reform for the tottering empire, a programme which he began immediately to put into effect in 'German Austria' in early 1919, when he was put in charge of the ministry of education of the new republic.

Changes of so radical a nature would probably not have been possible were it not that, if for only a brief moment, the chaos attending the fall of the Austro-Hungarian bureaucracy brought about a considerable reduction in the usual red tape. At any rate, experimental changes were already in effect as early as the school year 1919–1920. Only a few need to be mentioned here. The examination procedures according to which a child's future academic career virtually had to be decided at the age of 11 were radically modified; military academies were abolished and their sites transformed into state boarding schools for exceptionally able youths from remote homes. Also abolished were some of the old types of finishing schools for girls, like the Offizierstöchterschule and the Mädchenpensionate. The areas in which girls were permitted to study were expanded, and married women were permitted to teach. Compulsory attendance at religious exercises was abolished; associations rather like the American Parent Teachers Association were formed; and instruction in the manual arts and crafts was instituted, in order – among other aims – to give middle-class children the opportunity to acquire more respect for hand labour by learning through experience that it was not so simple as it might seem.[7]

These are surface phenomena, of symbolic significance certainly, but of questionable depth. More important, particularly for philosophy and psychology, were certain other changes – in particular, the systematic effort to undermine the very methodology of the Drillschule and Lernschule. Take spelling for example. Prior to 1919, the rules of spelling had been dictated, written on the blackboard perhaps, and then force-fed to students. After the school reform, various experiments were tried – under the general rubric of 'Self-Activity' or 'Selbsttätigkeit' – to encourage youngsters to figure out rules for themselves (with little help from the teacher) through the use of word-lists. I believe it likely that the second and final book that Wittgenstein published during his lifetime: namely, his *Wörterbuch für Volksschulen*, published as an official school text in 1926, was intended in part for this purpose. (Wittgenstein also had other purposes, not relevant here, though perhaps relevant to his philosophy: for example, the elimination of words of foreign origin, and the attempt to teach proper syntax through shrewd exploitation of dialect.) Possibly my suggestion here may prove helpful to those who have been bewildered to learn that a man of Wittgenstein's stature should have published a list of words, and who appear reluctant even to include this book in his bibliography.

Another significant reform experiment was dubbed 'integrated instruction'. Although this experiment, like the entire school reform programme, was later described as a Jewish concept, the word 'integration' has no racial or ethnic connotations here. It refers to an effort to allow individual teachers the latitude to determine how and when they would turn from one subject to another during the school day. No reading or spelling period as such, for example, was to be set aside. Although general goals were indicated, the interest of the children was supposed to determine how the day would be divided. Here again the implicit attack on the various associationist psychologies, with their emphases on 'unit ideas', is evident. That approach, so the school reformers believed, had led to an exaggeratedly compartmentalized and 'atomistic' approach to teaching and learning. As I shall explain below, this very practice of integrated instruction proves important in explaining certain episodes in Wittgenstein's life during the twenties.

Such experiments help explain the general name – so difficult to translate into English – that was given to the new type of school: namely, 'Arbeitsschule'. The word 'Arbeit', or 'work', referred not simply to the manual training and crafts taught but more importantly, in the context of the German phrase 'sich etwas erarbeiten', it referred to an active participation in lessons, aiming no longer simply at the storage of facts of the Drillschule or Lernschule, but at the development of capabilities. 'Sich etwas erarbeiten' suggests acquiring knowledge by working or puzzling something out for oneself. What was wanted was more independent and original thinking, activity, on the part of students – activity, as opposed to the *fatigue* that Herbart expected would set in when a pupil dared to play an active role in his education.

Under the impact of such reforms the Austrian school system was literally transformed between 1920 and 1926. But this was also a period of deep social division for the first republic, accompanied by serious economic difficulties; and gradually the country became polarized politically between the Social Democrats – those like Glöckel who had undertaken and carried out the school reform movement – and the much more conservative Christian Socialist Party, dominated by the Roman Catholic Church. For some seven years Social Democratic policies prevailed in the schools. By the middle twenties, however, the Christian Socialists and other even more right-wing groups began to grow rapidly stronger, resulting in a sharp retrenchment in the school reform – especially after the bitter political and religious strife of 1926. By 1927, when the Christian Socialists had obtained control almost everywhere in Austria except in the city of Vienna, the most interesting experimentation was forced to halt – except, that is, in Vienna, where the school reform movement continued under the leadership of Glöckel, who became administrative president of the Vienna School Council and remained in this office until 1934 – when the Dollfuss Dictatorship ended school reform, arrested many of its leaders, including Glöckel, and forced its chief publications – the journals *Die Quelle* and *Schulreform* – to cease publication. During the censorship that followed these journals were even locked up ('gesperrt') in the National Library, and thus made inaccessible to the general reader.

One other fact needs stressing here, since it too will prove useful below in understanding Wittgenstein. The compromises effected among the political parties in 1926 meant in practice that wherever the Social Democrats remained politically dominant – as in Vienna – Glöckel's essential programme (described by his opponents as 'school Bolshevism') could be maintained. But in the *country*, after 1926 the Christian Socialists had firm control of the schools.

Despite these upheavals, the general excellence that prevailed in Vienna until 1934 ought not to be underestimated. Robert Dottrens, of the Institut Jean-Jacques Rousseau in Geneva, reported: "At the conclusion of a tour through Czecho-Slovakia, Germany, Belgium, England and France, I do not hesitate to say that Vienna is ahead of all the other cities of Europe from the point of view of educational progress."[8] Dottrens continued: "It is to Vienna, the pedagogical Mecca...that the new pilgrims of the modern school must go to find the realization of their dreams and hopes."[9]

IV

One of the most important and famous shrines of this mecca was the Pedagogical Institute of the City of Vienna, the leading figures of which were Karl and Charlotte Bühler. As noted, an implicit psychology undergirded Glöckel's reforms: namely, a theory of the child as an active social being whose mind was far more than a bucket to be filled with appropriate information. In its attack on the Drillschule and Lernschule, the reform movement was essentially anti-Herbartian, anti-associationist, anti-elementarist in psychology.

In 1923 Glöckel wrote: "Die ganze innere Schulreform ist wesentlich auf die Ergebnisse der psychologischen Forschungen der Kindesseele aufgebaut."[10] Only a few months earlier, in 1922, Karl Bühler (27 May 1879–24 October 1963) the child psychologist, had been called to a chair of philosophy and psychology both at the University of Vienna and at the newly reorganized Pedagogical Institute. Bühler's career had begun in Würzburg, where, in 1906, he became assistant to Külpe, the critical realist and critic of Mach's positivism. Following Külpe, Bühler began in Würzburg to develop the theory of 'imageless thought'. This idea, as understood by Külpe and Bühler, was that in the intentional act of representation the particular image or model used, if any, need bear no imaginal resemblance to what is represented. Abstract words, used conventionally in this process, cannot be reduced to sense impressions. International attention was focused on Bühler's early ideas, as published in his habilitation thesis in 1907, as a consequence of a prolonged controversy with Wundt triggered by them. Bühler remained at Würzburg for only two years, following Külpe to Bonn in 1909 and to Munich in 1913. Whilst still at Würzburg, however, he became associated with Koffka, also an assistant to Külpe at this time, but who was soon thereafter to join Wertheimer and Köhler in forming the 'school' of Gestalt psychology. The association between Koffka and Bühler continued for many years, but it was not harmonious. For Bühler claimed priority in developing some of the basic laws and experiments of Gestalt psychology, and, as a result, bitter polemics were exchanged.

By 1920 Bühler had also made important contributions to the theory of language and to child and developmental psychology, fields which he cultivated in collaboration with his wife, Charlotte Bühler, herself an important psychologist. These subjects, which played important roles in his work at the Pedagogical Institute in Vienna and which were responsible for his being called to this post, provided the themes of his major works: *Die geistige Entwicklung des Kindes* (1918), an abridged version of which was translated into English as *The Mental Development of the Child* (1930), and of three other books, unfortunately still available only in German: *Die Krise der Psychologie* (1926), *Ausdruckstheorie* (1933), and *Sprachtheorie* (1934).

Whatever Bühler's differences with the Gestalt psychologists were, their theories were closely akin on many points – in particular in their opposition to associationist psychology, reductionism, behaviourism, positivism, psychological atomism. One need not rehearse here the main ideas of Gestalt psychology, but one might emphasize how minor some of the early Gestaltists considered the role of sense experience. Attacking the associationist and empiricist principles of Locke and Hume, the principles of contiguity and frequency, they sought to show that theory-making, organization, was a basic function of the human mind independent of associations of sense-impressions: structural properties of the human mind gave priority to the organizing and theorizing activity of the mind, which in turn determined the kinds of wholes with which we would deal as 'elements' of our thinking. The Gestalt psychologists did not doubt that their views were valid not only in psychology but also in epistemology. Köhler, in particular, stressed that his argument against psychological atomism affected epistemological atomism too. And Bühler rejected what he derisively called the view of language as *physiognomy* – that view, in short, which was variously known as 'the picture theory of language' and as 'logical atomism', and which Bertrand Russell had described in these words: "In a logically correct symbolism there will always be a certain fundamental identity of structure between a fact and the symbol for it; and ... the complexity of the symbol corresponds very closely with the complexity of the facts symbolized by it."[11]

During his sixteen years in Vienna, Bühler acquired many students and disciples who were later to attain distinction in their own right – among them Paul Lazarsfeld, Egon Brunswik, Else Frenkel-Brunswik, Konrad Lorenz, Karl Popper, Lotte Schenk-Danzinger, Albert Wellek, Edward Tolman. Egon Brunswik's name may appear surprising in such a list because of his own well-known association with logical positivism. This should not mislead one into thinking that Bühler looked favourably on the positivist movement. As a professor in the University of Vienna he naturally collaborated professionally with positivists; indeed, he and his wife were very good friends of Moritz Schlick and his wife. To give two pertinent examples of their professional cooperation, Schlick and Bühler were Popper's Ph.D. examiners; and they were the Ph.D. examiners as well for Thomas Stonborough, Ludwig Wittgenstein's nephew (who, incidentally, wote up his dissertation in Wittgenstein's hut in Norway). However cordial their social and professional relations, Schlick and Bühler were far removed philosophically. Indeed, Bühler regarded positivism with a combination of hostility and contempt. It is reported that he found it difficult to endure the thought that Brunswik had joined the Vienna Circle and had then gone on in America to advocate operationism and 'unitary science'. This Bühler is said to have regarded as a personal betrayal.[12]

V

I have just described Karl Popper as one of Bühler's disciples; and earlier I mentioned the personal participation of both Wittgenstein and Popper in the school reform movement. It is occasionally recalled that both Wittgenstein and Popper were schoolteachers in and near Vienna during the 1920's. But to my knowledge, no one has ever raised the question whether their activities as schoolteachers might not be relevant to their philosophies. I believe that there was a quite important relationship between these two activities.

Take Wittgenstein first. I am about to deal with the 'mystery years' in his life between the completion of the *Tractatus* and his return to Cambridge in 1929. The story I am about to relate makes Wittgenstein

appear a more rational and sympathetic figure than other accounts I
know. Writers often express some puzzlement about Wittgenstein's behaviour after the first war. George Pitcher, for instance, in his introduction to Wittgenstein's life and thought, muses as follows: "A man of
acknowledged genius who, after knowing next to nothing about logic and
philosophy, had made important contributions to both fields within a
remarkably short period of time, a man who could not help having a
brilliant future in one of the most sophisticated of all intellectual disciplines – this man turned his back on all that and devoted himself to the
humble task of teaching young children in remote villages."[13]

I do not want in any way to criticize the excellent studies of Pitcher,
W. D. Hudson or many other writers who have been similarly puzzled by
Wittgenstein's behaviour; but one cannot help being mildly amused by
the notion that only a rather eccentric person would pass up an opportunity for academic success in England – particularly to go into elementary education! Wittgenstein was, however, hardly of the stuff of which
the typical British don is made. He was not even British: he was a pa-
triotic Austrian who had during the war become a sort of socialist man-
darin of a Tolstoian stamp of mind. I suggest that he deliberately and for
good reason chose elementary school teaching as a career. It is unlikely
that Wittgenstein chose school teaching *because* of the Reform Movement;
such was hardly his style, and the slogans of school reform were frequently of a vulgarity which would have irritated him. But his family, including
the two sisters who took a particular interest in him, Hermine Wittgenstein (who herself ran a day-school for poor boys in Vienna) and Margarete Stonborough, knew Glöckel personally; and Wittgenstein could
hardly have been unaware of the opportunities available through the new
movement. Again, Ludwig Erik Tesar, an active school reformer, was
one of the beneficiaries of Wittgenstein's famous bequest to Ludwig von
Ficker, better known for its more famous beneficiaries: Rainer Maria
Rilke and Georg Trakl. It would, however, be wrong to overemphasize
any such 'rational' element in Wittgenstein's choice. Shattered both men-
tally and physically by the war, he certainly took up his school teaching
career at least in part as a means of 'Arbeitstherapie'. And his choice
was strongly influenced even before his return to Vienna by the persuasion of his fellow-prisoner in Monte Cassino, Dr. Ludwig Hänsel.

In any event, within ten days of his return to civilian life, Wittgenstein

was enrolled at a teacher training college – one of the first Lehrerbildungs-
anstalten operating under the general direction of Glöckel – and attended
the year-long course required for the certificate. By the autumn of 1920
he was teaching children in the third form, aged 9 and 10, in the tiny
village of Trattenbach in lower Austria.

About his day-to-day life as a schoolteacher the Wittgenstein literature
gives us comparatively few solid reports. We are told that Wittgenstein
was in constant friction with those around him, including his colleagues;
that he was unhappy in Trattenbach and was transferred first to Puchberg
(1922–1924) and then to Otterthal (1924–April 1926), also in lower Austria. Although he rarely saw his old English companions and admirers
during this period, he did not distantiate himself from the Austrian school
reform movement: he subscribed for a time to *Die Quelle* and to *Schulreform*; and in 1924 he produced his official wordbook.[14]

As a schoolteacher he is said to have been unusual in certain respects.
Pitcher reports:

The next class would often be kept waiting outside his door, and he regularly kept his
young charges one or two hours – sometimes longer – after the rest of the school had
been dismissed. The accepted teaching procedures held no interest for him; he was
always experimenting with new methods and devices of instruction. He dissected animal
corpses and assembled their skeletons, explained models of steam-engines, set up with
his students a potter's wheel on which they fashioned clay pots. ...If he happened by
chance to meet some of his youngsters in the evening, he might give them instruction
in astronomy on the spot. In mathematics, he had great success; he took his students
well beyond the ordinary requirements for their class, and introduced the older, more
gifted ones to advanced problems in algebra.

Despite such successes, a serious crisis flared up in Otterthal in 1926
having to do in part with complaints arising from his disciplining of a
child. This led to a trial and a compulsory psychiatric examination for
Wittgenstein. Although he was eventually acquitted, Wittgenstein volun-
tarily resigned his post in April 1926, thus bringing his career as schoolteacher to a close.

Taken in isolation, these events might appear odd but, apart from their
having happened to Wittgenstein, not terribly important. If we fit them
into the background of the school reform movement, however, we get a
more intelligible and significant picture. For example, Wittgenstein's
practice of disregarding the so-called 'usual' school periods, the division
of the school day into 'periods', was a policy entirely consistent with

Glöckel's principle of integrated instruction, which allowed individual teachers to determine how and when they would turn from one subject to another. Indeed every teacher was *encouraged* to experiment for himself, here as in other areas.

The claim that "the next class would often be kept waiting outside his door" by Wittgenstein is false. It suggests the changing of rooms procedure with which we are familiar in the upper grades of our elementary schools and in junior high and high schools. But as it happens this system was not in use in Wittgenstein's schools. Richard Meister reports of Austria in the 20's and 30's: "In the eight-year elementary schools in the country, as well as in the general elementary schools and final grades in the cities, there is a system of class teachers, where one master teaches all subjects except religion. The latter is given by special teachers of religion for each denomination separately. In the higher elementary schools there is a system of subject teachers, where each subject or group of subjects is taught by specially trained teachers."[15] Since Wittgenstein was a *lower* elementary schoolteacher there was no changing of classes in his school, a fact which I have confirmed by discussing these matters with Wittgenstein's former colleagues and students. Rather, it was not the students but the local priest (Father Alois Neururer, in Trattenbach), who was often kept waiting outside the door by Wittgenstein's experimental teaching!

In any case, the tale about Wittgenstein's resignation following complaints about his rough disciplining of certain students is incomplete as it stands. Since school discipline of the roughest sort had been common in the Habsburg domains, and was still common during the 20's and 30's (bamboo sticks often being used, so I am told), it is *prima facie* unlikely that his colleagues would not have supported him (as some of them indeed did) against a parent's complaint – even if the charge were true. Of course it is possible that Wittgenstein did mercilessly beat one of his charges – but such conduct does not accord with the memories of his former pupils in Trattenbach and Otterthal, who stressed that although Wittgenstein was strict his punishment was always fair.

Another explanation occurs to me. I conjecture that, as can often happen in schools, some small disciplinary quarrel was made the pretext for a more deep-seated complaint – very likely through some sort of collaboration (the details of which we are not likely ever to learn) among some of the townspeople, some of the local clerics (but not Neururer, who was a friend of Wittgenstein), and some of Wittgenstein's superiors and colleagues. For it will be recalled that the resignation came in 1926, an exceedingly turbulent year in schools throughout Austria. I suggest that conservative farmer forces in Otterthal lost no time in sending packing back to Vienna a man whom they must have felt threatening to them in many ways: Wittgenstein was thought to be rich (a story which is not strictly true: he made over most of his fortune – which was producing an annual income of 300000 Kronen in 1914 – to his sisters Hermine and Helene and his brother Paul. The fortune was not affected by the German and Austrian wartime and postwar inflation, neither during the war when still in Wittgenstein's name, nor later, when in the names of his siblings. For before his death in January 1913, Ludwig's father, Karl Wittgenstein, had invested virtually his entire fortune, apart from real estate, in American iron and steel stocks. That money which the family earned within Austria during the First World War with government bonds and similar investments was invested in real estate prior to the outbreak of serious inflation; and in the period between the wars the family fortune was kept largely in Holland, once again safe from inflation. Wittgenstein's siblings in effect held this money in trust for him, in case he should ever want it back, until the mid 1930's when, with the approach of war, the fortune was distributed amongst Wittgenstein's nephews and nieces and other family members.) He was a socialist; he was known not to be Roman Catholic; he was a proponent of progressive education, the author of a positivist tract, was known by his villagers to be homosexual and thought by some of them to be a misogynist to boot; and – what must have been hardest to bear – he was an extremely successful teacher for all that. His effect on his students was virtually magical.

To turn to Karl Popper, usually regarded as one of Wittgenstein's antagonists. Everybody knows that Popper's main formal background was in physics and mathematics. It so happens, however, that what everybody knows is false. In fact, Popper is an amateur physicist and mathematician, his formal training having been in education and in Gestalt psychology, under the supervision of Karl Bühler.[16] His thesis for his teacher's training certificate (1927) was entitled: 'Gewohnheit und Gesetzerlebnis'; and his doctoral dissertation, *Zur Methodenfrage der Denkpsychologie* (1928), was a defense of Bühler's ideas – as outlined for instance in *Die*

Krise der Psychologie – against the associationist physicalistic ideas of Schlick, which Popper vigorously attacked.[17]

Unlike Wittgenstein, Popper was not a recluse, but was for a time actively involved in socialist party activities in Vienna and in the School Reform Movement. One of the more important figures with whom Popper collaborated in his political and social activities was Alfred Adler. Throughout this period Adler worked as closely as he was permitted with the school reform movement in Vienna. Adlerians, including Adler himself, contributed to *Die Quelle* and *Schulreform*; Adler gave courses at the Volksheim and at the Pedagogical Institute; and some of his disciples – e.g., Birenbaum, Scharmer and Spiel – opened an Adlerian school for children of poor Viennese families in September 1931. An Adlerian child guidance clinic was opened as a pilot project at the Volksheim; and eventually twenty-eight such centres existed in Vienna, most of them situated within school buildings. Stressing his ideological kinship with the Bühlers, Adler argued that individual psychology showed many ideas in common with Gestalt psychology; and Wolfgang Köhler later agreed with this evaluation – although without reference to the Bühlers. In his book on Adler, Lewis Way writes that the viewpoint of Gestalt psychology "is as near as one could wish, given its different subject matter, to that of Individual Psychology."[18] Another social and psychological theory closely akin to Adler's in some respects is Popper's doctrine of 'the logic of the situation'.[19]

And this leads me from my digression on Adler back to Popper. It was not only through Adler that Popper was involved in the School Reform Movement. Indeed, Popper has often told the story of his personal break with Adler.[20] Popper also worked with Eduard Burger, the editor of *Die Quelle*, and contributed articles on pedagogy to both *Schulreform* and *Die Quelle*. These virtually unknown publications, in which some of Popper's later ideas are sketched, include two fairly substantial articles ('Zur Philosophie des Heimatgedankens', and 'Die Gedächtnispflege unter dem Gesichtspunkt der Selbsttätigkeit', published in *Die Quelle*), and a piece of juvenilia, 'Über die Stellung des Lehrers zu Schule und Schüler', published in *Schulreform*. Popper also published dozens of short reviews of books and of articles on psychological and educational topics. These reviews, sometimes only a few lines in length, are occasionally revealing about Popper's allegiances, betraying his extensive familiarity with the publications of Adler's school as well as his alliance with Bühler in the latter's quarrels with the Koffka group. For instance, in reviewing a work on Gestalt psychology published in 1931, Popper complains that it fails to consider the views of Külpe and Bühler.[21]

Considering the depth of Popper's involvement with both individual psychology and Gestalt psychology, it is curious that in his later writings he mentions them so rarely: some brief, favourable but mildly critical remarks about Gestalt psychology – without any reference to the individual founders of the school – are to be found in *The Poverty of Historicism*.[22] Popper's interest in education, on the other hand, can easily be seen in his later works, perhaps best in his essay 'Back to the Pre-Socratics', where he contrasts the dogmatic as opposed to critical traditions in education.

When one views Popper's thought against this background, it is hard, surprising as it may seem to some, to find much of strikingly novelty in his philosophy. His methodology turns out to be in effect a kind of critical continuation of the theories of Külpe, Bühler, and Koffka – one that also bears a close resemblance on many points with the work of Piaget. I mention Piaget because it is sometimes said that his theory differs from that of the Gestalt psychologists in that he – by contrast to the Gestaltists – thinks that learning occurs not only in the elaboration of intellectual structures but also in perception. Some Gestaltists have indeed written as if perceptual constancies of shape and size belong to the object perceived and are not modifiable by the observer. Whether or not this is a misreading of the intentions of most of the Gestalt writers, it is clear that for Piaget perception of such things as figures is not only gradually built up but also gradually corrected. Clearly Bühler, as well as the teachers of art in Vienna with whom he collaborated, held such a view. And Popper's view of conjecture and refutation – or, as expressed by his associate E. H. Gombrich, 'making and matching' – is also close to Piaget.

Popper's attacks on the positivists may, then, be construed as direct *applications* of the attacks already mounted by Koffka and Bühler on the associationist psychologists. Even some of Popper's constructive ideas, including the emphasis on testability in connection with the hypothetico-deductive method, may be found in the work of his teachers: in particular, in that of Heinrich Gomperz.[23] Popper's views acquire their distinctive form and emphases from the fact that they were elaborated in dialogue with the logical positivists; but they acquire no originality from this circumstance. Popper's notorious disagreements with Otto Neurath (1882–1945), for example, if put into the context of Popper's real ideological background, no longer appear like the internal feuding of two positivists; they are the disagreements of men sharply opposed on basic issues. Neurath had also contributed to *Die Quelle*, outlining as his contribution to educational reform the so-called 'Viennese method', suggesting the use of pictures rather than words in tables of statistics, in order to avoid verbal misunderstandings arising from translation from one language to another. His tables, so Neurath maintained, could be used throughout the world without regard to language, and would, incidentally, contribute to the creation of a 'universal language.' [24] This was only part of his broad programme of visual education, which aimed for an international language of simplified pictures, or 'isotypes'. In essence, Neurath's 'Vienna Method' aimed to give an invariant and self-explanatory pictorial sign for any given thing.

It is hard to imagine a more striking contrast than that between this theory for educational reform through linguistic reform and Bühler's theory of language and of imageless thought. What was a programme for Neurath, was, in effect, a problem for Bühler. In particular, Bühler was puzzled by the remoteness of child art (which he treated as representational in this context) from our visual experience; and he found an explanation for this phenomenon in our use of language – but without in any way suggesting that our language need on *this* account to be reformed. Arguing that children do not draw what they see but what they know or remember, Bühler explains that this is due to their *mastery* of language which, he writes, "models the mind of man according to its requirements." [25]

"As soon as objects have received their names," Bühler continues, "the formation of concepts begins, *and these take the place of concrete images*. Conceptual knowledge, which is formulated in language, dominates the memory of the child. What happens when we try to impress some event on our own memories? As a rule the concrete images fade, but as far as the facts are capable of being expressed in language, we remember them. This development begins as early as the second year in the child and when it begins to draw – in its third or fourth year – its memory is by no means a storehouse of separate pictures, but an encyclopedia of knowledge. *The child draws from its knowledge, that is how its schematic drawings come about.* ...If it wants to draw a man, it does not look around for a model or copy, but cheerfully goes ahead with its task and puts into the drawing whatever it knows about a man and whatever comes to mind. The man must have two eyes, even in profile, the horseman two legs. Clothes are hung round him afterwards, as one would clothe a doll. One can see what is in his pockets and the coins in his purse, as in an X-ray photograph. Models and copies at most serve as suggestive impulses." [26]

Bühler continues: "The order of a story is not necessarily the spatial order of the objects it describes. If we read in a fairy tale: 'the dwarf had a huge head and two short little legs, snow white hands and a nose like a glowing coal,' we should certainly not criticize the style because of the irregular order adopted. If such a sentence were to guide the efforts of a child that does not see the picture in its mind as a whole, we might expect that the short legs will be drawn as growing straight from the head and the hands likewise. The nose, again, might be put in its proper place in the middle of the face. But that is exactly what we see in some of the earliest pictorial efforts of the child. Its drawings are, in a sense, *graphic accounts.* Looked at in this way, the irregular order found in its drawings becomes intelligible. We cannot simply assume that the mind of the child is in a state of chaos, because if these graphic accounts be translated into language, they might be found to be well ordered. The fault lies not so much in a chaotic mind, as in *errors of translating from knowledge – formulated in language – to the spatial order of pictorial representation....*" [27]

In effect, Bühler is suggesting that representational drawing is one 'language-game' (although he does not use this phrase) in which children and others may engage. This game or activity has its own rules which are not the same as those of descriptive or representational verbal language. Moreover, attempting to apply the rules of the one representational activity to the other activity – equally representational – may result in distortion. *If one wishes to learn to draw,* then one must *learn* how to translate knowledge formulated in language into accurate drawing. But inaccurate drawing, inaccurate pictorial rendering, is in itself no index whatsoever of linguistic or conceptual confusion. Bühler would reject Neurath's educational programme, as well as that of the Russell-Wittgenstein of the 'picture theory' as entirely unwarranted and bound to be self-defeating, being based on untenable psychological and linguistic premises.

Bühler's own ideas about teaching children how to draw were applied quite directly in Vienna by such craftsmen teachers as Franz Cizek. As Dottrens reports Cizek's method: "The class...is asked to draw some ordinary object, say a pair of scissors. No model is put before them; the pupils draw from memory this object which they have occasion to use every day. With how many errors and how much difficulty they produce a satisfactory picture of that little instrument which they have seen a hundred times, but never really observed! The drawing finished, the teacher and the class examine together a pair of scissors...Next each one makes another drawing, as before, trying to keep in mind the attentive observations he has just made. In this way Cizek's teaching leads to cultivation of the sense of observation, and a continual growth of the capacity of expression." [28] Here again we find an account virtually identical with Popper's's theory of conjecture and refutation and E. H. Gombrich's application of it to art in his book *Art and Illusion* and elsewhere, through his concept of 'making and matching'.

VI

I mentioned earlier the possibility of construing the later thought of Wittgenstein as that of an amateur but gifted child psychologist who turned, partly as the result of his experiences in school teaching during the twenties, from an essentially associationist psychology to a configurationism or contextualism closer to that of the Gestaltists. [29] It is to this theme that I now wish to return.

In *Zettel* 412, Wittgenstein asks: "Am I doing child psychology?", and answers: "I am making a connexion between the concept of teaching and the concept of meaning." [30]

Wittgenstein's question whether he is doing child psychology obviously does have to be answered affirmatively. *Zettel, The Philosophical Investigations*, and *The Blue and Brown Books* must be read in a number of different ways. But two of the necessary ways in which one must read them are: (1) as polemics on the atomism represented by the *Tractatus* or by Russell or Herbart; and (2) as attempts to develop a child psychology of language. After all, how does the *Investigations* open except as a critique of St. Augustine's account of how a child learns a language? Indeed, much of the first part of the *Investigations* focuses on the question of how children learn their native languages.

Moreover, since his child psychology is developed in part as a polemic against his earlier atomism, that atomism could not have been, as has been occasionally affirmed in recent years, *purely* formal, neutral with respect to psychological issues: it was also psychological. Whether it was *identical* with the particular sense-data theories of Russell or the positivists is of course quite another question.

My suggestion, then, is that there exists an important family resemblance between the views of the Gestalt psychologists, Popper's philosophy of science, and the views of the later Wittgenstein. To avoid misunderstanding, I am not chiefly interested here in making categorical claims about 'intellectual influences'. That Popper's thought was decisively moulded by that of Bühler, Külpe and the Gestalt psychologists is beyond dispute. Whether Wittgenstein was directly influenced by Bühler or other of the Gestalt theoreticians is uncertain. He definitely was familiar with Bühler's ideas. The connection here is more direct – and more complicated – than it would be simply because of Wittgenstein's participation in school reform. Wittgenstein knew Karl and Charlotte Bühler socially and personally: in fact they were present at the famous first encounter between Wittgenstein and Moritz Schlick, as the guests of Wittgenstein's sister, Frau Margarete Stonborough,[31] and had been invited at the suggestion of Wittgenstein's nephew, who was studying with Bühler at the University of Vienna. Whether Wittgenstein ever made any conscious connection between Bühler's psychology and his own later thought is, however, an open question. Friends and members of his family recall that Wittgenstein did not like the Bühlers personally, and that he occasionally referred to Karl Bühler as a 'charlatan'. This personal reaction, however, by no means precludes at least some positive intellectual influence.

As for the 'family resemblances' among these thinkers, I have room to mention only a few of the most striking features. Foremost, perhaps, is the common opposition found in these three ways of thinking to psychological and logical atomism.

Second, there is the contextualism or configurationsim shared by them. According to the Wittgenstein of the *Investigations*, there is no sense in talking of a one-to-one correspondence between the simples of language and the simples of reality (even assuming such simples exist). Wittgenstein reasons that simplicity is not a matter of absolutes, but is context-dependent: one *might* break down the visual image of a flower into all the different colours of which it is composed.[32] But as Wittgenstein shows in *Investigations* 47, the question of which properties are *more* simple makes little sense. Multi-colouredness is one kind of complexity; being composed of straight lines is yet another. Since, on Wittgenstein's view, we use the words 'composite' and 'simple' in a great many different ways, and ways that are also differently related, questions that presuppose *absolute* complexity and simplicity *apart from context* are not answerable and ought not to be asked.

If we turn to Popper – or to one of his disciples, like Paul Feyerabend, who was also influenced by Wittgenstein – we find an entirely different way of putting the matter. But the basic point, and the family resemblance, remains. For Popper and Feyerabend too, what is relevant in one's analysis of an object will depend upon the theory one is entertaining or testing. Like Wittgenstein, Popper also uses a series of geometrical shapes to illustrate his argument. In Appendix No. *10 of *The Logic of Scientific Discovery* (p. 421), he presents a series of shaded and plain circles, triangles, squares and rectangles to show that "similarity, and with it repetition, always presuppose the adoption of *a point of view*." As shown by this example, and indeed throughout Popper's writings, the problem on which a scientist may be working will determine which theories are relevant – which in turn determines which 'simples', or observation statements, are relevant. The network of problems, theory, and observation make up the context which determines relative simplicity and complexity.

As for Wittgenstein and Popper, so for the Gestalt psychologists. Take Külpe's famous experiment with cards (which also bears some resemblance to Wittgenstein's example of coloured boxes in *Investigations* 48). Külpe had contrived his experiment to combat Mach's claim that mental processes could be reduced to sensations; in it, Külpe presented his subjects with cards containing nonsense syllables of various colours and arrangement. Some subjects were asked to report on the *colour*, others on the *pattern*, others on the *number* of the items seen. In every case, the subject abstracted the features he had been instructed to report and made no mention of – and in many cases did not even remember! – other features of the card which could easily well have been taken as simples. Here again the answers depended on the question, on the *context*. Whereas for the associationist organization or theory arises from previous association, for Külpe and the other Gestaltists association depends on organization or theory.

Thirdly, there is the *conventionalism in respect to words* (not sentences or theories) shared by Wittgenstein, Popper, and many of the Gestaltists. That both Popper and Wittgenstein – rejecting the picture theory of language – regard words as tools is so familiar that it does not bear comment. In both cases, in Popper explicitly, in Wittgenstein implicitly (as Feyerabend has pointed out) [33] there is an attack on 'essentialism' in respect to words. The case of Külpe is less well known. For Külpe, abstract words, being 'impalpable', cannot be reduced to sensations – or otherwise reduced; they are used instrumentally.

Fourthly, and closely related to this conventionalism, is the idea of 'imageless thought'. I have already referred briefly to the views of Külpe and Bühler on this matter. Roughly the same idea occurs frequently in Wittgenstein: for example, in *Philosophical Investigations* 395, 396, and 397. One can do no better than quote Wittgenstein himself: "There is a lack of clarity about the role of imaginability in our investigation. Namely, about the extent to which it insures that a proposition makes sense. It is no more essential to the understanding of a proposition that one should imagine anything in connexion with it, than that one should make a sketch from it. Instead of 'imaginability' one can also say here: representability by a particular method of representation. And such a representation may indeed safely point a way to further use of a sentence. On the other hand a picture may obtrude itself upon us and be of no use at all."

Wittgenstein talks in a similar way in his *Lectures and Conversations*,[34] where he denies that when a Frenchman says 'Il pleut' and an Englishman says 'It is raining', something happens in both minds which is the real sense of 'It is raining'. Wittgenstein writes: "We imagine something like imagery, which is the international language.[35] Whereas in fact: (1) Thinking (or imagery) is not an accompaniment of the words as they are spoken or heard. (2) The sense in the thought 'It's raining' is not even the words *with* the accompaniment of some sort of imagery."

One could continue to sketch such family resemblances among the ideas of Wittgenstein, Popper, and the Gestaltists. For example, one could compare the accounts Bühler and Wittgenstein give of the relationship between naming and describing and Popper's critique of the 'causal theory of naming'. But enough has been said to indicate a broad family resemblance. A more detailed survey of the similarities *and* differences among these thinkers could be carried out on some other occasion.

And indeed, even if it does go without saying – and even if Wittgenstein would nonetheless be the first to say it – there are many *many* differences between the theories I have sketched. But one needs comparable theories at one's disposal before the important job of differentiation can be carried out: to do intellectual history one must compare *and* differentiate. Either alone is insufficient.

Indeed, Popper, Wittgenstein, and the Gestalt psychologists would have to agree that it is not only shapes and figures which are similar or different depending on their 'ground' or context. *People and their ideas are also similar or different in relationship to a background.* By providing some of the missing background of the 'very different' philosophies of Wittgenstein, Popper, and the Gestaltists, I have tried to bring out some of their similarities on basic issues.

But, to repeat, I have not solved the intriguing and complicated question, why the background was allowed to disappear in the first place. To answer that question would involve a vast programme of research – one that I hope will one day be carried out.

VII

One can hardly sum up such a paper as this except by remarking that 'there are things in Vienna undreamt of by our philosophies'. Wittgenstein, Popper, and the members of the Vienna Circle were not the only, nor even the most important philosophical thinkers to flourish in Central Europe in the 20's and 30's – even if they were the most important philosophical thinkers whose ideas *emerged* from Central Europe. Conceivably, the common practice of emphasizing their ideas may have some pedagogical value – just as teaching the history of philosophy in the 18th century as if it were the story of the development from Locke, through Berkeley, to Hume, may be pedagogically convenient. Personally, I doubt that this is so: such parochialism is rarely desirable, pedagogically or otherwise. Whatever the answer to the pedagogical question, these com-

mon prejudices regarding the highlights of our recent philosophical history bear little relationship to the facts.

California State University, Hayward

Written in Ascona, Switzerland
August 1968

Bibliographical postscript. Due to the delay in the publication of the present monograph
I add this note in proof. I have been studying central European intellectual history of
the period between the two World Wars for some ten years. A small part of my research,
which greatly amplifies the information on Wittgenstein's Austrian context, was published in 1973 in my book *Wittgenstein*. The present monograph overlaps in some
places with that book, but is for the greater part supplementary to it. Most of the
material on Gestalt psychology and all of the material on Sir Karl Popper is presented
here for the first time.

This monograph was read at Boston University in October 1968, and also served as
the basis of papers I read to the Western Pennsylvania Philosophical Association, to the
Philosophy Colloquium of Vanderbilt University, to California State University, Hayward, and to the Humanities Seminar of California Institute of Technology. Abbreviated forms have twice been published in German. After a meeting in Minneapolis in
1969 the essay was copied and freely circulated without my prior knowledge, and it has
since then been quoted both in periodical publications and in dissertations.[36]

Scholarly attention to Austrian intellectual history has significantly advanced during
the past decade. Among those works which have contributed to understanding of this
period and of Wittgenstein's Austrian context are: Paul Engelmann: *Letters from
Ludwig Wittgenstein with a Memoir*, 1967; Wilma Abeles Iggers: *Karl Kraus: A Vien-
nese Critic of the Twentieth Century*, 1967; Frank Field: *The Last Days of Mankind:
Karl Kraus and His Vienna*, 1967; and William M. Johnston: *The Austrian Mind*, 1972.
A less responsible work on these matters is: Allan Janik and Stephen Toulmin:
Wittgenstein's Vienna, 1973.

Unfortunately, there has been no comparable development in the understanding of
Popper and his thought in their Austrian context. His book *Objective Knowledge*, 1972,
fails to indicate the significance of Bühler's work or of his Viennese background. A
general hagiographic account of Popper's life and work by Bryan Magee (*Karl Popper*,
1973) almost completely neglects his education and early development, and does not
mention Bühler.

NOTES

[1] Wolfgang Köhler *Dynamics in Psychology*, Washington Square Press, New York,
1965, pp. 116 and 122.
[2] Acknowledgements: This work would have been impossible without the help of the
following persons. Needless to say, they are in no way responsible for my conclusions
and have in several cases disagreed with them. I also am indebted to several persons
who, for reasons of their own, do not wish their names to be mentioned. My most
sincere thanks go to: Frau Anny Eder, Vienna; Herr Helmut Kasper, Innsbruck;

Professor Russell Dancy, Cornell University; Professor I. M. Bochenski, University of Fribourg, Switzerland; the late Professor Rudolf Carnap; Professors Joseph Agassi, Marx Wartofsky, and R. S. Cohen, Boston University; Professor J. O. Wisdom, York University, Toronto; Professor Paul K. Feyerabend, University of California, Berkeley; Professor Richard H. Popkin, Washington University, St. Louis; Professor Bernard Kaplan, Clark University; Dr. W. D. Hudson, University of Exeter, Professor Sylvain Bromberger, Massachusetts Institute of Technology, Professors Walter Kaufmann and George Pitcher, Princeton University; Dr. George Steiner, Cambridge University; Professor Charlotte Bühler, Los Angeles; Professor Albert Wellek, University of Mainz; Professor Viktor Kraft, University of Vienna; Herr und Frau Rudolf Koder, Vienna; Herr Dr. F. Lenz, Vienna; Dr. Thomas H. W. Stonborough, Vienna; Frau Gerda Leber-Hagenau, Vienna; Dr. Gustav Lebzeltern, Graz; Dr. Günter Posch, Innsbruck; Mr. Wallace Nethery, Hoose Library, University of Southern California; Herr Karl Gruber, Vienna; Herr und Frau Konrad Gruber, Vienna; Dr. Hanns Jäger, Stadtbibliothek, Vienna; Herr Dir. Martin Scherleitner, Schwarzau; Herr Norbert Rosner, Puchberg am Schneeberg; Frau L. Hausmann, Kirchberg am Wechsel; Frau Dir. Georg Berger, Trattenbach; Professor Herbert Spiegelberg, Washington University; Professor Herbert Feigl, University of Minnesota; Frau Margarete Bicklmayer, Otterthal; Herr Franz Brenner, Trattenbach; Herr Johann Scheibenbauer, Trattenbach; Herr Bürgermeister Emmerich Koderhold, Trattenbach; Professor F. A. von Hayek, Salzburg.

I am also indebted to the staff of the following institutions for their cooperation and help: Landesarchiv für Niederösterreich, Vienna; Kriegsarchiv, Staatsarchiv, Vienna; Pädagogisches Institut der Stadt Wien; Pädagogische Zentralbücherei, Vienna; Stadtschulrat für Wien, Vienna; Stadtarchiv, Bielefeld; Stadtarchiv, Korbach.

I am deeply indebted to the Research Committees of the University of Pittsburgh; of Gonville and Caius College, Cambrige University; and of the University of California, which at various times supported my investigations into German and Austrian intellectual history.

[3] See Karl Strack, *Geschichte des deutschen Volksschulwesens*, Gütersloh 1872, pp. 327 and 329–30; and also Charles A. Gulick, *Austria from Habsburg to Hitler*, The University of California Press, Berkeley and Los Angeles, 1948, Vol. I, p. 546.

[4] H. Gomperz, 'Philosophy in Austria During the Last Sixty Years', *The Personalist* (1936) 307–311.

[5] *Die Quelle* **81** (1931) 607–619.

[6] J. F. Herbart, *Outlines of Educational Doctrine*, Macmillan, 1904, p. 165.

[7] There was of course nothing new in this. Already in 1911, at the Congress for the Education of Youth in Dresden, Kerschensteiner (who was later to serve as the model for Hagauer in Robert Musil's novel *Der Mann ohne Eigenschaften*) had maintained that a public school that neglected practical education did the public a disservice, and went on to demand that students be trained in manual labour as well.

[8] Robert Dottrens, *The New Education in Austria*, The John Day Co., New York, 1930, p. ix.

[9] *Ibid.*, p. 202. Similar testimony may be found in Richard Meister, 'Teacher Training in Austria', *Harvard Educational Review* **8** (1938) 112–121.

[10] Otto Glöckel, *Die Österreichische Schulreform*, Verlag der Wiener Volksbuchhandlung, Vienna, 1923, p.11.

[11] See Bertrand Russell, 'The Philosophy of Logical Atomism', *The Monist*, 1918–19.

[12] See also Bühler's critical comments on Carnap in *Das Gestaltprinzip im Leben des*

Menschen und der Tiere, Verlag Hans Huber, Bern and Stuttgart, 1960, pp. 96–99. And see Albert Wellek, 'Karl Bühler 1879–1963', *Archiv für die gesamte Psychologie* **116** (1964) 2–8.

[13] George Pitcher, *The Philosophy of Wittgenstein*, Prentice Hall, Englewood Cliffs, 1964, pp. 6–9. See also W.D. Hudson, *Ludwig Wittgenstein*, John Knox Press, Richmond, 1968.

[14] Ludwig Wittgenstein, *Wörterbuch für Volksschulen*, Holder-Pichler-Tempsky A. G., Vienna, 1926. The following appears on the title page: "Mit dem Erlasse des Bundesministeriums für Unterricht vom 12. Oktober 1925, Z. 15444/9, zum Unterrichtsgebrauch an allgemeinen Volksschulen und an Bürgerschulen allgemein zugelassen."

[15] Richard Meister, "Teacher Training in Austria', *Harvard Educational Review* **8** (1938) 112–121.

[16] Popper did of course have *some* formal training in physics, having attended Hans Thirring's lectures.

[17] *Zur Methodenfrage der Denkpsychologie* is available in the University of Vienna Library. *Gewohnheit und Gesetzerlebnis* is described by Popper in *Conjectures and Refutations*, p. 50.

[18] Lewis Way, *Adler's Place in Psychology*, Collier Books, New York, 1962, p. 318.

[19] See my paper: W. W. Bartley, III 'Theories of Demarcation between Science and Metaphysics', in *Problems in the Philosophy of Science* (ed. by I. Lakatos and A. E. Musgrave), North-Holland Publishing Co., Amsterdam, 1968.

[20] K. R. Popper, 'Science: Conjectures and Refutations', in *Conjectures and Refutations*, Basic Books, New York, 1962.

[21] Pädagogische Zeitschriftenschau', *Die Quelle* **82**, 301.

[22] Cp. pp. 76–8 and pp. 82–3. Von Üxküll's theories, praised by Popper in *Conjectures and Refutations*, p. 382, are close to some versions of Gestalt psychology – to Kurt Lewin's cognitive field theory for example.

[23] Heinrich Gomperz, 'Kann die Deduktion zu 'neuen' Ergebnissen führen?', *Kantstudien* (1930) 466–479.

[24] See Otto Neurath, 'Die Pädagogische Weltbedeutung der Bildstatistik nach Wiener Methode', *Die Quelle*, 1933. Another positivist to contribute to the reform movement publications was Edgar Zilsel, whose article, 'Kant als Erzieher', appeared in *Schulreform* (1924) 182ff. Rudolf Carnap has told me that on his arrival in Vienna in 1926 he heard much discussion about the school reform, and remarked on its many differences from the German school reform movement with which he was well acquainted, having known some of the Landerziehungsheime led by Lietz, Wyneken, and Bondui.

[25] Karl Bühler, *The Mental Development of the Child*, Harcourt, Brace and Co., New York, 1930, pp. 106ff. See also E. H. Gombrich, *Meditations on a Hobby Horse*, Phaidon Press, New York 1963, p. 8, and the criticisms of this view which Gombrich cites therein.

[26] Bühler *Mental Development*, *op. cit.*, pp. 113–115.

[27] *Ibid.*, pp. 116, 120.

[28] Dottrens, *op. cit.*, p. 64.

[29] That Wittgenstein was 'in essential respects' a Kantian philosopher has frequently been maintained. See Erik Stenius, *Wittgenstein's 'Tractatus'*, Basil Blackwell, Oxford, 1960, and also J. Fang, 'Wittgenstein vs. Kant in a Philosophy of Mathematics', *Akten des XIV. Internationalen Kongresses für Philosophie*, Herder Verlag Vienna, 1968, pp. 233–236. I take issue with such Kantian interpretations of Wittgenstein in my book *Wittgenstein*, J. B. Lippincott, New York, 1973.

[30] Ludwig Wittgenstein *Zettel* (ed. by G. E. M. Anscombe and G. H. von Wright), University of California Press, Berkeley and Los Angeles 1967, p. 74e.
[31] See Paul Engelmann, *Letters from Ludwig Wittgenstein with a Memoir*, Basil Blackwell, Oxford 1967, Chapter V, p. 118; and B. F. McGuinness, *Friedrich Waismann: Wittgenstein und der Wiener Kreis*, Basil Blackwell, Oxford 1967, p.15n. I am indebted to Joseph Agassi for calling this reference to my attention in October 1968, after the first draft of this paper had been completed.
[32] And even here, a rather sophisticated theory of colour absolutes is presupposed!
[33] P. K. Feyerabend, 'Wittgenstein's Philosophical Investigations', in K. T. Fann (ed.) *Ludwig Wittgenstein: The Man and His Philosophy*, Dell Publishing Co, New York, 1967, pp. 214–250.
[34] p. 30.
[35] Is this a veiled criticism of Neurath?
[36] See, for example, D. W. Harding's review article of Wolfgang Köhler's 'The Task of Gestalt Psychology', in *The New York Review of Books*, December 18, 1969, pp. 16–20.

BIBLIOGRAPHY

Achs, O., 'Die Schulreform in der Ersten Republik (1918–1927)', *Österreich in Geschichte und Literatur* **13** (1969) 223–235.
Bartley, W. W., III, 'Rationality versus the Theory of Rationality', in *The Critical Approach to Science and Philosophy*, Free Press, New York, 1964.
Bartley, W. W., III, 'Theories of Demarcation between Science and Metaphysics', in *Problems in the Philosophy of Science* (ed. by I. Lakatos and A. E. Musgrave), North-Holland Publ. Co., Amsterdam, 1968.
Bartley, W. W., III, 'Sprach- und Wissenschaftstheorie als Werkzeuge einer Schulreform: Wittgenstein und Popper als österreichische Schullehrer', *Conceptus* **3** (1969) 6–22, Innsbruck-München-Salzburg.
Bartley, W. W., III, 'Die österreichische Schulreform als die Wiege der modernen Philosophie', *Club Voltaire, Jahrbuch für kritische Aufklärung*, Vol. IV, Rowohlt Paperback, 1970.
Bartley, W. W., III, *Wittgenstein*, J. B. Lippincott, New York, 1973 and Quartet Books, London, 1974.
Basil, O., *Trakl*, Rowohlt, Hamburg, 1965.
Beiträge zur Problemgeschichte der Psychologie, Festschrift for Karl Bühler's 50th Birthday, 1929.
Belohoubek, V., 'Schulwesen', in Stepan (ed.), *Neu-Österreich*, Vienna 1923, p.253.
Benjamin, W., *Illuminations* (ed. with introduction by Hannah Arendt), Schocken Books, New York, 1969.
Bühler, K., *Die geistige Entwicklung des Kindes*, Gustav Fischer Verlag, Jena, 1918.
Bühler, K., 'Die Krise der Psychologie', *Kantstudien* **31** (1926) 455–526.
Bühler, K., *Die Krise der Psychologie*, Gustav Fischer Verlag, Jena, 1927.
Bühler, K., 'Displeasure and Pleasure in Relation to Activity', in *Feelings and Emotions*, Wittenberg Symposium, Clark University Press, Worcester, 1928.
Bühler, K., *The Mental Development of the Child*, Harcourt, Brace and Co., New York, 1930.
Bühler, K., *Ausdruckstheorie*, Gustav Fischer Verlag, Jena, 1933.
Bühler, K., *Sprachtheorie*, Gustav Fischer Verlag, Jena, 1934.

Bühler, K., 'Forschung zur Sprachtheorie: Einleitung', *Arch. ges. Psychol.* **94** (1935), 401–412.

Bühler, K., *Das Gestaltprinzip im Leben des Menschen und der Tiere*, Verlag Hans Huber, Bern und Stuttgart, 1960.

Bühler, K., *Abriß der geistigen Entwicklung des Kleinkindes*, Quelle und Meyer, Heidelberg, 1967.

Bühler, K., *Die Uhren der Lebewesen und Fragmente aus dem Nachlass*, herausgegeben und mit einer Biographie versehen von Gustav Lebzeltern unter Benützung von Vorarbeiten von Hubert Rzainger. Foreword by Hubert Rohracher, Hermann Böhlaus Nachf., Kommissionsverlag der Österreichischen Akademie der Wissenschaften, Vienna, 1969.

Bühler, K., *Die Zukunft der Psychologie und die Schule*, Heft 11, *Schriften des Pädagogischen Institutes der Stadt Wien*, Deutscher Verlag für Jugend und Volk, Wien–Leipzig.

Bullock, M., *Austria 1918–1938: A Study in Failure*, Macmillan and Co., London, 1939.

Burger, E., *Arbeitspädagogik*, Engelmann, Leipzig, 1923.

Carnap, R., 'Andere Seiten der Philosophie', in *Club Voltaire, Jahrbuch für kritische Aufklärung*, Vol. III, Rowohlt Paperback, Munich, 1967.

Cox, Philip W. L., 'Austrian Teachers in the Crisis', *Progressive Education* **11** (1934) 360–368.

The Critical Approach to Science and Philosophy (ed. by Mario Bunge), Free Press, New York, 1964.

Dengler, Paul L., 'Creative Personality and the New Education', *Progressive Education* **6** (1929), 132–135.

Diamant, A., *Austrian Catholics and the Social Question*, University of Florida Press, Gainesville, 1959.

Dottrens, R., *The New Education in Austria*, The John Day Co. New York, 1930.

'Educational Events: Summer School in Psychology at the University of Vienna', *School and Society* **41** (1935) 284.

Engelmann, P., *Letters from Ludwig Wittgenstein, with a Memoir*, Basil Blackwell, Oxford, 1967.

Fadrus, V., *Die Österreichische Schulreform*, Vienna, 1927.

Fadrus, V., 'Professor Dr. Karl Bühlers Wirken an der Wiener Universität', *Wiener Zeischrift für Philosophie, Psychologie, Pädagogik* **7** (1959) 3–25, Festgabe for Karl Bühler's 80th birthday.

Feigl, H., 'The Wiener Kreis in America', *Perspectives in American History* **2** (1968) 630–673.

von Ficker, L., 'Rilke und der unbekannte Freund: In Memoriam Ludwig Wittgenstein', *Der Brenner*, Innsbruck 1954, pp. 234–248.

Fischl, H., *Wesen und Werden der Schulreform in Österreich*, Vienna 1929.

Fogelin, R. J., 'Wittgenstein and Intuitionism', *American Philosophical Quarterly* **5** (1968).

Glöckel, O., *Das Tor der Zukunft*, Vienna 1917.

Glöckel, O., *Die Österreichische Schulreform*, Verlag der Wiener Volksbuchhandlung, Vienna, 1923.

Glöckel, O., *Die Wirksamkeit des Stadtschulrates*, Vienna 1925.

Glöckel, O., *Die Entwicklung des Wiener Schulwesens*, Vienna 1927.

Glöckel, O., *Drillschule, Lernschule, Arbeitsschule*, Verlag der Organisation Wien der Sozialdemokratischen Partei, Vienna, 1928.

334   W. W. BARTLEY, III

Gombrich, E. H. J., *The Story of Art*, Phaidon, London, 1958.
Gombrich, E. H. J., *Art and Illusion*, Phaidon, London, 1960.
Gomperz, H., *Über Sinn und Sinngebilde, Verstehen und Erklären*, J. C. B. Mohr, Tübingen, 1929.
Gomperz, H., 'Kann die Deduktion zu 'neuen' Ergebnissen führen?', *Kantstudien* **35** (1930) 466–479.
Gomperz, H., *Die Wissenschaft und die Tat*, Gerold und Co., Vienna, 1934.
Gomperz, H., 'Philosophy in Austria during the last sixty Years', *The Personalist* **17** (1936) 307–311.
Gomperz, H., 'Interpretation', *Erkenntnis* **7** (1936) 225–232.
Gomperz, H., *Interpretation: Logical Analysis of a Method of Historical Research*, Library of Unified Science, Monograph Series, No. 8–9, W. P. van Stockum en Zoon, The Hague, 1939.
Gomperz, H., 'The Meaning of 'Meaning' ', *Philosophy of Science* **8** (1941) 157–183.
Gomperz, H., 'Autobiographical Remarks', *The Personalist* **24** (1943) 254–270.
Gomperz, H., *Philosophical Studies* (ed. by Daniel S. Robinson, with foreword by Philip Merlan), Christopher Publishing House, Boston, 1953.
Gulick, Ch. A., *Austria from Habsburg to Hitler*, University of California Press, Berkeley and Los Angeles, 1948, two volumes.
Hänsel, L., *Die Jugend und die leibliche Liebe*: *Sexualpädagogische Betrachtungen*, Tyrolia Verlag, Innsbruck-Vienna-Munich, 1938.
Hänsel, L., 'Ludwig Wittgenstein (1889–1951)', *Wissenschaft und Weltbild, Monatsschrift für alle Gebiete der Forschung* **8** (1951) 315–323.
Hänsel, L., *Begegnungen und Auseinandersetzungen mit Denkern und Dichtern der Neuzeit*, Österreichischer Bundesverlag für Unterricht, Vienna, 1957, pp. 315–323.
Hausmann, L., *Wittgenstein als Volksschullehrer*, typescript.
Herbert, J. F., *Outlines of Educational Doctrine*, Macmillan, New York, 1904.
Hon, F., 'School Reform in Austria', *The Canadian Forum* **11** (1930) 13–14.
Hudson, W. D., *Ludwig Wittgenstein*, John Knox Press, Richmond, 1968.
'100 Jahre Reichsvolksschulgesetz', *Erziehung und Unterricht* (1969).
Hunter, J. F. M., ''Forms of Life' in Wittgenstein's *Philosophical Investigations*', *American Philosophical Quarterly* **5** (1968).
Kaplan, B. and Wapner, S., (eds.) *Perspectives in Psychological Theory*, International Universities Press, New York, 1960.
Katz, J. J., *The Philosophy of Language*, Harper and Row, New York, 1966.
Keyserling, A., *Geschichte der Denkstile*, Verlag der Palme, Vienna, 1968.
Koffka, K., *The Growth of the Mind*, Harcourt Brace and Co., New York, 1925.
Koffka, K., *Principles of Gestalt Psychology*, Harcourt Brace and World, New York, 1935.
Köhler, W., *Dynamics in Psychology*, Washington Square Press, New York, 1965.
Kraft, V., *The Vienna Circle*, New York, 1953. The Philosophical Library.
Külpe, O., *Introduction to Philosophy*, George Allen and Unwin Ltd., London, 1927.
Külpe, O., *Outlines of Psychology*, Swan Sonnenschein and Co., London, 1909.
Külpe, O., *Vorlesungen über Psychologie*, S. Hirzel, Leipzig, 1922.
Külpe, O., *Vorlesungen über Logik*, S. Hirzel, Leipzig, 1923.
Lazarsfeld, P. F., 'Amerikanische Beobachtungen eines Bühler-Schülers', *Zeitschrift exp. angew. Psychol.* **6** (1959) 69–76.
Leinfellner, E., 'Zur nominalistischen Begründung von Linguistik und Sprachphilosophie: Fritz Mauthner und Ludwig Wittgenstein', *Studium Generale* (1969) 209–251.
Lewin, K., *A Dynamic Theory of Personality*, McGraw-Hill, New York, 1935.

Lewin, K., *Principles of Topological Psychology*, McGraw-Hill paperbacks, New York, 1966.

London, K. L., 'Nazi Germany and the Austrian School System', *The Educational Record* (1941) 170–178.

Lyford, G. L., 'Education in Eight European Cities', *Childhood Education* 6 (1930) 259–260.

Meister, R., 'Teacher Training in Austria', *Harvard Educational Review* 8 (1938) 112–121.

Methlagl, W. and Rochelt, H., 'Das Porträt: Ludwig Wittgenstein zur 80. Widerkehr seines Geburtstages', broadcast over Radio Tirol, Österreichischer Rundfunk, 1969.

Meyer, A., *Modern European Educators*, Prentice Hall, New York, 1934, pp. 99–160.

Naess, A., *Four Modern Philosophers*, University of Chicago Press, Chicago, 1968.

'The Nazis and their Subjugated Neighbors', *School and Society* 52 (1940) 625.

Neurath, O., 'Die Pädagogische Weltbedeutung der Bildstatistik nach Wiener Methode', *Die Quelle*, 1933.

'Obituary for Ludwig Wittgenstein', in *Die Presse*, Vienna, 6 May 1951, p. 51.

Pears, D. F., 'The Development of Wittgenstein's Philosophy,' *The New York Review of Books* (January 16, 1969), 21–30.

Piaget, J., *Play, Dreams and Imitation in Childhood*, William Heinemann Ltd., London, 1951.

Piaget, J., *The Child's Conception of Number*, Routledge and Kegan Paul Ltd., London, 1952.

Piaget, J., *The Origins of Intelligence in Children*, International Universities Press, New York, 1956.

Pitcher, G., *The Philosophy of Wittgenstein*, Prentice-Hall, Englewood Cliffs, 1964, pp. 6–9.

Popper, Karl R., 'Über die Stellung des Lehrers zu Schule und Schüler', *Schulreform* 4 (1925) 204ff.

Popper, Karl R., 'Zur Philosophie des Heimatgedankens', *Die Quelle* 77 (1927) 899ff.

Popper, Karl R., 'Die Gedächtnispflege unter dem Gesichtspunkt der Selbsttätigkeit', *Die Quelle* 81 (1931) 607ff.

Popper, Karl R., 'Pädagogische Zeitschriftenschau', *Die Quelle* 82 (1932) 301, 580, 646, 712, 778, 846, 930.

Popper, Karl R., 'The Demarcation between Science and Metaphysics', in *Conjectures and Refutations*, Basic Books, New York, 1963.

Popper, Karl R., 'Language and the Body-Mind Problem', in *Conjectures and Refutations*, Basic Books, New York, 1963.

Popper, Karl R., *The Open Society and Its Enemies*, Harper Torchbooks, New York, 1963, two volumes.

Popper, Karl R., 'Philosophy of Science: A Personal Report', in *British Philosophy in the Mid-Century* (ed. by C. A. Mace), Allen and Unwin, London, 1957, and in *Conjectures and Refutations*, Basic Books, New York, 1963.

Popper, Karl R., 'Towards a Rational Theory of Tradition', in *Conjectures and Refutations*, Basic Books, New York, 1963.

Popper, Karl R., *The Logic of Scientific Discovery*, Harper Torchbooks, New York, 1964.

Popper, Karl R., *The Poverty of Historicism*, Harper Torchbooks, New York, 1964.

Popper, Karl R., *Conjectures and Refutations*, Harper Torchbooks, New York, 1968.

Scheibenreif, F., *Orts- und Hauschronik von Trattenbach*, Gemeindeamt Trattenbach: Im Selbstverlag und Kommissions-Verlag, Druck: K. Mühlberger, Neunkirchen, 1934.

Schneider, W., '100 Jahre Reichsvolksschulgesetz und Schulaufsicht im Schulbezirk Neunkirchen', *Erziehung und Unterricht*, May 1969.
Schwab, J. J., 'On the Corruption of Education by Psychology', *Ethics* **68** (1957).
Schwendt, J., *Schule im Wald: Trattenbach, Dorf am Wechsel*, Trattenbach 1966.
Siegl, M. H., *Reform of Elementary Education in Austria*, New York 1933.
Smital, O., 'Eine Stimme aus Wien: Ludwig Wittgenstein', Broadcast, Österreichischer Rundfunk, 19 September 1968.
Smoor, M. P., 'Impressions of the Schools of Austria', *The High School Teacher* (1930) 417–8.
Spiegelberg, H., 'The Puzzle of Wittgenstein's *Phänomenologie*', *American Philosophical Quarterly* **5** (1968).
Steiner, G., *Language and Silence*, Atheneum, New York, 1970.
Stern, J. P., *Lichtenberg: A Doctrine of Scattered Occasions*, Indiana University Press, Bloomington, 1959.
Tesar, L. E., 'Die Schulreform in Österreich', in *Handbuch der Pädagogik*, H. Nohl, L. Pallat, Vol. IV, 1928.
Tumlirz, O., 'Austrian Education in the Past Fifteen Years', *School and Society* **41** (1935) 197–203.
Vossler, K., *Sprachphilosophie*, Max Hueber, Munich, 1923.
Vygotsky, L. S., *Thought and Language*, John Wiley and Sons, Inc., New York, and the MIT Press 1962.
Waismann, F., *Wittgenstein und der Wiener Kreis* (ed. by B. F. McGuinness), Basil Blackwell, Oxford, 1967.
Walker, J., 'Wittgenstein's Earlier Ethics', *American Philosophical Quarterly* **5** (1968).
Way, L., *Adler's Place in Psychology*, Collier Books, New York, 1962.
Wein, H., *Sprachphilosophie der Gegenwart*, Martinus Nijhoff, The Hague, 1963.
Wellek, A., 'Ein Dritteljahrhundert nach Bühler's 'Krise der Psychologie'', *Zeitschrift für experimentelle und angewandte Psychologie* 6 (1959) 109–117.
Wellek, A., 'Karl Bühler 1879–1963', *Archiv für die gesamte Psychologie* **116** (1964).
Wellek, A., 'The Impact of German Immigration on the Development of American. Psychology', *Journal of the History of the Behavioural Sciences* 4 (1968) 207–229.
Wellmer, A., *Methodologie als Erkenntnistheorie: Zur Wissenschaftslehre Karl R. Poppers*, Suhrkamp Verlag, Frankfurt a. M., 1967.
Werner, H. *Comparative Psychology of Mental Development*, Follett Publishing Company, Chicago, 1948.
Werner, H., (ed.) *On Expressive Language*, Clark University Press, Worcester, 1955.
Wissenschaftliche Weltauffassung: Der Wiener Kreis, Arthur Wolf Verlag, Vienna, 1929.
'Wittgenstein's Lectures on Ethics', *The Philosophical Review* **74** (1965) 3–26. Also printed in *Neue Zürcher Zeitung*, and in *Conceptus* 2 (1968).
Ludwig Wittgenstein The Philosophical Investigations (ed. by George Pitcher), Doubleday New York 1966.
Ludwig Wittgenstein: The Man and His Philosophy (ed. by K. T. Fann), Dell Publishing Co., New York, 1967.
'A Wittgenstein Miscellany', in *The American Philosophical Quarterly* (ed. by Nicholas Rescher) **5** (1968).
Wittgenstein, L., *Tractatus Logico-Philosophicus*.
Wittgenstein, L., *Wörterbuch für Volksschulen*, Hölder-Pichler-Tempsky A. G., Vienna 1926.
Wittgenstein, L., *The Blue and Brown Books*, Basil Blackwell, Oxford, 1958.

Wittgenstein, L., *Philosophical Investigations*, Basil Blackwell, Oxford, 1958.
Wittgenstein, L., *Lectures and Conversations* (ed. by Cyril Barrett), University of California Press, Berkeley and Los Angeles, 1967.
Wittgenstein, L., *Zettel*, University of California Press, Berkeley and Los Angeles, 1967.
Wuchterl, K., *Struktur and Sprachspiel bei Wittgenstein*, Suhrkamp Verlag, Frankfurt, 1969.
Zilsel, E., 'Kant als Erzieher', *Schulreform* **3** (1924) 182ff.

Periodicals

Amtlicher Bericht des Stadtschulrates für Wien.
Deutsche Erziehung (ed. by Max Fritz), Vienna.
Der Neue Weg (ed. by Leopold Lang and Ludwig Battista).
Pädagogisches Jahrbuch (ed. by Wiener Pädagogischen Gesellschaft).
Die Quelle (ed. by Eduard Burger), Vienna.
Schulreform, Haase Verlag, Vienna.
Die Volkserziehung, Nachrichten des Bundesministeriums für Unterricht.
Volksbildung, Zeitschrift für die Förderung des Volksbildungswesens in Österreich (ed. by Max Mayer).

RICHARD POPKIN

BIBLE CRITICISM AND SOCIAL SCIENCE*

In this paper I will deal with the crucial role played by the Bible critics of the 17th century in providing some of the framework in which modern social science developed. Besides providing some of the methodology of the social science, in terms of the research approaches of the Bible critics, and posing some of the basic problems in the area,[1] the Bible critics contributed greatly to forming the ideology of what Hume called 'the science of man'. It is this latter contribution that I shall treat here. In so doing, I certainly do not wish to claim that there were not other important streams of influence, such as the theories of the dynamics of man and society presented by Machiavelli and Hobbes.

In this paper I will deal with four of the Bible critics, Isaac La Peyrère, Baruch de Spinoza, Richard Simon and Pierre Bayle. Although it is often claimed that the modern higher criticism of the Bible began with Hobbes' questioning of the Mosaic authorship of the Pentateuch in *Leviathan*,[2] various problems about what constitutes the correct text of Scripture, and whether Moses wrote the entire Pentateuch had been raised earlier. Renaissance humanistic scholarship had indicated many difficulties.[3] However, I think it is fair to say that, by-and-large, Bible scholarship up to 1640 was trying to reconcile difficulties *within* an accepted framework in which the Bible was seen as the record of the development of human history, and that human history is Providential, directed by God from Creation, to the Fall, to the Flood, to the Exodus, to Sinai, to the Prophets, to the Incarnation, to the Resurrection, to the Second (or First) Coming. In the early 17th century, works of men like Rabbi Menasseh ben Israel and Archbishop Ussher, were still devoted to making all known information about the human world fit within such a framework. Sir Walter Ralegh, in his *History of the World* (1614), traced human history from the Creation of man in 4004 B.C. up to Roman history. Menasseh ben Israel, in his *Conciliador* (1633), tried to show how the seemingly conflicting passages in the Old Testament could be interpreted and reconciled into one harmonious whole. In his *Hope of Israel* (1650), he sought to show how modern European history followed from Biblical history, and that the American Indians could be included by recognizing that they came from China, and were the Lost Tribes of Israel. Archbishop Ussher placed all the events in all of the known cultures in a consistent chronology from 4004 B.C. onward.

And, at the same time that so much ingenuity was being employed to make the Biblical world and the known human world all fit togeher, a radical bombshell that launched modern Bible criticism appeared, La Peyrère's theory of the Pre-Adamites.

Isaac La Peyrère (1596–1676), is a scholar's dream. He was a man who played a crucial role in the development of our intellectual world, and yet he is practically unknown today. No book has been written on him (except for the inordinate number of refutations that appeared in the 17th century). A few articles and a few chapters of books have been devoted to him. He has been described as 'l'incorrigible hérésiarque' by René Pintard,[4] as a Socinian by Leo Strauss,[5] as an unbeliever by Don Cameron Allen,[6] as a precursor of the Deists by David R. McKee,[7] and as a Marrano precursor of Zionism by Hans Joachim Schoeps.[8]

La Peyrère was an amazing figure and an amazing mentality, whose over-all theory has not, I believe, been fully understood. He came from Bordeaux from a family engaged in government business.[9] He was raised as a Calvinist, though there was and is ample suspicion that he was actually Jewish, from a family of Portuguese refugees. His friends in later life assumed he was Jewish, and he, when asked said he was in the sense St. Paul used the term, presumably as a convert.[10] There is a document in Paris from 1626 indicating that he had had theological difficulties with the Reformed Church at that time.[11] Around 1640 La Peyrère started working for the Prince of Condé in Paris, and shortly afterwards became his secretary. In Paris he was an associate of many of the leading intellectuals of the time – Grotius, Mersenne, Gassendi, the Du Puys, La Mothe le Vayer, and probably Hobbes. In 1642–3 La Peyrère wrote his two most important works, the *PraeAdamitae* and *Du Rappel des Juifs*, the first rasing the most serious and basic problems about the Bible and whether it is an accurate account of the world's history, and the latter analyzing the the role of the Jews, past, present and future, in which he foresaw the imminent coming of the Jewish Messiah, the union of the Jews and Christians, and the establishment of a new age through the alliance of the King of France and the Messiah, and the salvation of everybody. The two books were originally one continuous presentation of La Peyrère's picture of the world.[12] From his youth onward, he was worried about inconsistencies he had found in the Bible, especially in the story in *Genesis*. He also early developed his Messianic vision. Around 1642–3, he united it all into a revolutionary reinterpretation of Scripture and the course of Judaism, in which the Bible only dealt with Jewish history. The world of the gentiles had been going on aimlessly in the state of nature before Adam. God decided to save mankind and created the Jews, first Adam, and then the first Jewish momma, Eve. Through Jewish history the Divine drama of salvation is being played out. Its culmination is about to occur, and though its participants are only the Jews, the rest of mankind has been, and will be saved through mystical imputation.[13]

He excitedly dedicated his works to Cardinal Richelieu and presented them to him. La Peyrère may have felt Richelieu would be sympathetic, since Richelieu had been the secretary of the Jewish leader of France, Concino Concini, le Maréchal de l'Ancre, who was apparently a part of a Cabalist circle.[14] Richelieu, a clever politician, and an adept in-fighter in the religious battles of the time, immediately banned the work. La Peyrère anonymously published the second part, *Du Rappel des Juifs* (1643), and passed the other part around among his friends. Grotius saw the radical implications of La Peyrère's pre-Adamite theory, and wrote an attack on it in the second edition of his work on the origins of the American Indians in order to save Christianity. Grotius argued that the Indians came from the Adamic world, and were really Norwegians, descended from the Norse explorations.[15] Various letters of the time, including some from the Vatican, indicate that La Peyrère's theory was well known, considered amusing by some, shocking by others.[16] The author went off to Copenhagen with the French ambassador, and there wrote his two most scientific works, *Relation de Groenland* and *Relation d'Islande*, both landmarks in the development of anthropology.[17] Though he never visited either Greenland or Iceland, he studied the Danish records. On the basis of these, he became convinced that the Eskimos were not descended from Adam and Noah, and that they pre-dated the Norse invasions, a point he used later on to refute Grotius.[18]

Around 1647 La Peyrère returned to the service of the Prince of Condé, stopping off to discuss his criticisms of the Bible with various Dutch

scholars. He travelled widely for Condé, eventually getting as far as Spain
and England before his troubles began. He went with the Prince to Brus-
sels, where he lived next door to another religious mystic, Queen Christina
of Sweden, who had just given up her throne, and was about to turn
Catholic. Christina read La Peyrère's manuscript, and persuaded him to
have it published, either by agreeing to pay for it, or encouraging the
author to seek a publisher in Holland.[19]

La Peyrère's story of how his book got printed was that while he was
in Amsterdam for Condé, he did not want to leave his manuscript
lying around. When he got to Amsterdam, the Dutch publishers tried
to grab it from him. Finally he consented, and in 1655 it appeared in
three editions by Elsevier, another in Basel, and another unidentified,
plus an English translation in 1656, and a Dutch one in 1661.[20] Also
immediately refutations and condemnations appeared everywhere.[21] The
author went to Spain, then returned to Belgium, and spent a quiet few months
while the book was being banned and burned.[22] He was denounced at
Namur as a Jew and a heretic. Finally he was arrested by order of the Archbishop of Malines.[23] Condé tried to rescue him, but to no avail. Then it was
indicated that if La Peyrère turned Catholic and apologized to the Pope, he
would be freed. He did so, and went to Rome.[24] His abjuration, prepared
with much Papal assistance, consists of blaming his views on his Calvinist
upbringing. La Peyrère insisted that as long as he were a Calvinist, he
could only use his own reason and his reading of Scripture to guide him,
and so could only come to the conclusions that there were men before
Adam, that the Biblical text is not by Moses, that it is confused and contradictory –

I need not trouble the Reader much further, to prove a thing in it self sufficiently evident, that the first five books of the Bible were not written by *Moses*, as it is thought.
Nor need any one wonder after this, when he reads many things confused and out of
order, obscure, deficient, many things omitted and misplaced, when they shall consider
themselves that they are a heap of Copie confusedly taken[25]

– that the Flood was a local event in Palestine, and that the Bible only
describes the history of the Jews, not the history of the world.[26]

When La Peyrère was attacked, and the only evidence brought against
him was to point out the minor fact (minor to the author), that he disagreed with the *entire* Christian and Jewish traditions, he then claimed
that he had to find some authority other than himself to tell if he was right.

And who was a better authority than the Pope? La Peyrère's *Apologie* reeks with hypocrisy, but it worked. Not only was the Pope pleased with him, but offered him a post in the Vatican. La Peyrère refused and returned to Paris and to the Prince of Condé, as his librarian. He later retired as a lay brother to the pious order of the Oratorians.[27]

In all of his later works, published and unpublished, La Peyrère was busy gathering further evidence for his theories, establishing his pre-Adamite hypothesis, and arguing that the culmination of Jewish history was imminent. The six letters of the great Bible scholar, Father Richard Simon (who will be discussed shortly) to him in 1670–72 indicate that La Peyrère was busy in his old age arguing his case, and rewriting his *Rappel des Juifs* (which Simon assured him would be banned, as it was).[28] In 1671 La Peyrère tried publishing a French Bible with notes, which was suppressed by the time he got to *Leviticus*, since the notes kept presenting the evidence for his pre-Adamite views, which he put forth as a crank theory, condemmed by the Church, but which can be supported by such and such facts.[29] He insisted to Simon up to the end of his life that there was no evidence against his theory either in the Bible or in reason. And he died without giving up the fight (though, in a last attempt to get his manuscript published, he took out the pre-Adamite theory).[30] A friend wrote as his epitaph

'Here lies La Peyrère, that good Israelite
Hugenot, Catholic and finally Pre-Adamite
Four religions pleased him at the same time
And his indifference was so uncommon
That after 80 years when he had to make a choice
The Good Man departed and did not choose any of them'[31]

La Peyrère's evidence for his revolutionary theory was first of all an interpretation of St. Paul's *Romans* 5:12–15 that sin began with Adam. Hence there must have been a sinless state of nature before this. Next La Peyrère explored the evidence in *Genesis* that there must have been another source of people besides Adam and Eve. Who was Cain's wife? Who was in the city Cain moved to? Where did Lilith come from? Was Noah's Flood universal, or just a local occurrence in Palestine? Lastly La Peyrère introduced the evidence that was to carry the day – that the recent discoveries in Asia, America and the South Pacific could not be reconciled with the Biblical picture of the world. Cultures from antiquity to the present had chronologies, monuments and histories that could not be fitted into a chronology starting in 4004 B.C., and a lineal descendence from Adam and Noah.[32]

La Peyrère's point was to separate Jewish history from world history, to insist that Adam was *only* the first Jew, that the Bible only dealt with Jewish events, and that the rest of the world had independent origins. For La Peyrère Jewish history was and is Providential to the nth degree. The *Prae-Adamitae* is dedicated "To all the Synagogues to the Jews, dispersed over the face of the EARTH. People holy and elect! Son of *Adam*, who was the Son of God. Son of God also your selves. One, I know not who, wishes you all happiness; and himself to be one of you."[33] The theory in the work and its sequel, *Du Rappel des Juifs*, seems to me to be that of a Marrano, seeking to show first the all-importance of Jewish history in world affairs, and in the present the importance of Marrano history. It is the Marranos who will lead to the Messianic age and rule the world with the Messiah and the King of France. Everyone else is essentially just a bystander in the course of theological history.[34]

La Peyrère's theory provided several ingredients in the development of social science. His denial of the Adamic-Noaic origin of most peoples and cultures opened the way to studying their cultures and their histories apart from the Providential context of Judeo-Christianity. Secular, naturalistic examination of most of human society was now possible. Second, he opened the door to the study of the Bible as a historical document that developed over the course of time and within human historical and cultural conditions. And third, he was one of the first to begin the anthropological study of religion, and to use anthropological evidence to interpret religion. La Peyrère's pre-Adamite theory, his critique of the Bible, and his introduction of anthropology led to several lines of development of importance for the social sciences – Vico's formulation of a science of human history, Spinoza's naturalistic metaphysics within which to study Biblical religion, and Father Simon's presentation of the scientific study of the Bible, revealing the epistemological problems in the study of past.

Though Vico was the last of these, I shall treat him first. Vico's *Scienza nuova* starts out in Book I with a fundamental problem initiated by La Peyrère, that of the apparent conflict of the chronologies of Egypt, China, etc., and that of the Bible. Vico quickly decided "the Hebrews were the first people of the world" "whose prince was Adam," and that the chronologies of all other ancient peoples are fables or mythology. The world of the gentiles can be studied scientifically out of their mythologies, but Jewish history is known only by Revelation. In trying to reject La Peyrère's pre-Adamism, Vico paradoxically advanced a somewhat similar framework. Jewish history is to be completely separated from gentile or national history. In separating them, and giving Jewish history chronological priority, the history of all groups *except* the Jews can be studied in scientific and secular terms. La Peyrère had initiated the separation. Vico was unwilling to accept La Peyrère's heresies, but did follow out one of the major consequences of the separation, that the scientific study of history was possible if the one case of Providential history were omitted.[35]

Others like Pascal[36] and the Protestant scholar, Pierre Jurieu tried to write off La Peyrère as a crank with an implausible theory (though Jurieu, orthodox fanatic that he was, still felt the theory interesting enough to expound it at length for his readers who might not have read it.)[37] The learned Bishop Stillingfleet tried to show that the reasonable man, operating on common sense standards, would find the Biblical account more plausible than La Peyrère's polygenetic interpretation of pagan history and the explorer data.[38] Bishop Huet tried to trivialize La Peyrère's point by contending that various cultures had independent origins and developments, but that they all expressed the same religious truth. Through the comparative study of religion, one could find the same Moses, Mary and Jesus figures, but only Christianity spelled their names correctly.[39]

Lots of refutations of La Peyrère occurred. His *Prae-Adamitae* is probably the most refuted work of the 17th century. But, as the author contended, the critics had no evidence to offer except to report that *all* the authorities, Jewish and Christian, from ancient times to the present, insisted that Adam was the first man, and that the Bible was correct. The two Bible critics who took La Peyrère most seriously, and who built on some of his claims were Spinoza and Father Richard Simon.

Spinoza probably came in contact with La Peyrère's ideas shortly before his excommunication. La Peyrère was in Amsterdam for six months in 1655. He was a friend (and opponent) of Spinoza's teacher, Manesseh ben Israel, and was involved in discussions in the Millenarian circles around Menasseh ben Israel and Felgenhauer.[40] Spinoza owned a copy of the *Prae-Adamitae*, and the *Tractatus–Theologicao Politicus* shows lots of influence of La Peyrère's theories and specific claims.[41] Recent studies by I. S. Révah indicate that there were other people in the Jewish community highly critical of the Bible, who were also punished by the Synagogue. One, Juan de Prado, who was expelled along with Spinoza, had written a work claiming that the Law of Nature took precedence over the Law of Moses. In the charges against Prado, he was accused of holding the pre-Adamite theory, and of giving the same evidence for it as La Peyrère had.[42] It is reported in the literature from the mid 17th century onward that right after La Peyrère's book appeared, there was a short-lived sect of pre-Adamites in Amsterdam.[43] Considering all of this, it is possible, even likely, that La Peyrère and Spinoza were brought into personal contact by the currents they were involved in. La Peyrère's burning interest in Judaism, and his probable family connections with the Pereiras of Amsterdam, should have brought him into contact with the Jewish community. His relations with Manesseh ben Israel may have put him in contact with some of Manesseh's students. His radical ideas would have made him sympathetic to the young radical Bible critics in the community.

From Spinoza's writings, he apparently only knew the *Prae-Adamitae*. Unlike La Peyrère who could not read Greek or Hebrew, and who seems to have only known Maimonides, and maybe Judah Halevi, among the Jewish authors, Spinoza was quite learned in Hebraic materials.[44] Spinoza pressed the difficulties raised by La Peyrère and Rabbi Ibn Ezra. He explored the problem of Biblical authorship, and the many contradictory or conflicting elements in the Bible, offering evidence that appears to come from both Ibn Ezra and La Peyrère. But much more than the details of his Biblical scholarship, Spinoza contributed an entirely new outlook to the evaluation of the Biblical material. "I may sum up the matter by saying that the method of interpreting Scripture does not widely differ from the method of interpreting nature – in fact, it is almost the same."[45]

For Spinoza the Bible and the Biblical world could be interpreted naturalistically. It could be treated not as Revelation or Providential history, but as a primitive stage of human development. Using the psychology and sociology of his day (primarily Hobbesian analysis), Spinoza interpreted the Biblical world, and later in the *Ethics* and in his letters, the entire religious world, in terms of human fears and superstitions. He offered entirely naturalistic evaluations of the whole religious dimension of man's development and conduct. Unlike La Peyrère, for whom the entire Gentile world could be studied in secular and natural terms, but for whom the Jewish world still remained Providential, Spinoza provided a metaphysics in which the Judeo-Christian world became a part of the human scene to be understood, as everything else, in terms of the application of the 'new science' to man.

Spinoza's world left no room for Divine intervention and direction. Human activities were no longer seen as part of an unfolding Divine drama, but as necessary features of the logical order of Nature (or God). La Peyrère's secularization of most of the human world was due to his mysticism about Judaism. Spinoza's naturalization of *all* of the human world was the outcome of his metaphysics, designed, apparently, originally to justify his Bible criticism. The *Tractatus* is supposed to be a revised version of his original answer to the rabbis who excommunicated him. In it the metaphysics of the *Ethics* emerges as the justification of his denial of miracles, and of his interpretation of nature and God.[46] This metaphysics provides a basis for social science by making all of human behavior intelligible in a naturalistic framework, and by making Judeo-Christianity, in its message and its history, part of the natural world to be examined scientifically.

Following Spinoza, the next stage in the development of Bible criticism was the work of Father Richard Simon, 1638–1712. Simon, the greatest Bible scholar of his day, and The Father of Higher Criticism, was heavily influenced by both La Peyrère and Spinoza. He knew La Peyrère personally, coming in contact with him when the latter was in his 70's, when La Peyrère was living with the Oratorians. Simon studied and taught at the Oratory in the 1660's. The six letters of Simon to La Peyrère, and a letter giving La Peyrère's biography,[47] indicate that they were pretty close. They argued about the pre-Adamites, whom Simon did not believe in. Simon read the last manuscript of La Peyrère on the *Rappel des Juifs*, and advised against its publication. He sent his Jewish crony, Jona Salvador, a Cabalist and Messianist, out to see La Peyrère to tell him of the Messiah, Sabbatai Zevi. Simon's Bible criticism indicates that he had absorbed the fundamental points of La Peyrère's approach. Simon also studied Spinoza's *Tractatus* when it appeared in 1670, and unlike the many theologians who were horrified by it, Simon accepted many of

Spinoza's maxims, but denied or disputed the irreligious conclusions Spinoza drew from them.[48]

Father Richard Simon is a fascinating mind, who has been too neglected in studies of the 17th century.[49] He was fabulously erudite in Near Eastern languages, in Jewish and early Christian history, in the manuscripts of the Bible, in the published and unpublished materials about Judaism and early Christianity. He was a critical, humanistic scholar, concerned to establish the best text of the Bible, and the best interpretation of its Message. His first writings dealt with Judaism and early church practices.

His most famous work, *The Critical History of the Old Testament* (1678), was originally intended to refute the Calvinist claim that one could find the Truth by reading Scripture. Simon sought to show the monumental difficulties involved in ascertaining what the Bible said and what it meant. To tell what it said, one first had to establish an accurate text, not only of the present Bible, but of the actual message from God.

Simon debated at great length with those who contended that the present Bible, the Vulgate, was what had actually been dictated by God to Moses. He pointed out the variation between the Vulgate and the Septuagint and the Massoretic text, and the variations in different manuscripts. The oldest known manuscript in his day dated from many centuries, even millenia, after the original Revelation. The text, itself, building on points raised by La Peyrère and Spinoza, could not be by Moses, and Simon proposed as an explanation of the internal problems, that it was written down by public scribes over an 800 year period. What has come down to us is copies of copies of copies, with all sorts of alterations, glosses, mistakes and variants, creeping in. The study of the history of the Jews and the early Christians, of their development, their customs, their beliefs, would aid in unravelling the mess, and in approaching the actual Message. However, each level of this study also indicated more problems in gaining any certainty about what God had revealed to man. The more one delved into the study of ancient languages, manuscripts, history, etc., the more one realized the impossibility of establishing a past fact with complete certainty. Simon, who was actually a modest fellow, thought he knew Hebrew better than anyone else alive at his time. Most Jews only learned it by usage.[50] The rabbis did not seem to know the grammar, the etymology, the development of Hebrew. But Simon realized that for all of his study of manuscripts, for all of his analysis of the structure of the Hebrew language, for all of his knowledge of its history and development, he was still restricted to a 17th century interpretation of what ancient Hebrew was like and what it meant. The application of what he called the 'critical method' to the Bible revealed the insuperable problems in recapturing the context of a past event. In his works from the *Critical History of the Old Testament*, suppressed before publication, to his last work, the French translation of the New Testament – (all of his later works were banned) – Father Richard Simon showed the difficulties in *really* ascertaining what God had said or what He had meant. The more one did research, the more one showed that one had somehow to cut through a miasma of human interpretations and human events to get to the Message, and all that one had to work with were present evaluations of past human happenings. Critical study could show what were the most probable conclusions that human beings at this moment could come to.

Simon's researches were condemned by his Protestant and Catholic opponents as leading to an 'historical Pyrrhonism', a doubt about all past facts, and to an undermining of confidence in all of the doctrines of Christianity. Simon, with a deft touch and with incredible erudition, drowned his opponents in learning and in a sea of problems. The Catholic Church rejected his achievements resisting the turn toward Higher Criticism for a couple of centuries, while insisting they were standing fast on their doctrines and dogmas, and that they would not let their case rest on the latest findings of the Bible scholars. Some Protestant scholars followed out some of Simon's leads, and developed the scientific study of the Bible. Astruc a generation later proposed a solution to the Biblical authorship problem by arguing that there were two different kinds of authors.[51] In Germany starting with Reimarus a wholesale reconsideration of the Bible and its Message began, leading to the 19th and 20th century Higher critics.

Although Simon's work had the effect of unsettling many intellectuals' confidence in the truth or the content of Judeo-Christianity, Simon himself seems to have remained an indefatigible truth-seeker all of his life, convinced that there was a Message that somehow, sometime, human beings would recapture. He did not accept La Peyrère's pre-Adamite hypothesis and its attendant consequence that Biblical history was not world history. He did not accept Spinoza's metaphysical rejection of the supernatural import of the Bible. He agreed that there were problems, far more, and far more troublesome ones than his brave predecessors ever thought of, since they had never done the research Simon had. Every claim about the Bible, about the content of Judeo-Christianity, was fraught with difficulties. The more one knew about the Bible, and about Judaism, Jewish history and early Christian developments, and about the evidence on which this knowledge was based, the less sure one could be of any of the doctrinal claims made by the churches. Although he was accused of being an unbeliever (and he burned all his notes and books to prevent them being seized and inspected by the authorities), I think he remained seriously within the religious context. He was seeking for some common religious truth behind Christianity and Judaism, and his scholarship led him to reject the doctrinal points at which particular religious groups had stopped. There is a strong suggestion of modernism in his views, approaching the spirit of Vatican II. This may account for the attempt to revive him as a theologian and a religious spirit by the late Abbé Jean Steinmann, one of the most liberal Catholics in recent French thought.[52]

Simon's contributions to the modern world are many. In terms of the theme of this paper, I should just like to single out a few. He is in part part responsible for the development of modern methods for studying the past. He revealed the epistemological problems involved, and worked out in great detail the 'critical method' of historical research for gaining the best account and evaluation at a given time of past events. Using scholarly tools far beyond La Peyrère's and Spinoza's, he showed how this could illuminate a crucial part of man's behaviour – his religious dimension. He showed how present investigations, into the practices and customs of Jews, the Greek Orthodox, the Nestorians, the Cairites (with whom Simon was fascinated), the Moslems, etc. could throw light on understanding the Bible and its Message. He provided many tools and methods, and an approach that was not only to make religion amenable to social scientific investigation, but was to open the door to a kind of science of man.

One of those who saw some of the potentialities of what Simon had done or started was the great sceptic, Pierre Bayle, 1647–1706. Bayle, a son of a Protestant pastor, became a Catholic when a student at Toulouse, then relapsed to Calvinism and had to flee to Geneva where he studied philosophy. He secretly returned to France, became a teacher at the last

Protestant academy in France, Sedan, and had to flee to Holland when the school was closed. He lived the last 25 years of his life in Rotterdam, fighting against all sorts of theologians, philosophers, historians and scientists.[53] From his earliest unpublished writings to his monumental *Historical and Critical Dictionary* (1697–1702) (which Voltaire labelled 'the Arsenal of the Enlightenment'), Bayle was concerned with uncovering the mistakes and errors that had crept into the historical record. An unpublished notebook of his now in Copenhagen, indicates he had started once to list all the errors.[54] Then he turned to writing a correct life of Jesus. Before he wrote the *Dictionary*, an immense work of 7–8 000 000 words, he had proposed to publish just a compilation of all the errors, but no one was interested. The *Dictionary* was a compromise. It includes masses of his historical erudition on questions like whether there ever a Papesse Jeanne, whether Calvin was a sodomist, whether Pope Leo X was an atheist, whether George Buchanan, Montaigne's teacher, was arrested partaking of a Passover Seder, etc. It also includes endless digressions of sceptical attacks on all kinds of philosophical and theological theories, developing Bayle's basic contention, that all intellectual effort is 'big with contradiction and absurdity' and that all one can do when one realizes this, is give up reason and turn to faith.

One side of Bayle's efforts to undermine confidence in the rational world was his attack on religious practices and on the Bible. Though he rejected Spinoza's metaphysical rationalism, in the longest article in the *Dictionary*, and treated La Peyrère as a pleasant amusing crank, he did imbibe much of the attitude of the Bible critics. He appeals often to Simon's work. Bayle was not a Hebraist or a Greek scholar, but he read inordinately, and corresponded with experts to find out what the facts were. In a long series of articles on Old Testament heroes and on post-Biblical saints and religious leaders, Bayle plunged into a new dimension of Bible and religious scholarship, namely the secular evaluation of the personnages and the events. Starting with article 'Abimelech', through the notorious one on 'David' to the hilariously ribald one on 'Ham' down to 'Sarah' and her sex life, and on to the end, Bayle analyzed the Biblical stories in terms of their plausibility in ordinary natural terms, and evaluated them in terms of ordinary moral standards. In construing them in this way, the foibles, immoralities, humanity of the Biblical characters was developed. They were seen not in a religious framework, but a human one, that extends all through man's history. Bayle, with his perverse sense of humor, made the most of the sexual motifs and the signs of human weakness and failure amongst the heroes of the faith. (He did such a thorough job on King David, that George the First of England was scandalized when he was compared to David.) The fact that Bayle could do this already showed how much the world was changing so that the sacred could be treated in the same way as the profane. But more, Bayle was using his form of Bible criticism as a kind of social scientific investigation of man.

In his earliest book, the *Thoughts on the Comet* (1681), Bayle had startled the learned world with his claim that a society of atheists could be more moral and harmonious than a society of Christians. Human behaviour was not a function of what one believed, or claimed to believe. And, if human behaviour was to be understood, one had to examine how men functioned regardless of what the alleged religious context of their acts was. The study of the Biblical characters, the study of the saints, the religious fanatics, the Reformers, all revealed that they were human, all too human. Bayle's study of man encompassed not only religious man, but also secular man, kings, queens, generals, politicians, literati, etc. His Bible criticism eliminated the need to consider man within a religious framework, since the same sorts of behaviour took place among the religious and among the secular. Coursing through human history, Bayle found all sorts of clues developed by later philosophers and social scientists – the importance of climate on human behavior, the all-importance of sex (in fact Bayle's favorite view seems to be that events are caused by what is between the navel and the knee), the influence of greed, vanity, fear, etc. Eighteenth century philosophers and social scientists built on these various insights. Bayle's secularization of the Bible, and his reduction of man at all times and all places to an object of study made a kind of social science possible. For Bayle, unlike the Enlightenment optimists who followed him, "Properly speaking, history is nothing but the crimes and misfortunes of the human race." [55] Bayle could see no progress, no hope, for man. But he could see that if one examined man in his human context, from Biblical days to the present, one could find the same kinds of behaviour and the same kinds of explanations for this behaviour.

From Bayle's analyses of the human scene, Voltaire and Hume were able to go on to a much more thorough and naturalistic account of the human situation. Hume probably best illustrates the legacy of the Bible critics from La Peyrère to Bayle. Hume was an avid reader of Bayle and we know that when Hume entered 'his new scene of thought' when he was going to develop his 'science of man', he went to France lugging with him his sets of the folio volumes of Bayle's *Dictionary* and his *Oeuvres diverses*.[56] Hume was going to apply the experimental method of reasoning of Isaac Newton to the moral subjects. For Hume, the laboratory study for this new science that would explain man, and through him, his intellectual achievements, was to be history – gleaning observations over the course of man's career.[57] When he came to do this, he chose English history rather than the Bible. Bayle had secularized the whole history of man, so that any area would do. (Voltaire still wandered as Bayle had done through the Bible, the secular heroes and the present.) Using far more sophisticated psychological, economic and political tools than Bayle had, Hume was able to give a more 'scientific' account of the human scene. The religious dimension was no longer a framework at all, but had become just part of the data. Hence Hume could write a book on the *Natural History of Religion*, applying his social science to this area as well as any other.

The Bible critics of the 17th century contributed greatly, I believe, to making the modern social scientific evaluation of man possible. They provided some of its crucial methods in developing the techniques of historical investigation and in introducing anthropological methods. Because religious questions were in the forefront in the 17th century, much creative and scholarly energy was poured into Biblical studies.[58] The critics I have been discussing contributed much more, namely removing the study of man from a religious framework. La Peyrère raised the possibility that most of humanity (that is, all, except the Jews) could be studied outside of Judeo-Christianity, and could be studied anthropologically. Spinoza provided a metaphysical framework for seeing man only in scientific terms. Simon revealed what happened to the quest for religious knowledge if this became a scientific matter – Bayle secularized mankind from Biblical man to contemporary man, showed that religious beliefs did not determine his behaviour, and that he could be studied in terms of the natural factors that did account for his actions. The removal of man from a religious framework to a purely secular one was in part, accomplished by the Bible critics. La Peyrère had said that his theory was like Copernicus'.

It did not change any of the events in the world. It *just* involved interpreting them differently.[59] The different interpretation led to the possibility of studying human behaviour outside of Judeo-Christianity, in terms of geographical, economic, political, sociological and psychological factors.

A Dr. Thomas Smyth of Charleston, South Carolina, said in the early 19th century, "When, however in modern times, infidelity sought to erect its dominion upon the ruins of Christianity, Voltaire, Rousseau, Peyrère and their followers introduced the theory of an original diversity of human races, in order thereby to overthrow the truth and inspiration of the Sacred Scriptures."[60] La Peyrère may not have had such intentions, but the effect of his theory was to change the study of man from a religious to a social scientific endeavor.

If the Bible critics from La Peyrère onward had this effect, one can ask in conclusion, was it good or bad? As part of the 17th century drama of the Bible critics, the Western world lost its religious convictions. Some of the participants did too, like Spinoza. Others like La Peyrère held on to their religious fantasies. Simon kept looking for the Message, and Bayle retreated into an avowal of blind faith. Their evidence certainly made it difficult, if not impossible, to hold on to the religions delineated at the time. The Enlightenment rushed beyond them to throw out all versions of Judeo-Christianity, and to embrace the new secular faiths of scientism, nationalism, progress, etc.

Since these secular faiths have failed to be sufficiently convincing or satisfying, it may well be worth rethinking what happened in the 17th century. By so doing, we can search for new hope, now that our innocence is lost, in what may still remain vital in our religious and intellectual heritage. I believe that a reconsideration of, and a reliving of the intellectual and spiritual drama that runs from La Peyrère to Bayle, will reveal where the crucial steps were made that led to the abandonment of our previous heritage. Like Father Richard Simon, I think that one has to accept some of the maxims of Spinoza (as well as some of La Peyrère's, Simon's and Bayle's), and some of his interpretations. But does one have to draw Spinoza's conclusion that there is no Providential element in human affairs, and that there is no encounter between man and the Divine?

The attempts to explain human affairs in a totally naturalistic fashion have produced a series of ideologies from the Enlightenment until the present. The ideologies do not seem to me, at least, to have succeeded in

providing any satisfactory or satisfying answer to why we are here, and
what we are supposed to do about it. In re-examining and re-evaluating
the steps by which we passed from a religious to a secular orientation in
the 17th century, we *may* be able to find some guidance for today and
tomorrow. We *may* be able to see if too much was conceded too fast or if
alternative routes could have been taken.

Like Father Simon, I am convinced that there is some Message for us
to find. The only evidence I could offer for this conviction is my own religious experience, which probably would not convince any one else. Like
La Peyrère, I think this Message is somehow revealed at least in part in
the history of the Jews, and again I could only appeal to my own religious experience to buttress this claim. Unlike La Peyrère and his precursor, Judah Halevi, I do not think that this Message can only be found
in the Jewish tradition, but may be in many other sources and histories
as well.

When the task of finding the Message was turned over to the social and
biological scientists in the Enlightenment and in the 19th century, whatever supernatural clues there were blotted out. The methods and assumptions of these scientists assured that they would only find naturalistic data
and interpretations. If we now turn back to the 17th century context, to
the period when the study of the Bible spawned a modern social scientific
outlook, we may be able to reconsider in a radically new and modern
context whether there are any clues of the continuity of a viable tradition
liking our religious past with our secular present. If anything viable remains from our past, it may provide some hope for our future.

Washington University
St. Louis

NOTES

* Some of the research for this paper was done under grants from the American Council of Learned Societies and the Guggenheim Foundation. I should like to thank them
both for their generosity.
1 I have dealt with this aspect in my study 'Scepticism and the Study of History', in
David Hume, Philosophical Historian (ed. by R. H. Popkin and D. F. Norton), Indianapolis 1965, pp. ix–xxxi.
2 cf. Thomas Hobbes, *Leviathan*, Part III, Chapter XXXIII.
3 See the recent study, Klaus Scholder, *Ursprünge und Probleme der Bibelkritik im 17.*

Jahrhundert, Munich 1966. Also see Adolphe Lods, 'Astruc et la critique biblique de son temps', *Revue d'Histoire et de Philosophie religieuses* (1924), 109–39 and 210–227.

[4] René Pintard, *Le Libertinage érudit*, Paris 1943, pp. 423. La Peyrère is discussed on pp. 355–61, 379, 399, 420–24, and 430.

[5] Leo Strauss, *Spinoza's Critique of Religion* (transl. by E. M. Sinclair), New York 1965, Chapter III.

[6] Don Cameron Allen, *The Legend of Noah*, Urbana 1963, pp. 86–90 and 130–37.

[7] David R. McKee, 'Isaac de la Peyrère, a Precursor of the 18th Century English Critical Deists', *Publications of the Modern Languages Assn.* **59** (1944), 456–85.

[8] Hans Joachim Schoeps, *Philosemitismus im Barock*, Tübingen 1952, pp. 3–18, and *Barocke Juden Christen Juden Christen*, Bern and München 1965, pp. 15–24.

[9] For information about La Peyrère's family, Philippe Tamizey de Larroque and A. Communes, 'Isaac de la Peyrère et sa famille', *Revue critique d'histoire et de littérature* **19** (1885), 136–37.

[10] Guy Patin, in his letter to Charles Spon, 18 November, 1656, indicated he and others suspected La Peyrère was Jewish; Patin *Lettres de Gui Patin* (ed by J. H. Reveillé-Parise), Paris 1846, Vol. II, p. 263. Richard Simon, in his letter to La Peyrère, 27 May, 1670 said that La Peyrère must be a Marrano, and that this was also said at l'hôtel de Condé; Simon, *Lettres choisies*, Rotterdam 1702, Vol. II, p. 16.

La Peyrère himself seemed to make the matter clear in his dispute with a Protestant scholar in *Recueil de Lettres escrites à Monsieur le Comte de la Suze, pour l'obliger par raison à se faire Catholique*, Paris 1661. The Protestant had accused La Peyrère of being a Jew or demi-Jew. La Peyrère answered that he was not a Jew in the sense the Protestant meant, "Mais il fay gloire de l'estre comme S. Paul l'a entendu et l'a écrit". (Vol. I, p. 75): presumably as a Jewish Christian, a convert.

[11] Cf. Bibliothèque Nationale Ms. Français 15827, fols. 149–62.

[12] Cf. Pintard, *op. cit.*, p. 360. The evidence that it was originally one work is not only that both works were written in 1642–43, but also that the *Prae-Adamitae* contains a summary of the *Rappel des Juifs*, and also it ends with the clear indication that is the introduction to another work beginning with the rejection of the Jews. The English version, *Men before Adam* (1656) concludes 'The End of the First Part'.

[13] This is the gist of the theory in *Prae-Adamitae*, plus the imminent Messianism of *Du Rappel des Juifs*. Both of these works will soon appear in photoreproduction editions in the Johnson Reprint Corp. series on 'Texts in Early Modern Philosophy'.

[14] On Concini and his Cabalistic affairs see Pierre Bayle's articles on Concini and his wife Leonora Galigai, and the study by Fernand Hayem, *Le Maréchal d'Ancre et Leonora Galigai*, Paris 1910.

[15] Cf. Hugo Grotius, *Dissertatio altera de Origini Gentium Americanarum adversus obtrectatorem*, n.p. 1643.

[16] Pintard, *op. cit.*, p. 361; letter of Guy Patin to Charles Spon, 14 September, 1643, in *Lettres de Gui Patin* (ed. by J. H. Reveillé-Parise), Paris 1846, Vol. I, p. 296; Letters of André Rivet to Claude Sarrau, Bibliothèque Nationale Ms. Français 2390; Letters to Claude Saumaise, Bibliothèque Nationale, Ms. Français 3930; Claude Sarrau, *Epistolae*, Orange 1654, pp. 74–75; Marin Mersenne, Letter to Rivet, 7 November, 1643 to be published in Vol. XII of the Mersenne correspondence; La Peyrère's letters to De La Mare, Bibliothèque Nationale Ms. coll. Moreau, 846; Christian Du Puy to Boulliau, Bibliothèque Nationale Ms. Français 9778; Bibliothèque Nationale Coll. Dupuy 730, fol. 154ff. There are no doubt many more indications of the discussion and reading of La Peyrère's ideas in this period.

17 These works were written as letters to La Mothe le Vayer in 1644 and 1646. The *Relation du Groenland* was published in Paris in 1647, and the *Relation d'Islande* in 1663. The latter has been reprinted and translated and appears in the Hakluyt Society's publications, First Series, No. XVIII, 1855.

While La Peyrère was in Denmark, he discussed his theories with Thomas Bang, who reported in his *Coelum Orientis*, Copenhagen 1657, that in 1645 La Peyrère showed him the manuscript about the pre-Adamites. La Peyrère's correspondence with Ole Worm shows they also discussed the matter. Cf. *Olai Wormii et ad eum Doctorum Vivorum Epistolae*, Copenhagen 1754, Vol. II, esp. letters 860–897.

18 This appears in *Prae-Adamitae*, Vol. IV, Chap. XIV.

19 Cf. Pintard, *op. cit.*, pp. 399 and 420. See also Sven Stolpe, *Christina of Sweden* (ed. by Sir Alec Randall), New York 1966, p. 130.

La Peyrère's relations with Queen Christina deserve more investigation. He apparently met her when he was in Scandinavia, and he dedicated his work, *La Bataille de Lents*, Paris 1649, to her. They probably were in contact again in Rome after they both converted to Catholicism.

20 This account of the publication of the work appears in La Peyrère's *Lettre de la Peyrère à Philotime*, Paris 1658, p. 118.

21 The condemnations seem to have begun with one in Holland, dated November 25, 1655 and then spread over Belgium and France. The number of refutations is quite large. Not only were many books specifically written against La Peyrère's theories, but he is discussed and answered by theologians and scholars throughout the rest of the century and beyond.

22 Cf. La Peyrère's letter to Condé, 8 October, 1655, Condé Archives, Chantilly, p. XV, fols. 347–48.

23 Cf. La Peyrère, *Lettre à Philotime*, pp. 123–26.

24 On La Peyrère's conversion, see Pintard, *op. cit.*, pp. 421–22, and Condé's letter, Bibliothèque Nationale Ms. Français 6728, Papiers de Pierre Lanet, Documents divers, no. 20.

Much more detail on La Peyrère's career appears in my introduction to the forthcoming photoreproduction edition of *Men before Adam*, Johnson Reprint Corp.

25 La Peyrère, *Men before Adam*, p. 208.

26 Cf. La Peyrère's *Apologie de la Peyrère*, Paris 1663, and *Lettre à Philotime*.

27 Cf. the brief biography of La Peyrère written by Richard Simon in his letter to Z.S., Paris 1688, in Simon, *Lettres Choisies*, Rotterdam 1702, Photoreproduction by Minerva G.M.B.H., Frankfurt/Main 1967, Vol. II, pp. 23–28.

28 Simon, *Lettres Choisies*, Vol. II, Lettres I-III, Vol. III, Lettres VII–IX.

29 (Michel de Marolles) *La Livre de Genese*, n.p. 1671, notes by La Peyrère.

30 This work is in the Condé archives at Chantilly.

31 There is some dispute as to who wrote this. It is usually attributed to de la Monnoye or to Claude Ménage. The original appears in *Ménagiana*, Paris 1729, Vol. III, p. 69.

Pierre Bayle, in his article on La Peyrère, in the *Dictionnaire historique et critique*, received a report from one of La Peyrère's friends, Morin du Sandat, that "La Peirere étoit le meilleur homme du monde, le plus doux le qui tranquillement croyoit font peu de chose."

32 All of this is developed in the first four books of *Men before Adam*.

33 La Peyrère, *Men before Adam*, p. A4. In some editions this very pro-Jewish dedication appears at the end of the volume.

³⁴ My interpretation is close to that suggested by Schoeps in the two works cited in note 8. In a future study, I will show in detail the Marrano character of the theory. Leo Strauss, at pp. 82ff., *op. cit.*, also suggests the Marrano interpretation of La Peyrère's theory. I believe the aspect of the theory about the role of the King of France derives from Guillanme Postel. At various stages in his career, La Peyrère had Condé, then Pope Alexander VII, and finally Louis XIV cast for the role.

³⁵ Cf. Giambattista Vico, *Principi di Scienza Nuova*, in Opere (ed. by F. Nicolini), Milan, Naples 1953, libro primo. La Peyrère is specifically mentioned on p. 402, as well as the refutation of his work by Martin Schoock.

³⁶ In the *Pensées* (Lafuma ed.) pp. 575–651, Pascal refers to "extravagances des Apocalyptiques et préadamites, millénaristes, etc." In Lafuma's note on La Peyrère in Pascal *Oeuvres complètes*, Paris and New York 1963, p. 657, he cites the report La Peyrère gave Huygens about his visit to the Pope. The General of the Jesuits told La Peyrère that "I and the Very Holy Father laughed and laughed in reading your book."

³⁷ Cf. Pierre Jurieu, *Histoire critique des Dogmes et des Cultes*, Amsterdam 1704, preface and Chapter XXV. In the preface the pre-Adamite theory is called 'la rêverie du Juif la Peyrère'. In Chapter XXV a brief biography of La Peyrère is given and an outline of his theory. Jurieu, p. 176, said the theory is not worth refuting. The outline is given without comment, and then on p. 181 the theory is called 'une suite des réveries', and La Peyrère's character attacked, especially for becoming a Catholic. Considering how vehement Jurieu usually was towards heretics, the section is very moderate.

³⁸ A good deal of Edward Stillingfleet's *Origines Sacrae*, first published in 1662, is devoted to showing "that there is no ground of credibility in the account of ancient times given by any Heathen Nations different from the Scriptures, which I have with so much care and diligence inquired into, that from thence we may hope to hear no more of Men before Adam...," p.beᵛ.

³⁹ Cf. Pierre-Daniel Huet, *Demonstratio Evangelica*, Paris 1679.
A. Dupront, in his *Pierre-Daniel Huet et l'exégèse comparatiste au XVIIe siècle*, Paris 1930, pp. 45–46 discussed Huet's view on La Peyrère's theory.

⁴⁰ A letter of Manesseh ben Israel to Paul Felgenhauer, dated Amsterdam, February 1, 1655, cites *Du Rappel des Juifs* as evidence the Messiah is coming. Cf. Felgenhauer's *Bonum Nunciam Israeli*, Amsterdam 1655, p. 90. The *Rappel* is also cited in Manesseh ben Israel's *Vindiciae Judaeorum*, London 1656, p. 18, where it is attributed to "a most learned Christian of our time".
Felgenhauer reported, in his *Anti-Prae-Adamita*, Amsterdam 1659, that La Peyrère was in Amsterdam in late 1654 and early 1655, and that he was discussing his theory with Manesseh ben Israel, and that the latter gave Felgenhauer a copy of *Prae-Adamitae*, p. 89. In Manesseh ben Israel's list in *Vindiciae* of books he had ready for the press is a lost opus, *Refutatio libri cui titulus Praeadamitae*, p. 41. Cecil Roth, in his *A Life of Menasseh ben Israel*, Philadelphia 1934, p. 161 said that Manesseh "apparently never met him". This seems unlikely in view of Felgenhauer's attempt to get Manesseh to arrange a debate between himself and La Peyrère, and in view of Manesseh's visit to Queen Christina.

⁴¹ It is not possible to tell when Spinoza came into possession of La Peyrère's book. Strauss, in his *Spinoza's Critique of the Bible*, pp. 264 and 327 shows specific points Spinoza drew from La Peyrère in the *Tractatus*. In the list of Spinoza's library given from the original inventory when he died in Jacob Freudenthal, *Die Lebensgeschichte Spinoza's*, Leipzig 1899, item (54) is *Praeadamitae*. 1655.

⁴² Cf. I. S. Révah, *Spinoza et Juan de Prado*, The Hague 1959; and 'Aux Origines de

la rupture spinozienne: Nouveaux documents sur l'incroyance d'Amsterdam à l'époque de l'excommunication de Spinoza', *Revue des études juifs* **3** (CXXIII) (1964), 359-431. Révah does not mention La Peyrère as a possible source of the developments then going on. However, in his article, pp. 378 and 393, he shows Prado held that Chinese history was 10000 years old. Orobio de Castro, in his answer to Prado attacked those who thought the Creation took place long before the one described in the Bible. *Spinoza et Juan de Prado*, p. 43.

[43] This claim appears in encyclopedias well into the 20th century. I have found no evidence that there was such a sect, unless it refers to Spinoza and his friends. La Peyrère was sensitive to the charge that he was trying to start a new sect, and in his *Lettre à Philotime*, p. 132, specifically denied having any such intention.

[44] On Spinoza's sources, Jewish, Christian and ancient pagan, see Leo Strauss, *op. cit.*, Appendix, pp. 311-27.

[45] Benedict de Spinoza, *Theologico-Political Tractatus*, chapter VII, p. 99 in *The Chief Works of Spinoza*, New York 1955 (transl. by R. H. M. Elwes).

[46] In Lucas's life of Spinoza, reprinted in Freudenthal, *op. cit.*, the catalogue of the works of Spinoza, p. 25, includes "apologie de Benoit de Spinoza, où il justifie sa sortie de la Synagogue". Cette Apologie est écrite en Espagnol, et n'a jamais été imprimée. Sir Frederick Pollock, in *Spinoza, his life and Philosophy*, 2nd ed., London 1899, p. 19, pointed out that this work is supposed "to have contained some foreshadowing of the *Tractatus Theologico-Politicus*".

In Chapters IV and VI of the *Tractatus*, Spinoza's general theory starts to emerge. A Spanish Inquisition report discovered by Révah from 1659, a report by some one who had dined with Spinoza and Prado, indicates that at this early date they both held that God only exists philosophically, and were well on the way to the theory of the *Tractatus* and the *Ethics*. Cf., Révah, *Spinoza et Juan de Prado*, pp. 32, 53 and 64.

[47] All the information about Simon's relations with La Peyrère come from these letters, cited in note 27-28.

[48] His usual view was that "Spinoza a pû avancer dans son livre plusieurs choses véritables, et qu'il aura même prises de nos Auteurs; mais il en aura tiré des conséquences fausses et impies". Richard Simon, *De l'Inspiration des Livres Sacres*, Rotterdam 1687, p. 43. One reason for Spinoza's bad results was "Il ne paroît pas même qu'il ait fait beaucoup de reflexion sur la matière qu'il traitoît, s'étant contenté souvent de suivre le Système mal digeré de la Peyrère Auteur des Préadamites", p. 48. But as he wrote M. Dallo of the Sorbonne in 1682, "Ne m'objectez-point, que ce langage est de l'impie Spinoza, qui nie absolument les miracles dont il est fait mention dans l'Ecriture. Defaites-vous de ce préjugé dont plusieurs abusent aujourd'hui. Il faut condamner les consequences impies que Spinosa à tiré de certaines maximes il suppose. Mais ces maximes ne sont pas toujours fausses d'elles-memes, ni à rejetter". Simon, *Lettres choisies*, Amsterdam 1730, p. 80.

[49] On Simon, see A. Bernus, *Richard Simon et son histoire critique du Vieux Testament*, Lausanne 1869; Louis I. Bredvold, *The Intellectual Milieu of John Dryden*, Ann Arbor 1959, esp. pp. 98-107; Paul Hazard, *La Crise de la conscience européenne*, Paris 1935, Deuxième partie, Chap. III, pp. 184-202; Henri Margival, *Essai sur Richard Simon et la critique biblique en France au XVIIe siècle*, Paris 1900; and Jean Steinmann, *Richard Simon et les origines de l'exégèse biblique*, Paris 1960.

[50] See, for instance, his comments at the end of his letter to the abbé D.L.R., 10 Mais 1679, *Lettres choisies*, 1702 ed., Vol. I, p. 26.

[51] Jean Astruc, in his *Conjectures sur les Memoires originaux dont il paroit que Moyse*

s'est servi pour composer le livre de la Genese, Bruxelles 1753, claimed he was fighting back against La Peyrère, Hobbes and Spinoza, and following out Simon's theory, pp. 452–55. Astruc claimed his work "anéanit le vain triomple de Spinosa", p. 452.

[52] The abbé Steinmann, who had enough difficulties with the Church before his tragic death in Jordan, tried in the final chapters of his book on Simon and in the appendix to make him the apostle of modernism.

[53] The best recent studies of Bayle are Elisabeth Labrousse, *Pierre Bayle, Vol. I, Du Pays de Foix à la cité d'Érasme*, The Hague 1963 and Pierre Bayle, Vol. II, *Héterodoxie et Rigorisme*, The Hague 1964. See also Paul Dibon (ed.), *Pierre Bayle, le Philosophe de Rotterdam*, Amsterdam 1959, and R. H. Popkin, *Pierre Bayle, Historical and Critical Dictionary, Selections*, Indianapolis 1965.

[54] Cf. Leif Nedergaard, 'La genèse du Dictionnaire historique et critique de Pierre Bayle', *Orbis Litterarium* **13** (1958), 210–27.

[55] Bayle, art. Manicheans, Rem. D., Popkin ed. p. 147.

[56] Cf. my article 'Bayle and Hume', *Transactions of the 13th International Congress of Philosophy*, Mexico City 1963, and in *Felsefe Arkivi* (1970), 19–38.

[57] See David Hume, Introduction to *A Treatise of Human Nature* (Selby-Bigge, ed.), Oxford 1951, esp. xxii–xxiii.

[58] On this, see my 'Scepticism, Theology and the Scientific Revolution in the Seventeenth Century', in *Problems in the Philosophy of Science* (ed. by I. Lakatos and A. Musgrave), Amsterdam 1968, pp. 1–39.

[59] Cf. La Peyrère's *Lettre à Philothème*, p. 107; and *Apologie*, pp. 21–22.

[60] Thomas Smyth, D. D., *The Unity of the Human Races proved to be the Doctrine of Scripture, Reason and Science*, Edinburgh 1851. Smyth was not just a reactionary, religious type, but was part of a movement, including Alexander von Humboldt, fighting the revival of La Peyrère's polygentic theory by those using it to justify racism, with inferior races (the pre-Adamites) and superior (the Caucasians–the Adamites). The climax of this movement appears in Alexander Winchell's *Preadamites; or a Demonstration of the Existence of Men before Adam*, Chicago 1880, which comes with photographs of the pre-Adamites (!), photos of a Dravidian, a Mongoloid, a Negro, an Eskimo, a Hottentot, a Papuan, and an Australian aborigine.

JINDŘICH ZELENÝ

KANT, MARX AND THE MODERN RATIONALITY

I

In this paper I wish to consider some questions relating to the philosophical foundations of contemporary Marxism. In order to qualify the aspects to be discussed I must briefly indicate the circumstances which gave rise to our questioning the relationship between Kant, Marx and the modern rationality.

In recent years, an interdisciplinary group, headed by Radislav Richta, was constituted in Prague to study the problems of the scientific and technological revolution. All members of the team have been engaged – in some manner or other – in the revolutionary socialist movement in our country. We were proud of the positive achievements of the socialist social reconstruction. We became, however, increasingly worried about accumulating obstacles to socialist progress, especially from the beginning of the sixties. Being confronted with the facts of stagnation in the economy and some other fields, we were forced to reexamine some theoretical positions underlying our practical behavior. An investigation was made of the scientific and technological revolution from a socialist viewpoint, and some results were published.[1] Thus our main effort was to make clear our perspectives. In doing so we were sometimes compelled to go back in theoretical coverage: *reculer pour mieux sauter*. This was especially the case with the philosophical aspects of the inquiry.

Starting the analysis of actual social development with inherited Marxist-Leninist concepts and methods, we have been gradually forced to adjust them to the new forms of reality under investigation. Hence in continuous interplay of social analysis and metatheoretical reflection we have been stimulated to pose the problem of the *foundations* of Marx's and contemporary Marxist theory, and to ask ourselves: What does Marx's and contemporary Marxist theory represent, as far as the *type of theory*, the *type of rationality* is concerned?

We came to the conclusion that some new light might perhaps be thrown on the relationship between Marx's and the traditional concept of theory. In this connection, the clarification of the *relationship between Kant and Marx* seemed to us to be of relevance and to offer some necessary *prolegomena* for inquiring into the foundations of contemporary Marxism.

Before I turn to this point, I should like to make three brief preliminary remarks.

Many terms are used here without presenting an elaboration. They are to be understood with regard to our approach outlined above, which is, of course, spatio-temporally conditioned. E.g. the word 'science' is used in this paper in a broader sense, relating more to the German term 'Wissenschaft' than to the narrower usage that is customary in this country. This remark is made not to anticipate agreement with our usage, but to make communication easier, in order to be understood before being criticized.

Some of the conceptions mentioned in this paper could not be argued in adequate form and extent here – because of shortage of time. I hope we might be permitted to refer to other publications containing a more detailed argument.[2]

While I am very much indebted to the interdisciplinary team-work mentioned above, the team is not, of course, responsible for philosophical ideas presented here. Moreover, they are still controversial even among the team members.

II

Kant's view that every theory of nature is a science in so far as it contains mathematics,[3] presupposes a certain notion of 'nature' and at the same time a complementing notion of 'free human self-determination'. The reduction of Kant's opinion into a statement that scientific theory exists only where there is mathematics, is therefore a misconception of Kant's view which was, in fact, the first great attempt to give an explicit account of the fundamental problem of modern rationality: i.e. the problem of the unification of the notions of nature and freedom.

Kant presented a new formulation of the problem of determination and self-determination.

His chief problem, in relation to which even the well known initial question of his *Critique of Pure Reason* about the possibility of synthetic propositions *a priori* appears to be derived, may be formulated as follows:

If the autonomous beings are incomprehensible by Newtonian rationality (which has its irrevocable place in natural sciences) and if, at the
same time, there is no doubt that the autonomous beings (especially and
above all man acting as a human person) do exist, then what concept of
rationality is able to make their coexistence comprehensible?

We are concerned here with the relationship of the natural and of the
human; it is a question of their unity from a logical and ontological point
of view.

The *natural* is 'phenomenal', accessible 'within the limits of possible
experience' to science which is, according to its type, a Newtonian natural
science. Kant's criticism investigates what reality, experience, universality,
necessity, scientific truth, etc., are in the realm of this rationality. So far
Kant's investigation deals with the logical and ontological base of the
mathematical natural science of Newtonian type. The neo-Kantians paid
chief attention to this aspect. Here we are not interested in this aspect in
an isolated way, but in relation to problems emerging in a broader historical perspective concerning the foundation of science, expressed in the
question we called Kant's chief problem.

The *human* in its proper sense is something the existence of which cannot be defined in terms of time and space, because it is not 'phenomenal':
a human person acting morally, i.e. determining his own will according
to the moral law by which man is an end in himself (in Kant's terms:
acting 'freely'). To comprehend this self-determining reality and its mode
of being one cannot apply criteria valid for phenomena.[4] The acting subject, capable of 'free causality' ('Kausalität aus Freiheit')[5] is in a special
way real, differently from natural phenomena, and this reality *sui generis*
opens up certain new possibilities of knowledge, not in the form of theoretical statements of science, but nevertheless uncovering at least partially,
at least in certain limited area, the 'being-in itself'.

Now a problem arises of relation and unity of 'natural' and 'human'
thus comprehended. Kant's solution in a sense of dualistic coexistence
is known: the natural mechanism is valid in the realm of 'phenomena',
whereas freedom exist within the intelligible realm. It is theoretically un-
knowable, we only see its possibility, i.e. logical non-controversiality, and
this is sufficient. The contradiction would arise only if the phenomena and
the things-in-themselves were not discriminated. Then Spinozism would
become an unavoidable alternative, i.e. the system that, according to

Kant, does not know self-determination in its proper meaning and lets freedom loose in a universal fatal necessity.

The unity of natural mechanism and freedom is analogous to the dualist unity of the phenomenal world and the world of the things-in-themselves. The consciousness of moral law ('or, and it means the same, the consciousness of freedom')[6] is a given fact as original as that given as phenomenal; it is impossible to explain further the possibility of the consciousness of freedom.

Important for the development of philosophic inquiry after Kant was the way in which Kant brought together the problem of freedom as self-determination of will according to the moral law and the problem of so-called transcendental freedom in its absolute meaning, i.e. the problem of the unconditionable, of the absolute as the basis of phenomenal continuities. Both are often subsumed by Kant under one problem of 'free causality' ('Kausalität aus Freiheit'. 'eine sich von selbst bestimmende Kausalität') and this is generally defined as a capability of absolute spontaneity.

Right at the beginning of *The Critique of Practical Reason* we read that "the speculative reason needed freedom in its absolute meaning, using the notion of causality, in order to avoid the antinomy in which it inevitably falls if it wishes to think the unconditioned...".[7]

Let us especially notice that the notion of freedom was to Kant 'the staple of the whole structure of pure reason'. Here the proto-philosophical investigations begin in which the primary role is played by the self-determining structures analogous to the autonomous Ego; the non-self-determining structures are interpreted on the basis of self-determination's fundamental role.

Self-determination is something different from self-motion in the sense of automatic movement. The latter belongs fully to the realm of natural mechanism and it is possible to conceive of it either as an automaton materiale (Newton) or as an automaton spirituale (Leibniz).

Kant initiated an important deed in the development of philosophical metatheoretical reflexions. He assigned general philosophical priority to the problem of self-determination as it appears in human activity. Thus he humanized the problem of creativity which had formerly been handled by Christian metaphysics in alienated form, putting in the first place the question of what the divine *creatio* is and in what relation man stands to it.

An important inspiration in this direction was offered also by the *Critique of Judgement* where Kant paid attention to the regions of reality which appeared to him as connecting links between the natural and the humanly autonomous, i.e. natural organic products and human artifacts, characterized by the notion of 'inner purpose' uniting 'nature' and 'freedom'.

III

If we notice the *center* of Marx's critique of Hegel and the Young Hegelians, to which all other problems can be reduced, we then find that it also was a question of the relation of human freedom and natural necessity, of the relation and unity of man and nature, comprising now of course *second* nature too, objectified human work.

The question was: How is it possible to comprehend the coincidence of self-transformation and transformation of circumstances? What forms of rationality are necessary for this?

Marx considered the practical revolutionary movement a background for the solution of social and – what he understood as one aspect of it – theoretical problems of his time. This movement tends to replace the bourgeois forms of human activity with the new, communist forms of life process. An unavoidable part of this revolutionary process is to him scientific thought activity, in the first place the 'comprehension of practice'[8] as a 'positive science' concerning 'the practical activity' and 'the practical process of development of men'[9] There is no doubt that Marx meant his criticism of the bourgeois political economy, presented in *Capital*, only as an initiating elaboration of this positive-critical science fundamentally open and never finished, for from the standpoint of practical materialism real human practice itself necessarily appeared as open and giving birth to new contents and forms.

In Marx we do not find the Kantian question of the possibility and foundation of such science, since he would obviously consider the original criticist questions as uncritical, as a return to speculative philosophizing.

Even though Marx, as far as we know, did not explicitly deal with Kant's *Critique of Pure Reason*, we can rather easily – on the background of the *German Ideology* – reconstruct Marx's reasons for his principal refusal of Kant's way of putting questions. Human science – and there is no other – is a specific form of activity of real people, i.e. of the people as they produce both materially and spiritually under certain conditions. Thought, including scientific thought, is a part of the practical life process of men as individuals in society. If the investigation of the forms of thought, including the problem of the foundation of science, wants to avoid being abstract (in Marx's sense), speculative and 'ideological',[10] it must take into account from the very beginning the fact that human consciousness and thought have substantially *only* this special mode of being, i.e. they exist only as an aspect of practical human life processes. Since Kant does not take this into account from the beginning, it is necessary to reject his way of posing questions concerning the fundamentals of science.

For Marx an analysis and critical theory of bourgeois political economy represents the background for the analysis and criticism of forms of scientific thought belonging to the capitalist era, the background of the rational comprehension of the corresponding type of rationality.

In this sense Marx's approach to the fundamentals of science is a substantial breach not only with Kant's solutions, but also with his way of posing questions. At the same time, in the first thesis on Feuerbach, there is already expressed not only a breach with the German classical philosophy, but also the appreciation of its role in the preparing of a scientific approach which would be able to 'comprehend practice'. Marx marks out the positive meaning of German transcendental philosophy – overlooked by Feuerbach – for the preparing of rationality able to comprehend critically the essence and historically transitive nature of capitalist forms of life.

A more detailed interpretation of the historical connection of Marx's new materialism with the elaboration of the 'active side'[11] in German idealism might start from Kant's 'transcendental deduction of the pure concepts of the understanding', where experience and empirical reality is essentially being grasped as an act of the productivity of the understanding, i.e. as a certain form of action. One would have to trace how Fichte radically advanced the onset of Kant's transcendentalism by a refusal of the 'thing-in-itself' and opened the way for the conception of the subject-object relation and hence of being, as production activity in general; how Schelling enlarged Kant's idea of an 'intellectus archetypus' and enriched transcendentalism by means of a social and historical dimension; how Hegel endeavoured to produce a more consistent theory of experience and of freedom than that of Kant, Fichte and Schelling. Marx in his *Economic and Philosophical Manuscripts* starts from Hegel's philosophy of self-production of self-consciousness in his attempts to elucidate the philosophical presuppositions of his critique of bourgeois political economy and of his theory of communism as of 1844, still greatly under the influence of Feuerbach. The self-production of philosophical self-consciousness, as depicted in Hegel's *Phenomenology*, is explained as a speculative expression of the historical process of Man's self-production. This concept is then further reconstructed in the *German Ideology*, after an elimination of eschatological and 'ideological' elements of Feuerbachian and Hegelian origins, into a practical conception of reality in the sense of a new materialism.

Marx seems to be closer to the starting point of German transcendental philosophy, i.e. to Kant, rather than to the idealistic complementation of the transcendental philosophy in Hegel's dialectics, especially in three respects:

(a) In Hegel's view it was a flaw in Kantian criticism that its 'absolute standpoint' was, after all, just 'man and mankind'. Hegel says: "...And so the object of that kind of philosophy is not the cognition of God, but, so to say, of Man...."[12] Marx comes back to Kant on that point on a new level, in the sense that he sees the Alpha and Omega in real men, as they act at a given time, under historically changeable social and natural conditions.

(b) We may conceive the relation of Marx to the idea of mathematism as a rejection of the respective Hegelian critique of Leibniz and Kant and see in this a certain comeback to Kant. Even if Marx equally refuses the idea of mathematism in its absolutistic postulates, he does not conceive, like Hegel, mathematical cognition as a subordinate and an inferior one, having no pretension to be a 'fully scientific' knowledge. He takes pains towards a maximum and potentially increasing application of mathematics in cognition, and that also with respect to processes of a dialectic kind, as e.g. is proved in his letter to Engels (of May 1873) about the future feasibility of a mathematical ascertainment of the basic rules of economic crises. In the *German Ideology* Marx rejects impetuously the "belletristic Phillipics against numbers, connected with Hegelian traditions" on the part of the 'true' socialists.[13]

(c) Third, Marx seems to us to be closer to Kant than to Hegel in his

acknowledgement of the bounds and limits of human reason. even if this
non-absoluteness of human cognition is differently conceived by both
thinkers – in the case of Kant in connection with his differentiation between empirical knowledge and the 'thing-in-itself', in the case of Marx
as a consequence of his practical and historical conception of reality.

Let us now make a few comments on some milestones of the development from Kant to Marx.

IV

In his *Critique of Pure Reason* Kant is interested not only in the possibility of theoretical knowledge, but also in the possibility of practical
freedom. He brings arguments for their coexistence and formulates as a
demand the program notion of the unification of theoretical and practical
reason.[14]

From the transformation of Kant's original synthetic unity of apper-
ception and of his postulate of the unity of speculative and practical
reason a new concept of 'reason' arose, aimed against the traditional
ontology and epistemology. This concept was a starting point for Fichte
('Ich', 'Vernünftigkeit', 'Intelligenz'), Schelling ('absolutes Ich', 'Intelligenz'), and Hegel ('Selbstbewusstsein', 'Geist', 'Vernunft'). The unity of
being, action, and reason is expressed, in the development of German
transcendentalism after Kant, above all in these metaphysical concepts.

Fichte attempted to present a more consistent philosophy of freedom
than was that of Kant. He was the first to arrive at the logical and ontological concepts which represent a substantial breach with the traditional
concept of being and reason. The problem of being became fully and
substantially the problem of 'practice', in the sense that the whole reality
appeared to be identical with the process of the self-production of the
absolute Ego, of 'reason'.

Fichte is conscious of the fact that he thrusts new thoughts in a region
which used to be traditionally the domain of metaphysical ontology, and
therefore he struggles painfully for expression. According to him, the only
absolute entity is a 'pure activity',[15] being the action of Ego. It possesses
a subject-objectal structure in which both poles are the same: Ego, reason.[16] Ego is at the same time its act and its own product.[17] It is necessary
to comprehend the self *in the identity of subject and object of activity*.[18]

The 'absolute action' has a specific structure: it is self-propounding by contraposition ('Sichselbstsetzen als Entgegensetzen'), and contraposition as the self-propounding. It is an 'activity returning to itself',[19] an identity of subject and object. Contradiction ('Widerspruch') appears here for the first time as a basic principle of rationality. Negation, and thus also the negation of negation are factors of the absolute identity of Ego. As an activity returning to its point of departure it has a structure analogous to organic process with an inner purposiveness, treated by Kant in his *Critique of Judgement*.

Fichte's basic writings on the theory of science ('Wissenschaftslehre') explain separately the fundamentals of theoretical knowledge and of 'science of the practical' ('Wissenschaft des Praktischen'). But *before* this differentiation of theoretical and practical knowlege Fichte sets forth the principles of the whole theory of science. The theoretical and practical philosophy are united in it. Moreover, according to Fichte's own evaluation, the 'praxeological' part of the theory of science is "by far the most important," giving a firm background to the first section which deals with merely theoretical knowledge.[20]

In Fichte's writings on the legal and moral problems we find another elaboration of the idea concerning the preeminence of practice. While in his *Wissenschaftslehre* the practical approach was treated mostly in the considerations of endeavour, instincts and sentiments, in the *Grundlage des Naturrechts* and in *Das System der Sittenlehre* it is a subject of practice in its richer social and legal forms. Since Fichte thinks that the notion of right and the notion of individuum are a precondition for self-consciousness,[21] he integrates the practical social and legal problems with the proto-philosophical investigation, among others with the clarification of the character of 'rationality'.

The traditional ontology and epistemology are transformed by Fichte into metaphysics of practical reason. The fundamental questions of logic (of theoretical approach) are here inseparably united with certain – as we could say today – existential practical approaches which cannot be deduced in a theoretical way only and which cannot be reduced to any form of merely theoretical approach. The preeminence of the non-theoretical behavior proclaimed by Fichte is immediately – *sit venia verbo*: prenatally – rationalized and transformed into a condition which the self-realizing absolute reason creates and propounds to itself.

Fichte's unification of practice and reason by 'practical reason' whose

primary mode of being is an 'absolute action' (i.e. self-realizing through
contraposition) seems to us to be a key to the comprehension of the further
development of German transcendental philosophy, as well as of Marx's
evaluation of the role of German idealism in preparing the new practical
materialism.[22] The central problem of the post-Kantian German transcendental philosophy was human practice in a deforming metaphysical
reduction to 'practical reason' as an absolute entity.

Fichte arrived at the opinion that the notion of individuum and the
notion of right was a precondition for self-consciousness. Schelling had
this to add to it: *history* is then the presupposition of self-consciousness,
and thus also of 'reason'.[23]

This effort is continued by Hegel who elaborated the idea of 'absolute
action' and the principle of 'practical reason' into a universal philosophic
system. Everything that is or that we wish to be 'reasonable' (=free,
=autonomous), must possess a structure of 'absolute action' (self-propounding through contraposition, negation of negation). Then everything
possessing analogous structure is 'reasonable', *vernünftig*, e.g. especially
the bourgeois constitutional state. Making the bourgeois forms of life in
a reflected juristic sense absolute seems to us to be a secret of Hegel's
unification of theoretical and practical philosophies.

V

To Fichte, Schelling and Hegel, the philosophy of right had a special importance for the clarification of the character of 'reason'. For Marx, it is
the critique of political economy which plays a special role in this respect.

It was the critique of bourgeois political economy which enabled Marx
to get a deeper insight into the philosophy of Hegel as a culmination of
traditional metaphysics, and therefore to break with the whole traditional 'ideological' philosophy (including the Young Hegelian and Feuerbachian anthropology). In other words: the beginning of Marxist criticism
of the traditional ontology and epistemology, as it is outlined in the *Theses
on Feuerbach* and in *The German Ideology*, presupposes a critical attitude to the bourgeois political economy and a critical understanding of the
relation between the bourgeois forms of practice and modern metaphysics.

Marx did not return to the viewpoint of the pre-Kantian ontology that
attempted to comprehend knowledge as a direct reflection of the objec-

tive world not mediated by practice. He came long after the difficulties and 'ἀπορίαι of this pre-Kantian approach had been stated. German classical philosophy took an opposite turn. It sought to explain the subject-object unity, i.e. the possibility of knowledge by regarding the objective reality as a product of consciousness. Here, of course, new aporiae appeared – how is it possible to 'deduce' external being existing outside consciousness; without its recognition, at least in the form of Fichtean impulse, the free practical Ego cannot be 'deduced'. It was Feuerbach who began to uncover and criticize the aporiae of this latter transcendental idealistic attempt when he pointed out that transcendental idealism does not know the real objects outside thought and is therefore imprisoned within the immanence of consciousness. Marx considered the relationship between being, practice and reason having experienced the aporetic results of both attempts mentioned above. He endeavoured to find a new solution. In certain respects, he continued the work of transcendental philosophy, in so far as the latter was a theory of freedom in the sense of self-production and self-determination, and in so far as it reflected structures analogous to those of human practice. At the same time Marx broke with the fundamental circle of transcendental philosophy, discarded the immanence of consciousness and returned on a new level to empiricism. It was a new, materialistic empiricism in the sense of non-identity of theory and practice, based on a newly conceived (and experienced) unity of both. Instead of the traditionally reflected contraposition of being and knowledge (ontology and epistemology) a more complicated structure appeared: being – practice – knowledge with human practice holding the key role.

Marxism thus opened up a new approach for the investigation of questions pertaining to the foundations of science. According to its content this approach may be characterized as *ontopraxeological*.

Applying this approach to the history of scientific thinking, we can, I take it, distinguish at least *three basic types of rational theory*, or three *types of rationality*.

The term 'type of rational theory' or 'type of rationality' will here be understood to refer to the entire human theoretico-practical attitude, not only restrictively to the technological characteristic of calculable action.

The type of rationality can be characterized:

(a) By the structure of the why-, what-, and how-questions. The structure of questions, of course, implies the structure of answers and *vice versa*. Different types show not only variations, but also essential invariants in this respect.

(b) By the specific conception of theory – practice relation.

(c) By the specific conception of the relation between fact-statements and value-judgements.

In Marx's scientific work we find the initial stage of the third basic historical type of theory.

The first basic type, as represented e.g. by Aristotle, produced the contemplative concept of theory as the highest human attitude which has purpose and aim in itself.

The second main type. the concept of rational theory as presented e.g. by Descartes, may be characterized as non-historical and technical, founded on metaphysics. Its aim was: "nous rendre comme maitres et possesseurs de la nature." If we take Descartes' concept as an example of a type, we may add that the metaphysical foundation appears here either in the form of theological (mostly non-ecclesiastical) elements or in the form of Spinozist absolute 'nature' as a single substance.

Marx proposed a *historical* and *practical* concept of theory, based on the *negation of metaphysics*. Both criticism of political economy and of Hegel's philosophy are inseparably united in the Marxian formulation of the anti-metaphysical concept of rationality.

The term 'metaphysics' is here used in a sense derived from *The German Ideology*. Every conception is 'metaphysical' which incorporates supernatural and superhuman entities or which converts general abstract ideas into independent superhuman powers. So, even Feuerbach's atheistic anthropology, based on the belief in the fundamental role of the capitalized *Man* in general, is a kind of metaphysics. For Marx, not Man in general, but "real individuals, their activity and the material conditions under which they live, both those which they find already existing and those produced by their activity,"[24] are the only starting-point of all empirical (antimetaphysical) theory. Negating of metaphysics in Marx's sense does not mean refraining from inquiry into some very abstract problems of modes of being, i.e. problems treated previously traditionally under the title of ontology as a part of 'metaphysics'. On the basis of the negation of metaphysics, these problems do not disappear. They are rephrased and investigated in new contexts.

From the ontopraxeological point of view the concept of rationality in Marx's works can be characterized as corresponding to the first phase of the socialist critique of bourgeois society. At that time the revolutionary negating process was essentially confined to those forms of human practice which were shaped in a decisive way by the circulation of capital, the mightiest 'subject' of history of that time.

The concept of theoretical thinking initiated by Marx implies that the the clarification of the ontopraxeological problems will necessarily undergo profound transformation under conditions where human practice will become essentially different from that created under classical capitalism.

Let us assume that the profundity of the changes in practical being which result from the present-day scientific and technological revolution are not being overestimated. It may then be asked whether the contemporary Marxist theory is not facing – in respect to the main logical and ontological elements of the theory – such a change that would perhaps be even more radical than the change that came about with the *appearance* of Marxism in the 19th century.

In his *Poverty of Philosophy* Marx labelled Hegel's logic as an abstract speculative expression of the *industrial movement*. Marx's 'comprehension of practice' ('das Begreifen der Praxis') was a theory of the process by which industrial capitalist development will inevitably breed its own doom. The fundamental forms of practice, which in one case (in Hegel's philosophy) were apologetically sanctioned and in the other case (in Marx's revolutionary theory) presented as transitional and as producing their own negation, were of course *identical*: the basis of both theories was the 'industrial movement'. This explains a certain affinity – despite their substantial break – of Hegel's and Marx's dialectics as applied, for instance, in *Capital*.

If the inquiry into the foundation of contemporary science tries consistently (in an 'orthodox' way) to apply the approach discovered by Marx, it enters, today, a new context. It becomes a part of the critical analysis of the transition of the capitalist-industrial and socialist-industrial forms of practice into the post-industrial ones.

In this sense, I believe, it is possible today to differentiate the two principal developmental stages in the evolution of the third type of theoretical thinking, initiated by Marx.

Let me now take just one example to illustrate this differentiation.

E.g. the question arises. if and how the notion of 'objective theory' and the notion of the 'intellectual appropriation of the world by the reproduction of reality by thought' is different in the first and second stage, i.e. in Marx's work and in contemporary Marxism.

Though the rationality of the 'comprehension of practice' was in the first phase (in the work of Marx) historical and not naturalistic, still the historical process was, at that time, similar to the development of the natural biological organism.

The historical movement today loses some characteristics analogous to the development of the natural biological organism (while others remain). The character of historicity of human life is changing. This mainly concerns the changes in pre-determination by materialized products and in temporality. Knowledge acquired by experience begins to be reinforced by a new kind of experience. There is certainly a difference if:

(a) the present is controlled by the past and the reality in fact precedes its intellectual appropriation;

(b) the intellectual appropriation of the world (with the predominance not of the reproductive but of the anticipatory, alternative and similar forms of thought) becomes a concurrent aspect of the production of practical reality in new forms. In this case, the experience has a structure different from that corresponding to the social movement analysed by Marx.

We can speak of the two stages of Marxist criticism of a purely objectivist concept of theory. To be sure, Marx already explained for his time and theoretical task that the theory of social process supposes treating men not merely as puppets, but as the producers of their own world, as both actors and playwrights of their own drama – even though the 'comprehension of practice' was then attainable by theoretical forms, analogous formally to the development theories of biological formations.

The idea we proposed here about the two principal developmental stages of the Marxist type of theoretical thinking has, of course, but a hypothetical value. It has to be proved or disproved by elaborating a Marxist conception of the foundations of contemporary science, an elaboration which remains so far, in the main, a theoretical task to be done. What we were able to present here, are nothing more than *prolegomena*.

Institute of Philosophy, Prague

NOTES

¹ R. Richta *et al.*, *Civilization at the Crossroads* (2nd Czech ed.), Svoboda, Praha, 1968. [*Ed. note:* See English translation, 3rd edition, International Arts and Sciences Press, White Plains, N.Y., 1969].

² J. Zelený, *Die Wissenschaftslogik bei Marx und 'Das Kapital'*, Akademie-Verlag, Berlin, 1968.

J. Zelený, 'Kant und Marx als Kritiker der Vernunft', *Kant-Studien* **56** (1966) Italian translation: *Il Pensiero* **11** (1966).

J. Zelený, 'Zu Marx' Auffassung der Begründung eines Satzes' (co-author: P. Materna), in *The Foundation of Statements and Decisions*, PWN, Warsaw, 1965.

J. Zelený, 'Hegels Logik und die Integrationstendenzen in der gegenwärtigen Grundlagenforschung', in *Hegel-Jahrbuch* 1961, Vol. II, Dobeck-Verlag, München. Italian translation: *Differenze* **6** (1965), Argalia, Urbino.

³ Im. Kant, *Metaphysische Anfangsgründe der Naturwissenschaft*, Vorrede, *Kants Werke*, Vol. IV, Akademieausgabe, Berlin, 1911, p. 470.

⁴ Im. Kant, *The Critique of Practical Reason*, The University of Chicago Press, Chicago, Illinois, 1949, p. 163ff.

⁵ Im. Kant, *Werke*, Akademieausgabe, Vol. V, p. 16.

⁶ Im. Kant, *Kritika praktickeho rozumu*, Praha 1944, pp. 63–64.

⁷ Im. Kant, *The Critique of Practical Reason*, l.c., p. 118.

⁸ Karl Marx and Friedrich Engels, *The German Ideology*, International Publishers Inc., New York, 1939, p. 199.

⁹ *Ibid.* p. 15.

¹⁰ 'Ideological' in the sense exposed in *The German Ideology*. Cf. *MEGA I.*, Vol. V, pp. 16, 535.

¹¹ Cf. the first Marx's thesis on Feuerbach in *The German Ideology*, l.c., p. 197.

¹² G. W. F. Hegel, *Werke*, Glockner, Vol. I, p. 291.

¹³ Cf. Marx-Engels, *Briefwechsel* IV, Berlin 1950, p. 478; *MEGA I.*, Vol. V, p. 498.

¹⁴ Im. Kant, *Werke*, Akademieausgabe, Vol. IV, p. 391.

¹⁵ J. G. Fichte, *Werke* II. (ed. Medicus), p. 406: "Das einzige Absolute, worauf alles Bewusstsein und alles Sein sich gründet, ist reine Tätigkeit".

¹⁶ Cf. *ibid.*, p. 5: "Der Charakter der Vernünftigkeit besteht darin, dass das Handelnde und das Behandelte eins sei und ebendasselbe; und durch diese Beschreibung ist der Umkreis der Vernunft, als solcher, erschöpft".

¹⁷ *Ibid.*, p. 26.

¹⁸ *Ibid.*, p. 27: "Sich selbst in dieser Identität des Handelns und Behandeltwerdens, nicht im Handeln, nicht im Behandeltwerden, sondern in der *Identität* beider ergreifen, ... sich des Gesichtspunktes aller transzendentalen Philosophie bemächtigen".

¹⁹ *Ibid.*, p. 21.

²⁰ Cf. *Werke*, l.c., Vol. I., pp. 213, 317, 321, 411, 478, 486, 511.

²¹ *Werke* II., pp. 56–57.

²² E.g. in the Marx's first thesis on Feuerbach. Cf. *The German Ideology*, l.c., p. 197.

²³ Cf. Schelling, *System des transzendentalen Idealismus*, Cotta, Tübingen, 1800.

²⁴ Marx and Engels, *The German Ideology*, l.c., p. 7.

QUENTIN LAUER

THE MARXIST CONCEPTION OF SCIENCE

With the disappearance of Josef Stalin and, hopefully, of the Stalinist dogmatic interpretation of Marxist thinking, both the Marxist and the non-
Marxist worlds have been given an opportunity – previously either not
available or not considered fruitful – of reexamining the sources of Marxism and coming to a more nuanced understanding of its subtleties.
Not least significant among the reinterpretations thus made possible
is that of the 'scientific' character of the Marxist theory of history as an
explanation of socio-economic development.

The problem arises out of an extremely naive conception of the very
meaning of science. If we accept as knowledge only that which has been
scientifically established, and if we accept as scientifically established only
which has been apodictically verified, then, strictly speaking, we are denying
the possibility of a history of science. There can be a history of the process
of arriving at scientific verification, but, according to the naive conception, at no point along the way can we justifiably speak of 'science', since
only at the end of the process does it make any sense to speak of apodictic
verification; if there is a history at all it is of that which is not yet science.
Logically, however, such a conception should lead one to deny that there
is any science at all, since what is still corrigible has not been apodictically
verified, and it is questionable whether we would want to call any scientific theory incorrigible. Scientific 'laws' are constantly being modified,
corrected, or replaced, but they do not by that fact cease to have been
scientific – unless, of course, one wants to look upon science as the 'possession' of knowledge rather than as the process of approaching the ideal of
knowledge.

The problem of Stalinist dogmatism, then, is not that it ignored or
suppressed the earlier, more philosophical – and, hence, more tentative –
writings of Marx, but rather that it refuses to admit that the 'discoveries'
of later Marxist 'science' are corrigible. It agrees that there can be further
advances in science but denies that they can be such as to require a new
look at what has already been scientifically established.[1] There is, inci-

dentally, a certain naïveté in thinking that the late discovery (1932) of
Marx's early philosophical manuscripts has contributed disproportionate-
ly to our understanding of Marx. This is particularly true with regard to the
concept of science, which is nowhere more clearly expressed than in the
Contribution to a Critique of Political Economy (1859) and in the Preface
to the first volume of *Capital*. The same might be said of Engels' elucida-
tions of Marx's thought in *Anti-Dühring* and in *Ludwig Feuerbach and
the Decline of Classical German Philosophy*. What has happened, then,
since the inauguration of a policy of de-Stalinization in 1956 is not so
much that new evidence has become available; raher it has become possible to examine the old evidence with new glasses.

Marx's primary aim was not to *know* but to do, i.e. to change the
world[2] and, thereby, the human situation in accord with what history
tells him that situation is to be.[3] His secondary aim was to accomplish this
transformation on the basis of 'science' or of a 'scientific method for the
construction of this society'[4] wherein man is to become what history has
destined him to be. In this framework, then, it is 'historical' science which
enables Marx to know both what man is destined to be and how to bring
this about.[5]

Without going into the question of what meanings or connotations the
term 'science' has taken on in our contemporary – particularly English-
speaking – world, it might be well before going on, to look at the meaning
which the term *Wissenschaft* had for a German in Marx's day. Originally
the term was quite vague and indicated either something so indefinite as
'information' or the subjective state of having knowledge in this or that
particular area. Until late in the 18th century it was difficult to distinguish
this meaning of the term from that of 'art'. Only gradually did the term
begin to take on the meaning of a body of knowledge in the theoretical
sense, and only at the end of the 18th century did it appear primarily in
the singular as designating the sum total of what is objectively known, in
the strictest sense of the word 'know', which in German is *Wissen*. Thus,
to know 'scientifically' was to know in the best possible sense of that
term. With Hegel, however, who in this area had the strongest influence
on Marx, *Wissenschaft* became the whole process of coming to knowledge
in the strict sense, such that any stage in the process was knowing in a
relative sense, and was thus in connection with the whole process 'scientific'.[6]

It should be pointed out here that in the tradition of the Enlightenment and of early 19th century philosophical thinking, the concept of 'science' was largely negative in the sense that stressing the autonomy of human reason in 'scientific' thinking involved denying the validity of authority, tradition, the supernatural, etc. Thus, a natural explanation of what had been previously explained supernaturally, even though it might on our terms be considered only plausible, was in the eyes of the 'enlightened' a scientific explanation. It is true that they demanded for science that what it claimed to be true also be seen as *necessarily* true, but the concept of 'necessity' was at best a vague one. This is particularly the case where the science in question is historical and where the 'objective necessity' of scientific laws is thought to govern historical process.[7]

Against this background, we can perhaps, understand what Marx was trying to do in his endeavor to study man and society 'scientifically'. From the Enlightenment he had inherited the demand that reality be explained on the basis of rationally discovered 'objective laws'. From Kant, Fichte, and Hegel he had inherited the realization that these laws are not contained in nature but are the product of human thinking – as a human response to facts in an endeavor to account for them rationally. From Feuerbach he had inherited the conviction that there is no going beyond the evidence presented to the senses in doing this. His own unique contribution was to synthesize all three elements of this inheritance, basing his synthesis on a concept of man as truly human only in the context of society, which society was constantly the product of human material activity governed by dialectical laws intrinsic to the process itself.[8]

To understand Marx, then, we must begin with the socio-economic situation which he observed in 19th century industrial Europe. This in turn demands an awareness of a number of factors – political, economic, social, and religious – which characterized the world in which Marx lived.

Political: The French Revolution (although it was not unique in this) had given the impetus – however vague – to the ideal of political self-determination. The best German thinkers were fascinated by the attempt to realize a society based on a rational concept of man. By the time Marx came on the scene, however, the rise and defeat of Napoleon had apparently manifested the weaknesses of doctrinaire liberty, equality, and fraternity. Finally, the dominance of the Prussian state had reinstated the framework of political absolutism, not merely by restoring the *status quo* but also by seeking to bolster the absolute state with the support of both reason and religion.[9]

Economic: The Industrial Revolution, which, although most advanced in England, had a strong influence on European life in general, was in the process of intensifying a movement which had begun at the end of the Middle Ages: The power of wealth was supplanting that of the feudal aristocracy, and the impersonal power of the national state was replacing the relationship of individuals to a personal monarch. The obvious cleavage between those who produce and those who profit by what is produced was rendered more acute by the private ownership of machines which were becoming an increasingly important factor in production. With this an extremely important 'economic fact' imposed itself: with a quantitative increase in production the margin of profit on each individual item became smaller, and the need of weighing carefully the costs of production became paramount. From this it followed that labor– the human work force – was considered one of the 'costs. of production and, thus, a commodity governed by the 'law' of supply and demand, The result of all this was that the very logic of production produced a 'class' of people whose only asset was work itself (an asset only while they were working) but for whom the absolute necessity of subsisting made a favorable bargaining position impossible.

Social: The massive industrialization of society resulted in a clearly-defined division in relation to the process of production between those who owned the necessary instruments of production and those who were forced to sell their labor to such owners. One could still speak of various *levels* on the social scale, but more and more there were only two *classes* cutting across the whole of world-society.[10] With increasing industrialization (worldwide) the structure of society became hardened in such a way that political, economic, and social inequalities were given a permanence impossible to breach.

Religious: To top it all off religion was turning into a supernatural and supra-temporal justification of the *status quo*. No matter what the purposes of religion may be, the political powers were employing it as a support of their own domination. At the same time, to all intents and purposes religious teaching, with its emphasis on safeguarding the rights of individuals, was serving simply to solidify the socio-economic structure with its inequalities – when rich and poor are face to face as individuals, equality of rights is the grossest of inequalities. In addition, the emphasis which Christian religion put on the provisional character of the earthly pilgrimage was influential in persuading those on both sides of the socio-economic divide to accept the situation with its inequalities – the hereafter could take care of equalizing.

Against this background we can now, perhaps, see more clearly Marx's task as he saw it himself. In his own age he observes a situation in which man has become dehumanized, 'alienated' from what is truly human in in him.[11] This is true not only of the vast majority, the working class which is cut off from the product of its own labor, but also of the owner class which has turned what it owns into a vast impersonal 'capital', which it too serves without knowing it. In the spirit of his age Marx cannot be satisfied with observing a set of facts or even with criticizing the inhumanity of the situation he observes. Nor can he simply look for a program of reform which would ameliorate the situation. He had to have a 'scientific' explanation of the situation as he saw it, and he had to account 'scientifically' for the only possible way to remedy the situation. To do this he had to elaborate a practico-scientific explanation of social development. As Lenin said of him, he combined in his thinking "all that was best in German philosophy, British political economy, and French Socialism."[12]

To change Lenin's order, Marx was influenced strongly by the effort of the economists to explain 'scientifically' economic (and, hence, social) structures. What he found significant in their work was the endeavor to determine the 'laws' which govern economic processes. They were not at all unaware of the ambiguity introduced into the overall picture by the arbitrariness of human behavior, and yet they were able to predict with considerable accuracy the movements of the economy by observing and describing the interrelations of the major factors in the productive process. Where they failed, Marx felt, was in their unphilosophical – and unscientific – acceptance of the contemporary structure of society as a given, as the necessary framework within which economic laws were operative.[13] Marx simply could not accept the Enlightenment presupposition – which was that of the British economists – that any factual situation is a 'given' in nature. What is 'given' has been produced, and the task is to understand its production.[14] This is where "the best in German philosophy" comes in. The gigantic philosophical endeavor, which began with Kant and – in a very real sense – ended with Hegel, had highlighted two essential features of the relationship between human thought and concrete reality: (1) the 'objectivity' of thought can be meaningful only if it can be reconciled with the subjectivity of thinking; and (2) the objective world which thought thinks is a world in process, a science of which must be able to account for process. The first of these features made Marx reject the Enlightenment notion that the 'laws' which science 'discovers' are simply inscribed in nature independently of human thinking. Most important in Marx's thinking is his refusal to see in man or society an essential 'nature' which can be arrived at by abstract thinking.[15] The second convinced him that the only real science – science of the concrete –would be a science of development. If, then, science was to discover laws, they had to be laws which took into account the thinking subject, and they had to be historical, in the sense that they gave an account of the coming to be of what was. Such a science Marx found in the Hegelian 'dialectic', which sought to account both for the interrelatedness of thought and reality and for the developmental character of thought's object. Marx would go beyond Hegel in recognizing only what man had produced through action as the true object of knowledge. Thus, the object of knowledge is never a 'given, not even man himself, since he is only to the extent that he produces himself.[16]

All of this, however, could have remained empty and formalistic, had not Marx also included in his thinking "the best...in French socialism." He could not be satisfied with a thought which sought merely 'to understand the world'. He required one which would 'change it'. It was characteristic of the French social thought of his day that it sought to *do* something about the situation which thought analysed.

Here it seems necessary to indicate the pre-philosophical – or pre-scientific – character of the insight which enabled Marx to see in the socio-economic situation of his day a dehumanization or 'alienation' of man. It has often been objected that to speak of man's alienation involves an ideal, *a priori* decision as to what man *should* be. There is unquestionably much of the idealist in Marx. Marx's own answer to the objection, however, would be that even before the elaboration of a *theory* of history the observation of historical development has revealed an orientation which the present situation of man contradicts. To see this situation as one of alienation, then, can be ascribed to what we can call a 'historical insight', an insight into what the observed process demands. It will be important to recognize as we go along that without this insight the 'science' of history does not make much sense.

Out of this 'insight' comes a scientific theory of history which will back up the insight in two ways. It will explain on the basis of necessary 'objective laws' of history the development which has brought about the present situation, and it will provide the key – again based on the operation of these 'objective laws' – for a further development which will mean the reappropriation of what has been alienated. In the final analysis Marx will have combined the economic 'science' of the British, the 'philosophy' of the Germans, and the 'socialism' of the French in a science of history which reconciles theory and practice. The theory is Hegelian in its dialectical character; it is peculiarly Marxist in its materialistic character, not only in the sense that he sees only material development as susceptible of scientific investigation, but also in the sense that he sees the transforming activity of man as the labor which at one and the same time changes nature and makes man himself into what he is to become.[17] What Marx sought was 'concrete science', and, as he saw it this demanded that he reject Hegel's 'system' while retaining his method.[18]

Although it may be difficult for us to see what Marx can mean in insisting on a 'science' of historic' process or on 'laws' which govern overall social development – especially when this insistence is accompanied by a refusal to accept the given as simply given – we may be able to grasp the whole thing better if we approach it from a different angle. The very notion of 'science' – which we trace back to the Greeks, and which found its first articulate formulation with Aristotle – is based on two fundamental considerations: (1) the fact that we do observe certain regularities in events; and (2) the conviction we have that such regularities are susceptible of rational explanation in terms of 'laws' (however loosely that that term may be construed). Thus, as Aristotle says, when we observe that nature 'always or nearly always' acts in a certain way, we conclude that there is a reason in nature itself for its acting the way it does. This is what Archimedes was doing when he determined the specific weight of gold; it is what Galileo was doing when he determined the rate of acceleration of falling bodies (even if the only experiments he performed were thought experiments); it was what Boyle was doing when he determined the laws which govern the expansion of gases. By the same token, however, it is what Gresham was doing when he tried to determine the law which governs the economic flow of gold and silver, or what Marx was doing when he sought to establish the relationship between surplus value and the accumulation of capital. We know, of course, that scientific 'laws' are always corrigible; but they are always corrected in the form of other 'laws' – which might be interpreted as new and better ways of formulating the same 'laws' governing observed regularities. It is also significant that we are scientifically convinced that the laws which have been formulated will also be verified in events we have not observed – or have not yet observed – always remembering that further observation may demand modifications in the formulation of any law). Significant also is our basic conviction that, given other regularities which we have not (yet) observed, it will be possible to formulate laws which will explain them.

We are, it is true, not quite so naive as was Marx with regard to either the formulation or the operation of these 'laws', but we must remember that Marx was a child of the early 19th century (heir of 18th century scientific optimism) and that we are not. In any event Marx was convinced that the 'facts' he was able to observe exhibited the kind of regularity which demanded the formulation of laws governing their regularity. More than that, he was convinced that once these laws were formulated they enjoyed the kind of universality and necessity which made them operative throughout the entire process of history. At the same time, however, Marx was not so naïve as to think that the laws of history correspond exactly with the laws of nature – as conceived in his day – according to which the future is already written in the past. He sought to formulate laws which would always take into account human initiative in a process where novelty was constantly emerging.[19] In so doing he was aware that ongoing reality is inexhaustible and not reducible to the knowledge we have of it. He was aware, too, that any scientific model is provisional and that, therefore, we must always look for richer, more effective – and, in a sense, truer – models.[20]

Now, although it would not be mistaken to say that Marx would extend his concept of science – *mutatis mutandis* – to all science (he is not known to have raised any objections to Engels project of *Dialectics of Nature*) the fact is that his own aim was the development of a scientific socialism,

wherein he determines the laws that have governed the development of
human social relations and which will provide the possiblity of planning
the future of society – without predicting in detail what that future will
be and without constructing a univocal model to which all future societies
must in all circumstances correspond. He was convinced that class society
as he knew it had to go, that only revolution would put an end to it, and
that the rules for that revolution were to be found in the science of history
as he conceived it. He did not believe that the way to do this was to conceive of the ideal rational society – somewhat after the fashion of the
French Revolution – and then to institute the reforms which would bring
it about. As Engels put it: "The task no longer consisted in constructing
a social system perfect as possible but in studying the historical development of the economy which had necessarily engendered these classes and
their antagonism, and in discovering within the economic situation thus
created the means of resolving the conflict."[21] The key expression here is
'necessarily engendered'. For Marx the socialism he preaches is 'scientific' precisely because it contains the only possible and, therefore, the
'necessary' means of resolving the contradictions of society – which contradictions, incidentally, are dialectically necessary for the development
of society to the point where they need to be resolved. Methodologically
this consists of an analysis of bourgeois society – [22] a 'given' in the sense
that it has been produced by the forces of history – which reveals the scientific laws operative in that society's development and which will bring
about its destruction. Practically this science will channel the operation
of those real forces capable of realizing the necessary social transformation.[23]

   If, then, we are to understand Marx's concept of science (scientific
socialism) we must see the manner in which he derives its laws and then
the manner in which those laws operate to bring about the society of the
future. As in any science the procedure begins with an observation of
'facts' (regularities, of course, since no number of isolated facts can serve
as a basis for scientific formulation). These facts are then seen as the result
of a historical process, and the science consists in determining the necessary laws of whose operation these facts are the necessary result. It should
be pointed out, incidentally, that this position is based on a presupposition which is itself not scientific nor, properly speaking, philosophical. It
is that, if there is to be explanation (and explanation there must be, except

of how matter and motion, the only givens, came to be), the explanation will be natural (material) *and* historical.[24]

One would have to rewrite *Kapital* to describe all the 'facts' and to determine all the 'laws' operative in necessarily bringing them about.[25] Actually Marx begins with the contemporary 'fact' of human alienation and argues back to a complete explanation of human history, beginning with the production of food as that whereby man distinguishes himself from the animal, i.e. dialectically fulfills the need he has for the food which nature does not adequately supply, thus developing consciousness in order to plan, speech in order to communicate, division of labor in order to produce more efficiently, private property in order both to expand production and to satisfy the interests of the class which has become dominant as the result of the division of labor, etc.[26] We can, however, show in a more summary fashion how the thing operates by simply analysing the functions of capitalist society which is the necessary product of the unplanned operation of productive forces down through the ages.[27]

As a first observed fact, then, we can take the situation of the worker (the producer of value) who has sunk to the level of a commodity to be bought and sold on the market.[28] This fact is a regularity, such that it must be the necessary result of laws operating to produce it, and the laws are those of capitalist economy.which by simply following out its own logic must make the rich get richer and the poor poorer. As a second fact we might take the paradoxical observation that the worker becomes poorer the more wealth he produces and the more his production increases in power and extent.[29] This is true, because, as we saw before, a conditon for profitable mass-production is the lowering of the unit cost of production, a saving which is realized by lowering the share of the cost allotted to the labor force. This, too, is an 'objective law' of capitalist production, sometimes called the 'law of progressive pauperization'.[30]

A third fact (not necessarily in the order of importance) is that labor, as an activity (a quality) which distinguishes man as man, produces goods which take on a greater importance than the labor itself and, therefore, than man himself. Thus, labor, whose end becomes the product and not the producer, actually degrades rather than dignifies man. It is here that the concept of 'alienation' enters in; the product stands over against the producer and becomes his master, and man is master neither of the product nor of his labor – *a fortiori* not of himself. Simply by following the laws of capitalist economy the worker produces surplus-value, i.e. that part of the value he produces which does not accrue to himself and which is left over after his subsistence and his contribution to the ongoing labor force by reproduction has been paid for. By the same laws surplus value accumulates and becomes capital, a kind of impersonal personality to which the worker is subject, which means that he necessarily produces an alien force to which he is subject. Thus, the labor whose dialectic function is to transform nature and so to transform man has been diverted into producing an alien force which enslaves him; in exchange for his creative effort which produces use-value the worker receives money, which is exchange value and in no way proportionate to the value he has produced.

The purpose of this summary description of Marxist methodology in one area has not been, obviously, to give a complete picture of the vast synthesis of theory and practice contained in *Kapital* – to say nothing of Marx's earlier, more philosophical works. Nor has it been to criticize the adequacy of the analyses, which are unquestionably very keen. Rather it has been to give some idea of just what a 'scientific law' means in a Marxist context and to show briefly how such a law can be derived. As Marx himself says, "the ultimate purpose of my work is to discover the economic law of development of contemporary society,"[31] which is but a global way of expressing the various 'laws' which cannot but function, so long as capitalist society persists. It would defeat our purpose to give here (or elsewhere) a list of the dialectical laws of which Marx speaks. They are conceived of as emerging in the process itself and are not, except in general fashion, to be formulated definitively.

Here, however, we must go back to the more general notion of a science of history and of its laws. Even though it is true that Marx's primary concern from the theoretical point of view was to give a scientific explanation of *economic* development and from the practical to prepare an economic revolution, if all he did were to analyse the necessary laws governing *capitalist* economy he would not escape the criticism which he himself had directed against the British economists of not having pushed the science far enough and of having limited themselves to discovering the laws peculiar to a *given* situation.[32] If the revolution he proposes, whereby man is to reappropriate what has been alienated under the capitalist mode of production, is to have a genuinely scientific underpinning, the scope of his scientific contribution must be the whole of history and not merely the industrial revolution or the period of capitalist domination. To get this overall picture we can, I think, turn to Engels, not because he has expressed the scientific theory more clearly than Marx has, nor because he has made an original contribution which is not to be found in Marx's own work, but because he has concentrated into one book, his *Anti-Dühring* (cf. also *Utopian and Scientific Socialism*), a concise explanation of what 'scientific history' means to Marx. The book is, so to speak, the classic text-book of Marxism, and it was written for Marxists.

Marx himself had said, in a letter to Kowaleski, "It is impossible to think logically except on the basis of the dialectical method."[33] What this says is that the developing nature of reality is such that only dialectical thinking can give a scientific account of it. Engels, in his work, is trying to show that Marxism is not, properly speaking, a philosophy at all but rather a dialectical method which alone permits science to be scientific. Apart from Engels' rather naïve application of this principle to mathematics and to the natural sciences (which he repeats in the *Dialectics of Nature*), what he is saying is that a thinking which is other than dialectic must inevitably fail to see that the truths which science discovers are not eternal truths whose validity was always there to be discovered but rather are historical truths which are simply not true until historically they become true.[34] This is the 'real basis' without which socialism never could have become a science.[35] At the same time there is no philosophy distinct from or embracing the particular sciences: there is only the dialectical method, which is what gives the scientific character to any investigation, which is but a way of saying that reality in process can be grasped only in a thought which is dialectical.[36] Thus, there is only one law to which all conforms, the law of dialectical development. Actually the paradigm of science is the socio-economic science of Marx, which became possible only when Marx had discovered the principle of a non-mechanistic, i.e. dialectical, interpretation of history, whose key concept with regard to an understanding of capitalism was that of surplus value. "These two great discoveries – the materialistic interpretation of history and the uncovering of the mystery of capitalistic production by means of surplus-value – we owe to Marx. With them socialism became a science, and it is now a question of working it out in all its details and connections."[37]

"To work out this 'science' in all its details and connections," however is not a matter of deduction; rather it is a question of determining the (derivative) laws of the socio-economic process on the basis of the observed process of history.[38] This must be, since there is only one 'given' to which science as such can appeal, i.e. matter, whose manner of being is motion, and this will reveal its real possibilities only as in the process they become realities.

What comes out of all this is not science – or any particular science – but rather a criterion for determining whether what is called science is truly science. Be it physics, biology, or socialism, if it traces the real connections whereby reality becomes the reality it is, we have science, otherwise we do not. Marx, as we have said, had no interest, as did Engels, in showing that this was true of physics or biology (although the theory itself did manifest to him the necessity of evolutionary development): his sole interest is to trace the necessary development of social forms and economic structures, which are what they are through the working out of history's discoverable necessary laws. Science, then, is 'critical knowledge of historical movement',[39] Here there are no fixed natures of which static science would be a knowledge, but there is 'historical necessity',[40] which is exemplified in the "natural laws of capitalist production... working with iron necessity towards ineviable results."[41] Fortunately for Marx the revolutionary, to whom science would be meaningless did it not reveal the changes *to be* wrought in the socio-economic structure, an examination of the given situation reveals the 'necessity of the present order' which 'points to the necessity of a subsequent order.'[42]

Now, if all we had were the foregoing, the whole thing would remain extremely vague, precisely because we should have no criterion for determining what is and what is not an inexorable law of history. Marx is certainly not so naïve as to say that whatever happens necessarily happens; that would be the sheerest fatalism and totally incompatible with Marx's conviction that human initiative is operative in the working out of history. Marx was unalterably opposed to Hegel's dictum, "whatever is real is rational, and whatever is rational is real"[43] (at least to the common interpretation of it[44]). There has to be a sense in which somethings *should* happen and others *should not*, or talk about revolution is nonsense. It is meaningless to say that historical development is the criterion of truth, unless there is some sense in which development can be unhistorical and, hence, untrue. Marx was no proponent of the policy of 'folded hands', which simply awaits passively the working out of history's inexorable laws. Nor could he abide Hegel's 'trickery of reason', which simply carried out its pre-conceived plans despite men's illusion that they were acting freely.

There must, then, be a way in which history can be read such that it reveals not merely what *does* take place but also what *should* take place. Engels said with regard to the 'science' of history in general that, in it, the categories of good and bad, right and wrong, were no longer viable, only the categories of 'historical' and 'unhistorical'.[45] One could almost say that Hegel's identification of the real and the rational had been transformed into an identification of the 'historical' and the rational. Here it is, then, that the theory of dialectical 'contradiction' enters in. That the contradictions in question are not the abstract contradictions of formal logic, but the concrete, dynamic contradictions of history should be clear; they are 'surds' in the forward progress of historical development.[46]

The dialectic of need and fulfillment such as Marxism sees it working itself out in the historical process is one in which genuine development always means the resolution of contradictions in a given situation. The basic contradiction, of course, is found in the relationship of the forces of production to the product. Whether the stage of development be that of the most primitive production of food to fulfill the needs of purely biological life or that of capitalist production with its attendant division of the entire world into antagonistic classes, there arises a contradiction between the form of production which actually exists and the needs of man which require to be fulfilled. Contradiction, then, exists wherever a stage in the development of man has turned into – *the* 'chains' which impede further development.[47] Now, according to Marxism, where this contradiction arises it must be resolved, and it must be resolved by human activity. If the contradiction persists, the activity of man has by that very fact become unhistorical. The point is, however, that in the very contradiction itself is revealed the only true means of resolving it, and any other means reveals itself as untrue by either perpetuating the contradiction or being in itself contradictory. Thus, although there may seem to be many possibilities of solution, there is only one possibility of *real* solution – and that is the truly historical development. What this real possibility is, is not inscribed in history from all eternity but reveals itself only in the actual process of development,[48] and to look for change *before* the historical stage which calls for chage has arrived is as unhistorical as is resistance to change when it is called for.[49]

Because in the last analysis Marx has a kind of Hegelian confidence in the rationality of history he can look at the past and see all those developments which have contributed to the present structure of society as genuinely historical (he can also see that there have been deviations or retardations). Thus, history itself reveals what is and what is not truly historical. Only when the contradictions of history have been genuinely resolved, it is true, can he say what precise form the resolution *had to* take, but he does know that the resolution was the result of the working out of 'necessary laws'. By the same token he does *not* know the precise form which the resolution of the present contradiction must take, but he is able to say what forms would merely perpetuate the contradiction or simply introduce new contradictions. Thus, he is quite sure that the contradiction will not be resolved until the division of society into classes has been overcome. Moreover, he is equally sure that only revolution – communist revolution at that – will resolve the present contradiction. Despite all his efforts to avoid it there *is* an element of (somewhat illegitimate) prediction in this, but even this contradiction he resolves in a twofold way: (1) he sees the collapse of capitalism and the triumph of socialism as not an event in the future but as a development already in process;[50] and (2) by a process of elimination he has exhausted all other possibilities.[51]

Nowhere, perhaps, do we find expressed more clearly the Marxist notion of dialectical necessity in the historical process than in Engels' interpretation of Hegel's dictum, "Whatever is real is rational, and whatever is rational is *real*."[52] The proposition, he tells us, has been gratefully accepted by conservatives and angrily rejected by liberals, by both for the wrong reasons. What both failed to see, says Engels, is that, historically speaking, there is neither reality nor rationality without necessity. The real is rational only when it is historically necessary that it be; and the reality of the rational is the historical necessity of its coming into being. This means, according to the very rules of the Hegelian dialectic, that the real as given contains within itself the very irrationality which necessarily demands that it be superseded by what is truly rational and, thus, historically real. The proposition, then, becomes the foundation stone of revolutionary thinking (and action). Its meaning is, "Whatever *is* given deserves to disappear."[53] There are no absolute truths to be held on to dogmatically, precisely because truth lies in the process of developing knowledge which is science.

The truth which philosophy had as its function to grasp was with Hegel no longer a collection of completed dogmatic propositions, which, once they have been discovered, demand only to be learned by heart. Truth was now to be found in the very process of knowing, in the long historical development of science, which climbs from lower to ever higher levels of knowledge without ever, through the discovery of so-called absolute truth, arriving at a point where it can go no further, where there is nothing more for it to do but fold its hands in its lap and marvel at the absolute truth it has attained.[54]

The application, of course, is primarily to history, and it embraces not only the process of thinking but also the course of events and the human action necessary to bring about what is historically 'necessary'. This process has no ideally perfect human situation as its goal, but every stage in the process of development is rationally necessary in a process which has no term. The process is rationally necessary, and it is possible to know scientifically the inner dialectical laws of this process – in fact this is what science is.[55]

Admittedly this is not what Hegel himself said. The point is that it is a necessary consequence of his dialectical method, and it describes what in fact the Marxist dialectic is. The Hegelian 'system' is a dead end; but the method describes 'scientifically' the necessary march of history.[56]

Despite whatever weaknesses this theory may reveal we can see that they are not the weaknesses of Stalin's unequivocal interpretation of the form which the revolution had to take. I venture to say that the same is true of Lenin (at least in his most influential pronouncements) – and even of Engels– both of whom 'dogmatized' Marxism to an extent incompatible with the more tentative approach of Marx himself.[57] In any event the non-dogmatic interpretation makes more intelligible the efforts of Marxists in Czeckoslovakia, Jugoslavia, and Cuba, and in the non – (not yet) communist countries of France, Spain, Italy, etc. to realize 'scientific' socialism in a manner which is not dictated by the univocal interpretation of the most powerful communist nations.[58]

Although it is difficult to subscribe to the opinion of some contemporary interpreters – usually proponents of some sort of non-communist Marxism – who emphasize the differences between the early and the late Marx,[59] it is nevertheless true that the language of the later works is more susceptible of the sort of dogmatic interpretation which until recently was current. We are, therefore, fortunate to have available the earlier, more philosophical writings, against the background of which we can understand the later insistence on 'science' with its 'objective laws'. It is, of course, difficult to see what it can mean to speak of history's 'scientific' laws, if one does not assume in history a sort of built-in teleology, but it is clear enough – and this many recent commentators stress – that Marx himself intended a historical process which was neither teleological nor rigidly deterministic and yet susceptible of the kind of scientific knowledge which would enable him not only to criticize the aberrations of society but also to remedy them. The all-important point to remember – and here Marx harks back to the 'transcendental' viewpoint of Kant and Fichte – is that the necessary antecedent condition for transforming the present situation of society is that there be 'scientific' knowledge of the genesis of that situation. In the last analysis the reason why Marx calls this 'science', is that for his purposes it *had* to be science – otherwise 'revolution'[60] would be only 'reform'. What tends to make the whole thing suspect, precisely as 'science', is that it is difficult to conceive of what – for Marx or for a Marxist – could even hypothetically count as evidence *against* the theory. If we know antecedently that any evidence which can be discovered will count as evidence *for* and not *against* a theory, are we talking about a 'scientific' theory at all?

Fordham University

NOTES

[1] It is significant that some contemporary Marxist thinkers – outside the Soviet Union – are emphasizing the need of conceiving Marxist 'science' more along the lines of knowledge in process, constantly being modified. Cf., e.g., Gajo Petrović, *Marx in the Mid-Twentieth Century*, Doubleday, New York 1967, p. 11.
   There is no particular point here in distinguishing between scientific certitude and probability, since it was not a concern of 19th century thinkers, least of all Marxists.
[2] Cf. Marx, *Theses on Feuerbach*, No. 11.
[3] Cf. *Petrović, op. cit.*, pp. 22–23.
[4] Roger Garaudy, *Marxisme du XXe siècle*, La Palatine, Paris 1966, p. 8.
[5] Cf. Engels' remark, in the Preface to the *Communist Manifesto*, that Marx's science was "destined to do for history what Darwin's theory has done for biology".
[6] *Deutsches Wörterbuch*, by Jacob and Wilhelm Grimm, Vol. XV, Part II; Hirzel, Leipzig, 1960.
[7] For German philosophers of the 19th Century, the classical distinction between freedom and necessity was contributed by Kant in his *Critique of Practical Reason* and *Fundamental Principles of the Metaphysics of Morals*. Roughly, Kant would say that

man is spiritually free but materially determined. Thus, he can choose to act, but once he goes into action, his activities are governed by the necessary laws of material activity. Marx seeks to synthesize freedom and necessity in terms of 'laws' of history which govern material activity.

[8] Cf. *Theses on Feuerbach*, No. 1, where we are told that Marx's materialism differs from all other materialism (particularly the 'mechanistic' materialism of the 18th century) in giving a materialistic account of man's subjective activity – called 'praxis', i.e. purposeful activity. In *Capital I* (transl. by Samuel Moore and Edward Aveling), Kerr, Chicago 1908, Preface, p. 13, Marx distinguishes between the 'natural' development of capitalist production, which is manifested in Germany, with the situation in England, where at least an attempt had been made to curb operation of 'natural' laws by legislation. Cf. further, Marx, *Zur Kritik der politischen Oekonomie*, Dietz, Berlin 1951, p. 237, where he assures us that individuals are what they are only in society which develops according to its own dialectical laws.

[9] Cf. Franz Schnabel, *Deutsche Geschichte im neunzehnten Jahrhundert*, Herder, Freiburg 1964. Vol. I, *Die Grundlagen der neueren Geschichte*; Vol. II, *Der Aufstieg der Nationen*.

[10] Cf. *Manifest der kommunistischen Partei*, Dietz, Berlin 1951, p. 7, where we are told that modern bourgeois society has "simplified class antagonisms".

[11] The concept of 'Alienation' is certainly not original with Marx (or with Hegel and Feuerbach). It is a religious concept of ancient vintage, indicating man's estrangement from God and, hence, from his own true humanity, through sin. It is from Hegel, however, that Marx derives the notion of 'alienation' as an essential negative movement in the historical process of human development.

[12] 'Three Sources and Three Component Parts of Marxism', *Selected Works*, Moscow 1936–39, XI, p. 3.

[13] "In so far as Political Economy remains within that horizon, i.e. as the capitalist regime is looked upon as the absolutely final form of social production, instead of as a passing historical phase of its evolution, Political Economy can remain a science only so long as the class-struggle is latent or manifests itself only in isolated and sporadic phenomena". *Capital* I, p. 19; cf. pp. 17–19.

It would seem that Marx himself is guilty of a similar 'unscientific' procedure. In deriving the 'necessary natural laws' of capitalist production from the British industrialist stage in the history of capitalist production he is taking it as the 'absolutely final form' of this sort of production 'instead of a passing historical phase of its evolution'.

[14] Cf. *Die deutsche Ideologie*, Dietz, Berlin 1953, pp. 40–41; *Theses on Feuerbach*, No. 10.

[15] Cf. *Zur Kritik der politischen Oekonomie*, pp. 12–13. There are not any 'eternal natures'; there is only historical process (cf. *ibid.*, p. 236).

[16] *Die heilige Familie*, Dietz, Berlin 1953, pp. 84–85.

[17] Cf. *Theses on Feuerbach*, No. 9.

[18] *Die heilige Familie*, pp. 14–15.

[19] Cf. *Theses on Feuerbach*, No. 2.

[20] We are not, of course, saying that the term 'model' was familiar to Marx.

[21] Engels, *Anti-Dühring*, Dietz, Berlin 1935, p. 31. Marx himself takes economico-historical development out of the realm of individual initiative into that of 'natural' process: "My standpoint, from which the evolution of the economic formation of society is viewed as a process of natural history, can less than any other make the individual responsible for relations whose creature he socially remains, however much he may subjectively raise himself above them". (*Capital* I, p. 15.)

22 *Zur Kritik der politischen Oekonomie*, p. 237.

23 *Ibid.*, p. 13.

24 Cf. Marx, 'Nationalökonomie und Philosophie', *Die Frühschriften*, Kröner, Stuttgart 1953, pp. 246–48.

25 It is not for the philosopher (not even the philosopher of science) to judge the accuracy with which Marx presents economic 'facts' – that is the task of the economist. What the philosopher can do is to judge whether the *de facto* presentation justifies a theory of 'necessary laws' – and to what extent.

26 For all this cf., *Die deutsche Ideologie*, pp. 17–30. It should be noted that each need-fulfillment leads at the same time to the engendering of another need to be fulfilled (*ibid.*, p. 25).

27 *Zur Kritik der politischen Oekonomie*, p. 14. Because up to this point development has been unplanned it is 'pre-historical'. This comes out strongly throughout *Capital*, where emphasis is constantly put on the 'necessary' character of development up to bourgeois society.

28 Cf. *Capital* I, Part VI, pp. 586–617.

29 Cf. *ibid.*, Part VII, pp. 671–711.

30 Roger Garaudy, *Karl Marx*, Seghers, Paris 1964, pp. 232–33. Cf. *Capital* I, pp. 686–87, 706.

31 *Capital* I, Preface to First Edition, p. 14.

32 Cf. supra, note 13, where we have pointed out a certain contradiction in Marx's treatment of the situation in England as the 'given' of capitalism.

33 Letter to Kowalewski, quoted by Roger Garaudy, *Karl Marx*, p. 184.

34 *Anti-Dühring*, pp. 20–22.

35 *Ibid.*, p. 22.

36 Cf. *ibid.*, p. 29.

37 *Ibid.*, p. 32.

38 Cf. *ibid.*, p. 40.

39 Marx, *Das Elend der Philosophie*, Dietz, Berlin 1952, p. 43.

40 Cf. *Theses on Feuerbach*, No. 6, where Marx accuses Feuerbach of having missed the whole point of 'real science' by looking on humanity as a 'natural' genus and thus failing to see that what makes humanity one is not logical classification but historical activity.

41 *Capital* I, Preface to First Edition, p. 13; cf. p. 15, where he calls the description of this process 'natural history'.

42 *Ibid.*, Preface to Second Edition, p. 23.

43 Hegel, *Grundlinien der Philosophie des Rechts* (ed. by Johannes Hoffmeister), Meiner, Hamburg 1955, p. 14.

44 Cf. Engels, *Ludwig Feuerbach und der Ausgang der klassischen deutschen Philosophie*, Dietz, Berlin 1951, pp. 7–10, for a nuanced Marxist interpretation of Hegel's dictum.

45 *Das Elend der Philosophie*, Engels' Preface, p. 27.

46 In the Preface to the second edition of the first volume of *Capital*, p. 26, Marx tells that the 'contradictions' of which he speaks (in economic history) are 'cyclic crises'.

47 Cf. *Die deutsche Ideologie*, pp. 72ff; *Zur Kritik der politischen Oekonomie*, p. 13.

48 The similarites between the kind of verification envisaged here and that of Pragmatism are too obvious to be detailed.

49 Cf. *Zur Kritik der politischen Oekonomie*, p. 14.

50 Cf. *Capital* I, p. 14.

51 Cf. *Communist Manifesto*, Part III.

[52] *Ludwig Feuerbach und der Ausgang der klassischen deutschen Philosophie.*
[53] *Ibid.,* p. 7.
[54] *Ibid.,* pp. 7–8.
[55] *Ibid.,* p. 8.
[56] *Ibid.,* pp. 9–10.
[57] There is, of course, a suspicion – which applies more obviously to much contemporary Marxist writing (e.g. Garaudy's *Marxisme du XXe siècle,* pp. 62–63) – that 'non-dogmatic' is in itself a sufficient guarantee of 'scientific'.
[58] Petrović, *op. cit.,* p. 56. It is true, of course, that the present phenomenon of Soviet intervention to suppress 'counterrevolution' in Czechoslovakia and Pravda's saber-rattling with regard to 'neo-nazism' and 'militarism' in West Germany cast some doubt on the ability of the theory to work itself out, where the univocal interpretation has all the military power. But here we are concerned with the theory as such.
[59] Cf. *ibid.,* pp. 31–34.
[60] This does not mean, of course, that 'science' would give Marx absolute certainty that any particular revolution would be successful. It is not the function of science to eliminate risks.

WŁADYSŁAW KRAJEWSKI

THE IDEA OF STATISTICAL LAW
IN NINETEENTH CENTURY SCIENCE

I. INTRODUCTION: THE CONCEPT OF STATISTICAL LAW
(IN THE NARROWER AND THE WIDER SENSE)

It is now generally recognised that there are two main types of laws.
Laws of the first type establish a simple correspondence between two or
more parameters; they are fulfilled in each individual case (usually only
approximately in real phenomena but we shall not consider this question
here). There are different names for this type: 'dynamical', 'deterministic',
'causal', 'simple', 'nonstatistical' etc. Laws of the second type are fulfilled
only by large sets of events. They are called 'statistical', sometimes
'probabilistic' or 'stochastic'. Here we shall consider only the latter type.

The simplest statistical laws establish the probability of some event
under certain conditions. E.g., the probability of a new born child being a
boy in a given country is 0.51. In physics and many other sciences
'statistical distributions' are often used that establish the dependence of the
probability of different values of some parameter on these values themselves. A classic example is Maxwell's distributions of molecular velocities of a gas in a state of thermodynamic equilibrium:

$$\frac{\mathrm{d}N}{N} = Av^2 e^{-av^2}\,\mathrm{d}v$$

This formula gives the probability that a molecule chosen by chance
has a velocity between v and $v + \mathrm{d}v$.

There also are laws that establish the dependence of some macro-
parameter on the probability of some micro-parameter. The classic
example is Boltzmann's law that connects the entropy of a gas with the
probability of its microstate:

$$S = k \log W.$$

All these laws explicitly contain the probability of some events. Hence,
we shall call them 'statistical laws in the narrower sense'.

Boston Studies in the Philosophy of Science, XIV. All Rights Reserved.

But often in physics laws of quite another kind are also called 'statistical'. Namely, the laws of macro-parameters (called also 'phenomenological laws') that appear in the light of the atomic theory to be results of statistical equalizations of an immense number of fortuitous micro-processes. To this kind belong the gas laws (Pascal's, Boyle's, etc.), diffusion equations, laws of heat-conduction, and others. The molecular-kinetic theory shows that all these laws establish only a probable course of events although for the quantities of gas with which we have to do in ordinary life this probability is in practice equivalent to certainty.

Laws of this kind were considered in Max Planck's lecture *Dynamische und statistische Gesetzmässigkeit*[1], Planck stresses the difference between necessary and probable processes, giving as examples the flow of liquid from a higher level to a lower one, and the passage of heat from a warmer body to a colder one. Planck calls the law of processes of the first type 'dynamical', and of the second type, 'statistical'.

Planck's terminology is widely used in physics. But not always. E.g., the Russian physicist A. Samoylovitch writes that the diffusion equation expresses a 'typical dynamical law' although he stresses that this law is a result of the law of large numbers, i.e. the result of the averaged equalization of random movements of gas-molecules[2].

The terminology of Samoylovitch has some advantages. According to it the type of law depends only on its form and not on any additional information. In a previous paper I called gas laws and analogous ones 'dynamical laws with statistical background'[3]. But most physicists call them statistical. For this reason I now call them 'statistical in the wider sense'.

The statistical laws in the wider sense do not contain the probability, but rather they are fulfilled with a probability which is the greater the larger is the set of events considered. Here the probability appears 'on a higher level'. For small sets of events, the probability is low and the law loses its validity.

II. PRE-STATISTICAL STAGE OF SCIENCE

Until the middle of the nineteenth century, science did not know statistical laws. The notions of chance and probability were considered to be subjective.

The probability theory first arose in the seventeenth century as the theory of games of chance. In the same period there occurred the beginning of demographic and economic statistics. In the eighteenth century, probability theory was applied to different social phenomena: demographic (the probability of a birth of a boy or girl, the probability of death at certain age, etc.), economic (problems of insurance, of taxes, etc.), juridical (trustworthness of witnesses), political (the probability that a decision taken by a majority of voices should be correct). Probability calculations were even overused but nobody considered their results as laws.

Sometimes the theory of probability was also applied to natural sciences but only to the evaluation of hypotheses (e.g. a hypothesis of the origin of planets by Buffon). At the beginning of the nineteenth century, Gauss and other mathematicians applied probability theory to the investigation of the distribution of fortuitous measuring errors (the Gaussian curve) and to methods for establishing the most probable value of measured magnitude (the method of least squares). However the results of these investigations concerned measuring and other scientific procedures but not Nature itself. Nobody supposed that there are probabilistic laws of Nature.

Statements about the probability of social phenomena were also not considered as laws. In the eighteenth century the concept of social law was not used at all.

The application of the theory of probability to social phenomena even met with resistance from some philosophers. August Comte perceived in them a manifestation of 'indeterminism'; he wrote that statistics creates an appearance of laws in a domain in which genuine laws are not known[4]. Comte was the founder of sociology and the laws that he tried to discover were of course deterministic.

Other philosophers and scientists did not object to the application of probability theory to social life but they, like Comte, considered only deterministic laws to be 'genuine'.

III. SOCIAL SCIENCES: QUETELET, MARX AND ENGELS

Shortly after Comte's main works, two books of a Belgian statistician, Adolf Quetelet, were published. The first 'Sur l'homme...' in 1835[5], the second with the characteristic title *Du système sociale et des lois qui le régissent* in 1848[6]. In these books were cited the statistics of births, marriages, crimes, suicides etc., conclusions drawn about their probability, and predictions made about the numbers of such events in succeeding years. These predictions were afterwards verified with great precision. For Quetelet this was a confirmation of determinism and a falsification of the religious doctrine of free will. True, in this he overvalued the role of statistical laws.

The laws discovered by Quetelet are statistical in the narrower sense. They establish the probability of definite events: e.g., one of these laws states that the inclination of a Belgian townsman to marry equals 0.0884!

Many other authors continued the work of Quetelet. E.g., Henry Buckle in his *History of Civilization in England*[7] gives further statistical data confirming the conclusions of the Belgian scientist.

A new period in the social sciences began with Karl Marx and Friedrich Engels. They discovered the basic economic laws of capitalism, as well as general laws of social development. Most of these laws are statistical, of course in the wider sense. Marx and Engels often stressed this although they did not use the words 'statistical' or 'probabilistic'.

In his economic manuscripts of 1857–59 Marx wrote that the value determined by work-time is a 'mean magnitude'[8]. In the first volume of his main work *Das Kapital* (1867) Marx stressed that deviations between value and price are inevitable, that the rule that the price is equal to value is only an average-law, 'blindwirkendes Durchschnittsgesetz der Regellosigkeit'[9].

In essence almost all economic laws are statistical in the wider sense. In the preface to the first volume of *Das Kapital* Marx calls the laws of capitalism 'tendencies'.

The same idea is expressed more clearly in the third volume of *Das Kapital* (published in 1894 by Engels):

Es ist überhaupt bei der ganzen kapitalistischen Produktion immer nur in einer sehr verwickelten und annähernden Weise, ale nie festzustellender Durchschnitt ewiger Schwankungen, dass sich das allgemeine Gesetz als die beherrschende Tendenz durchsetzt.[10]

In another place in this volume we read:

...die Sphäre der Konkurrenz, die jeden einzelnen Fall betrachtet, vom Zufall beherrscht ist, wo also das innere Gesetz, das in diesen Zufällen sich durchsetzt und sie reguliert, nur sichtbar wird, sobald diese Zufälle in grossen Massen zusammengefasst worden.[10]

Engels in his paper 'Zufälligkeit und Notwendigkeit' included in his *Dialaktik der Natur*, citing Hegel, stresses the dialectic of chance and necessity. In the preface to the second edition of *Anti-Dühring* (1885) Engels writes:

...in der Natur dieselben dialektischen Bewegungsgesetze im Gewirr der zahllosen Veränderungen sich durchsetzen, die auch in der Geschichte die scheinbare Zufälligkeit der Ereignisse beherrschen.[11]

The law of chance, the average of fluctuations, the law of massphenomena is of course the statistical law.

We see how wrong Hans Reichenbach was[12] when he wrote that Marx did not understand the statistical nature of social laws. On the contrary, Marx was one of the first scholars who understood this.

IV. BIOLOGY: DARWIN

The idea of statistical law was introduced into biology by Darwin although he – like Marx and Engels – did not use the words 'statistical' or 'probabilistic'. The whole theory of the struggle for life and natural selection has an essentially statistical character. Darwin speaks much about indefinite variations of living organisms and about the role of chance in their life. It depends on chance whether a given organism will survive or will die in the early stage of life. But it is more probable that a fit individual will survive than one who is not fit. In other words, the law of natural selection deals only with fate of whole populations and not of single individuals. The effect of natural selection is revealed only on a mass scale and during a vast period of time. Darwin wrote that natural selection "always acts very slowly, often only at long intervals of time"[13].

Hence the law of natural selection is fulfilled only in large sets of individuals and shows only the general trend of evolution. It is a statistical law in the wider sense.

As is known, Darwin's theory aroused animated controversy. One subject of it was the nature of the laws of this theory. Certain biologists said that Darwin's laws are not 'genuine laws'. They blamed Darwin for introducing of 'blind chance' or 'indeterminism' into science. Other scientists and philosophers saw that Darwin had introduced a new type of laws into biology, which opened up new perspectives. E.g., Engels in the *Dialektik der Natur* stressed the great merit of Darwin as a scientist who had revealed the dialectics of necessity and chance in Nature.

At the end of the nineteenth century a new branch of biology arose: biometrics. A new journal appeared: *Biometrika. A journal for the Statistical Study of Biological Problems.* In the editorial to the first volume of this journal, preceded by a photograph of Darwin's statue in Oxford, we read that biologists formerly neglected the differences between individual members of a race or species whereas "the starting point of Darwin's theory of evolution is precisely the existence of these differences...."[14] In the second editorial we read that the problem of evolution is a problem of statistics because only mass-phenomena are effective factors of evolution. Although Darwin himself did not employ mathematics, every idea of his "seems at once to fit itself to mathematical definition and to demand statistical analysis."[14,15]

Francis Galton, Karl Pearson and other pioneers of biometrics discovered many empirical statistical regularities of biological phenomena. These were statistical laws in the narrower sense.

Gregor Mendel in the domain of genetics discovered the basic statistical laws in the narrower sense even earlier (in 1866) but his results remained unknown and were rediscovered in 1900.

V. PHYSICS: MAXWELL AND BOLTZMANN

The molecular-kinetic theory of gases was the first statistical theory in physics. The general ideas of this theory has already been expressed in the eighteenth century by D. Bernoulli, M. Lomonossov and others who interpreted gas-pressure as the effect of the impact of atoms on container-walls and in the light of this interpretation explained the Boyle's law. But they did not explicitly re-formulate the laws of physics as statistical laws.

That came only in the second half of the nineteenth century. In 1856–57 A. Krönig and R. Clausius created a theory according to which a gas is a set of molecules chaotically moving and colliding like elastic balls. Krönig even suggested applying probability theory to this gas but he did not develop this idea. Clausius supposed that all molecular velocities are equal; he said that the deviations from average velocity are not interesting because they cancel each other out.

Contrary to Clausius, J. C. Maxwell concentrated his attention on the differences among molecular velocities supposing that the collisions differentiate rather than equalise them. In a lecture delivered in 1859 (published in 1860) he formulated his equation for velocity distribution that we cited above (in Section I). This was the first statistical law in the narrower sense that was a law of purely physical phenomena, not depending on human action.

Maxwell realises that he was using quite new methods. In one of his lectures (1873) he said that statistical method establishes a new kind of regularity, 'the regularity of average'[16].

The ideas of Maxwell were developed by L. Boltzmann. In his papers of 1871–1877, he formulated his H-theorem connecting the entropy of a gas with its microstate; we cited it (in its contemporary form) in the first section above as an example of statistical law in the narrower sense. Boltzmann stressed that there occur deviations from average entropy, 'fluctuations'; Thus he revealed that the law of entropy (second law of thermodynamics) is a statistical law in the wider sense. He developed the 'fluctuation hypothesis' in cosmology in an attempt to show that the 'heat death' of the universe is not inevitable.

In a lecture in Vienna in 1886 Boltzmann spoke generally about statistical laws as 'laws of chance', which rule similarly in the world of molecules as in human society. Referring to Buckle, he cited the constancy of probability rates in marriages, crimes etc. and said that in molecular physics, too, we have to do with laws which establish the constancy of some averages. E. g., the average of the intensity of impacts of gas-molecules in a unit of piston-surface is constant, in spite of fortuitous deviations.[17]

Afterwards statistical methods in physics were developed further by J. W. Gibbs, who created statistical mechanics, by M. Smoluchowski, who discovered certain laws of molecular fluctuations and by others. Our century saw the beginning of quantum mechanics which has a fundamentally statistical nature (but this is not the subject of the present paper).

We see that Maxwell's and Boltzmann's attitude in molecular physics is very similar to Darwin's attitude in biology. Contrary to their predecessors, they concentrated their attention on contingent differences between individuals (organisms, molecules) and from this contingency passed to necessity (to statistical law). The contemporary French biologist F. Jacob stressing this similarity of attitudes (he speaks about Boltzmann and Gibbs in physics and about Darwin, Wallace and Mendel in biology), says that it was a turning point in science, "un changement dans la manière même de considérer les objets, le résultat d'une attitude radicalement nouvelle qui apparaît au milieu du XIXe siècle."[18]

VI. CONCLUSION

Statistical laws were discovered earlier in society (Quetelet) than in Nature. This state of affaire is easy to explain. Human individuals are well known and always interest us, therefore statistics deals primarily with them. Biological individuals are also well known but interest us less, so biological statistics arose later. Physical individuals (atoms and molecules) are known only indirectly (even the existence of atoms was for a long time a controversional issue); they cannot be observed. Empirical statistics is here impossible, and we speak about molecular statistics in a metaphorical way, using an analogy to social statistics. Statistical methods in social sciences probably influenced the application of statistical methods in molecular physics (cf. Boltzmann's reference to Buckle).

We see also that whereas in the social and biological sciences statistical laws in the narrower sense are empirical, in molecular physics they are purely theoretical, deduced from an idealized model of a gas (and now the social and biological sciences sometimes use theoretical models too).

Finally, we notice a very interesting coincidence. In 1859 Darwin published his revolutionary book and Maxwell delivered his lecture on kinetic theory. This was therefore the year when the statistical laws appeared in physics (in the narrower sense) and in biology (in the wider sense). In the same period Marx noted his first thoughts about the statistical nature (in the wider sense) of economic laws.

Warsaw University

NOTES

[1] Max Planck, 'Dynamische und statistische Gesetzmässigkeit', in *Wege zur physikalischen Erkenntnis*, Leipzig 1944.

2 A. G. Samoylovitch, *Thermodynamics and Statistical Physics* (in Russian), Moscow 1953; see p. 179.
3 Władysław Krajewski, 'Remarks on the Dynamical and Statistical Laws' (in Polish), *Myśl Filozoficzna* (1954), No. 3.
4 August Comte, *Discours sur l'esprit positif*, Paris 1835, pp. 47-48.
5 Adolf Quetelet, *Sur l'homme et le développement de ses facultés*, Paris 1835.
6 Adolf Quetelet, *Du système social et de lois qui le régissent*, Paris 1848.
7 Henry Thomas Buckle, *History of Civilization in England*, London 1857.
8 *Marx-Engels Archives* (in Russian), Vol. IV, Moscow 1935, p. 47.
9 Karl Marx, *Das Kapital*, Erster Band, Berlin 1947, p. 107.
10 Karl Marx, *Das Kapital*, Dritter Band, Berlin 1949, pp. 186 and 882.
11 Friedrich Engels, *Herrn Eugen Dührings Umwälzung der Wissenschaft*, Berlin 1948, p. 11.
12 Hans Reichenbach, *The Rise of Scientific Philosophy*, Berkeley 1951, Chapter 4.
13 Charles Darwin, *The Origin of Species by Means of Natural Selection*, London 1902, p. 98.
14 *Biometrika, A Journal for the Statistical Study of Biological Problems* (ed. in consultation with Francis Galton by W. F. R. Weldon, Karl Pearson and C. B. Davenport), (October 1901 to August 1902), Editorial (pp. 1 and 3–4).
15 O. B. Sheynin, 'Newton and the Classical Theory of Probability', *Archive for History of Exact Sciences* 7 (1971), No. 3.
16 James Clerk Maxwell, *Scientific Papers*, Vol. II (Paris ed. of 1927), p. 374.
17 Ludwig Boltzmann, *Populäre Schriften*, Leipzig 1905, p. 34.
18 François Jacob, *La logique du vivant*, Paris 1970, p. 190.

SYNTHESE HISTORICAL LIBRARY

Texts and Studies
in the History of Logic and Philosophy

Editors:

N. KRETZMANN (Cornell University)
G. NUCHELMANS (University of Leyden)
L. M. DE RIJK (University of Leyden)

1. M. T. BEONIO-BROCCHIERI FUMAGALLI, *The Logic of Abelard*. Translated from the Italian. 1969, IX + 101 pp.
2. GOTTFRIED WILHELM LEIBNITZ, *Philosophical Papers and Letters*. A selection translated and edited, with an introduction, by Leroy E. Loemker. 1969, XII + 736 pp.
3. ERNST MALLY, *Logische Schriften*, ed. by Karl Wolf and Paul Weingartner. 1971, X + 340 pp.
4. LEWIS WHITE BECK (ed.), *Proceedings of the Third International Kant Congress*. 1972, XI + 718 pp.
5. BERNARD BOLZANO, *Theory of Science*, ed. by Jan Berg. 1973, XV + 398 pp.
6. J. M. E. MORAVCSIK (ed.), *Patterns in Plato's Thought. Papers arising out of the 1971 West Coast Greek Philosophy Conference*. 1973, VIII + 212 pp.
7. NABIL SHEHABY, *The Propositional Logic of Avicenna: A Translation from al-Shifā': al-Qiyās*, with Introduction, Commentary and Glossary. 1973, XIII + 296 pp.
8. DESMOND PAUL HENRY, *Commentary on De Grammatico: The Historical-Logical Dimensions of a Dialogue of St. Anselm's*. 1974, IX + 345 pp.
9. JOHN CORCORAN, *Ancient Logic and Its Modern Interpretations*. 1974, X + 208 pp.

SYNTHESE LIBRARY

Monographs on Epistemology, Logic, Methodology,
Philosophy of Science, Sociology of Science and of Knowledge, and on the
Mathematical Methods of Social and Behavioral Sciences

Editors:

DONALD DAVIDSON (The Rockefeller University and Princeton University)
JAAKKO HINTIKKA (Academy of Finland and Stanford University)
GABRIËL NUCHELMANS (University of Leyden)
WESLEY C. SALMON (University of Arizona)

1. J. M. BOCHEŃSKI, *A Precis of Mathematical Logic.* 1959, X + 100 pp.
2. P. L. GUIRAUD, *Problèmes et méthodes de la statistique linguistique.* 1960, VI + 146 pp.
3. HANS FREUDENTHAL (ed.), *The Concept and the Role of the Model in Mathematics and Natural and Social Sciences. Proceedings of a Colloquium held at Utrecht, The Netherlands, January 1960.* 1961, VI + 194 pp.
4. EVERT W. BETH, *Formal Methods. An Introduction to Symbolic Logic and the Study of Effective Operations in Arithmetic and Logic.* 1962, XIV + 170 pp.
5. B. H. KAZEMIER and D. VUYSJE (eds.), *Logic and Language. Studies dedicated to Professor Rudolf Carnap on the Occasion of his Seventieth Birthday.* 1962, VI + 256 pp.
6. MARX W. WARTOFSKY (ed.), *Proceedings of the Boston Colloquium for the Philosophy of Science, 1961–1962,* Boston Studies in the Philosophy of Science (ed. by Robert S. Cohen and Marx W. Wartofsky), Volume I. 1963, VIII + 212 pp.
7. A. A. ZINOV'EV, *Philosophical Problems of Many-Valued Logic.* 1963, XIV + 155 pp.
8. GEORGES GURVITCH, *The Spectrum of Social Time.* 1964, XXVI + 152 pp.
9. PAUL LORENZEN, *Formal Logic.* 1965, VIII + 123 pp.
10. ROBERT S. COHEN and MARX W. WARTOFSKY (eds.), *In Honor of Philipp Frank,* Boston Studies in the Philosophy of Science (ed. by Robert S. Cohen and Marx W. WARTOFSKY), Volume II. 1965, XXXIV + 475 pp.
11. EVERT W. BETH, *Mathematical Thought. An Introduction to the Philosophy of Mathematics.* 1965, XII + 208 pp.
12. EVERT W. BETH and JEAN PIAGET, *Mathematical Epistemology and Psychology.* 1966, XXII + 326 pp.
13. GUIDO KÜNG, *Ontology and the Logistic Analysis of Language. An Enquiry into the Contemporary Views on Universals.* 1967, XI + 210 pp.
14. ROBERT S. COHEN and MARX W. WARTOFSKY (eds.), *Proceedings of the Boston Colloquium for the Philosophy of Science 1964–1966, in Memory of Norwood Russell Hanson,* Boston Studies in the Philosophy of Science (ed. by Robert S. Cohen and Marx W. Wartofsky), Volume III. 1967, XLIX + 489 pp.
15. C. D. BROAD, *Induction, Probability, and Causation. Selected Papers.* 1968, XI + 296 pp.

16. GÜNTHER PATZIG, *Aristotle's Theory of the Syllogism. A Logical-Philosophical Study of Book A of the Prior Analytics.* 1968, XVII + 215 pp.
17. NICHOLAS RESCHER, *Topics in Philosophical Logic.* 1968, XIV + 347 pp.
18. ROBERT S. COHEN and MARX W. WARTOFSKY (eds.), *Proceedings of the Boston Colloquium for the Philosophy of Science 1966–1968*, Boston Studies in the Philosophy of Science (ed. by Robert S. Cohen and Marx W. Wartofsky), Volume IV. 1969, VIII + 537 pp.
19. ROBERT S. COHEN and MARX W. WARTOFSKY (eds.), *Proceedings of the Boston Colloquium for the Philosophy of Science 1966–1968*, Boston Studies in the Philosophy of Science (ed. by Robert S. Cohen and Marx W. Wartofsky), Volume V. 1969, VIII + 482 pp.
20. J. W. DAVIS, D. J. HOCKNEY, and W. K. WILSON (eds.), *Philosophical Logic.* 1969, VIII + 277 pp.
21. D. DAVIDSON and J. HINTIKKA (eds.), *Words and Objections: Essays on the Work of W. V. Quine.* 1969, VIII + 366 pp.
22. PATRICK SUPPES, *Studies in the Methodology and Foundations of Science. Selected Papers from 1911 to 1969.* 1969, XII + 473 pp.
23. JAAKKO HINTIKKA, *Models for Modalities. Selected Essays.* 1969, IX + 220 pp.
24. NICHOLAS RESCHER et al. (eds.). *Essay in Honor of Carl G. Hempel. A Tribute on the Occasion of his Sixty-Fifth Birthday.* 1969, VII + 272 pp.
25. P. V. TAVANEC (ed.), *Problems of the Logic of Scientific Knowledge.* 1969, XII + 429 pp.
26. MARSHALL SWAIN (ed.), *Induction, Acceptance, and Rational Belief.* 1970, VII + 232 pp.
27. ROBERT S. COHEN and RAYMOND J. SEEGER (eds.), *Ernst Mach; Physicist and Philosopher*, Boston Studies in the Philosophy of Science (ed. by Robert S. Cohen and Marx W. Wartofsky), Volume VI. 1970, VIII + 295 pp.
28. JAAKKO HINTIKKA and PATRICK SUPPES, *Information and Inference.* 1970, X + 336 pp.
29. KAREL LAMBERT, *Philosophical Problems in Logic. Some Recent Developments.* 1970, VII + 176 pp.
30. ROLF A. EBERLE, *Nominalistic Systems.* 1970, IX + 217 pp.
31. PAUL WEINGARTNER and GERHARD ZECHA (eds.), *Induction, Physics, and Ethics, Proceedings and Discussions of the 1968 Salzburg Colloquium in the Philosophy of Science.* 1970, X + 382 pp.
32. EVERT W. BETH, *Aspects of Modern Logic.* 1970, XI + 176 pp.
33. RISTO HILPINEN (ed.), *Deontic Logic: Introductory and Systematic Readings.* 1971, VII + 182 pp.
34. JEAN-LOUIS KRIVINE, *Introduction to Axiomatic Set Theory.* 1971, VII + 98 pp.
35. JOSEPH D. SNEED, *The Logical Structure of Mathematical Physics.* 1971, XV + 311 pp.
36. CARL R. KORDIG, *The Justification of Scientific Change.* 1971, XIV + 119 pp.
37. MILIČ ČAPEK, *Bergson and Modern Physics*, Boston Studies in the Philosophy of Science (ed. by Robert S. Cohen and Marx W. Wartofsky), Volume VII. 1971, XV + 414 pp.
38. NORWOOD RUSSELL HANSON, *What I do not Believe, and other Essays*, ed. by Stephen Toulmin and Harry Woolf. 1971, XII + 390 pp.
39. ROGER C. BUCK and ROBERT S. COHEN (eds.), *PSA 1970. In Memory of Rudolf Carnap*, Boston Studies in the Philosophy of Science (ed. by Robert S. Cohen and

Marx W. Wartofsky), Volume VIII. 1971, LXVI + 615 pp. Also available as a paperback.
40. DONALD DAVIDSON and GILBERT HARMAN (eds.), *Semantics of Natural Language.* 1972, X + 769 pp. Also available as a paperback.
41. YEHOSUA BAR-HILLEL (ed.), *Pragmatics of Natural Languages.* 1971, VII + 231 pp.
42. SÖREN STENLUND, *Combinators, λ-Terms and Proof Theory.* 1972, 184 pp.
43. MARTIN STRAUSS, *Modern Physics and Its Philosophy. Selected Papers in the Logic, History, and Philosophy of Science.* 1972, X + 297 pp.
44. MARIO BUNGE, *Method, Model and Matter.* 1973, VII + 196 pp.
45. MARIO BUNGE, *Philosophy of Physics.* 1973, IX + 248 pp.
46. A. A. ZINOV'EV, *Foundations of the Logical Theory of Scientific Knowledge (Complex Logic),* Boston Studies in the Philosophy of Science (ed. by Robert S. Cohen and Marx W. Wartofsky), Volume IX. Revised and enlarged English edition with an appendix, by G. A. Smirnov, E. A. Sidorenka, A. M. Fedina, and L. A. Bobrova. 1973, XXII + 301 pp. Also available as a paperback.
47. LADISLAV TONDL, *Scientific Procedures,* Boston Studies in the Philosophy of Science (ed. by Robert S. Cohen and Marx W. Wartofsky), Volume X. 1973, XII + 268 pp. Also available as a paperback.
48. NORWOOD RUSSELL HANSON, *Constellations and Conjectures,* ed. by Willard C. Humphreys, Jr. 1973, X + 282 pp.
49. K. J. J. HINTIKKA, J. M. E. MORAVCSIK, and P. SUPPES (eds.), *Approaches to Natural Language. Proceedings of the 1970 Stanford Workshop on Grammar and Semantics.* 1973, VIII + 526 pp. Also available as a paperback.
50. MARIO BUNGE (ed.), *Exact Philosophy – Problems, Tools, and Goals.* 1973, X + 214 pp.
51. RADU J. BOGDAN and ILKKA NIINILUOTO (eds.), *Logic, Language, and Probability.* A selection of papers contributed to Sections IV, VI, and XI of the Fourth International Congress for Logic, Methodology, and Philosophy of Science, Bucharest, September 1971. 1973, X + 323 pp.
52. GLENN PEARCE and PATRICK MAYNARD (eds.), *Conceptual Change.* 1973, XII + 282 pp
53. ILKKA NIINILUOTO and RAIMO TUOMELA, *Theoretical Concepts and Hypothetico-Inductive Inference.* 1973, VII + 264 pp.
54. ROLAND FRAÏSSÉ, *Course of Mathematical Logic* – Volume I: *Relation and Logical Formula.* 1973, XVI + 186 pp. Also available as a paperback.
55. ADOLF GRÜNBAUM, *Philosophical Problems of Space and Time.* Second, enlarged edition, Boston Studies in the Philosophy of Science (ed. by Robert S. Cohen and Marx W. Wartofsky), Volume XII. 1973, XXIII + 884 pp. Also available as a paperback.
56. PATRICK SUPPES (ed.), *Space, Time, and Geometry.* 1973, XI + 424 pp.
57. HANS KELSEN, *Essays in Legal and Moral Philosophy,* selected and introduced by Ota Weinberger. 1973, XXVIII + 300 pp.
59. ROBERT S. COHEN and MARX W. WARTOFSKY (eds.), *Logical and Epistemological Studies in Contemporary Physics,* Boston Studies in the Philosophy of Science (ed. by Robert S. Cohen and Marx W. Wartofsky), Volume XIII. 1973, VIII + 462 pp. Also available as a paperback.
60. ROBERT S. COHEN and MARX W. WARTOFSKY (eds.), *Methodological and Historical Essays in the Natural and Social Sciences. Proceedings of the Boston Colloquium for the Philosophy of Science, 1969–1972,* Boston Studies in the Philosophy of Science

(ed. by Robert S. Cohen and Marx W. Wartofsky), Volume XIV. Also available as a paperback.

In Preparation

58. R. J. SEEGER and ROBERT S. COHEN (eds.), *Philosophical Foundations of Science. Proceedings of an AAAS Program, 1969.* Boston Studies in the Philosophy of Science (ed. by Robert S. Cohen and Marx W. Wartofsky), Volume XI.
61. ROBERT S. COHEN and MARX W. WARTOFSKY (eds.), *For Dirk Struik. Scientific, Historical, and Political Essays in Honor of Dirk J. Struik*, Boston Studies in the Philosophy of Science (ed. by Robert S. Cohen and Marx W. Wartofsky), Volume XV.
62. KAZIMIERZ AJDUKIEWICZ, *Pragmatic Logic*, transl. from the Polish by Olgierd Wojtasiewicz.
63. SÖREN STENLUND (ed.), *Logical Theory and Semantic Analysis. Essays Dedicated to Stig Kanger on his Fiftieth Birthday.*